Next-Generation Sequencing

This cutting-edge reference book compiles standard operating procedures, protocols, and applications of *next-generation sequencing* (NGS). It discusses genomic testing applications through NGS. It pays special focus on the protocols for cataloguing variants of uncertain significance. Over the years, NGS and advanced bioinformatics approaches have allowed the transition of genomic assays into translational practices. The book covers visualisation of NGS datasets, investigation of early development impairment, and metagenome protocols. It also discusses the challenges in NGS methods.

Key Points:

- Includes case studies of application of NGS in different taxa like humans, rodents, plants, and bacteria
- Compiles protocols from various reputed companies like Illumina, PacBio, and ThermoFisher
- Discusses the translational applications of NGS methods
- Reviews machine learning heuristics for NGS data interpretation
- Discusses emerging genomic assay technologies and characterising mechanisms of disease prevalence

The book is meant for researchers and industry experts in genomics, computational biology, and bioinformatics.

Next-Generation Sequencing
Standard Operating Procedures and Applications

Edited by
Prashanth N. Suravajhala and Jeffrey W. Bizzaro

CRC Press
Taylor & Francis Group
Boca Raton London New York

CRC Press is an imprint of the
Taylor & Francis Group, an **informa** business

Designed cover image: Jonathan A. Bizzaro

First edition published 2025
by CRC Press
2385 NW Executive Center Drive, Suite 320, Boca Raton FL 33431

and by CRC Press
4 Park Square, Milton Park, Abingdon, Oxon, OX14 4RN

CRC Press is an imprint of Taylor & Francis Group, LLC

Funding Body: Luxembourg National Research Fund (FNR) and EUROSTARS (EUREKA Network)

ISBN: 978-1-032-39262-2 (hbk)
ISBN: 978-1-032-40635-0 (pbk)
ISBN: 978-1-003-35406-2 (ebk)

DOI: 10.1201/9781003354062

Typeset in Times
by KnowledgeWorks Global Ltd.

To my revered Sadguru Dev Maa – Prashanth N. Suravajhala

To my parents, William and Louise, who recognised and encouraged my interest in science – Jeffrey W. Bizzaro

Contents

Foreword

It has been more than 50 years since the invention of the first technologies for nucleic acid sequencing and 25 years since the landmark sequencing of the human genome, but the impact of these advances continues to grow. We have gone from an era where sequencing a single gene was a monumental achievement to one where we routinely sequence whole genomes or transcriptomes of thousands of single cells. We now have a plethora of technologies for next-generation sequencing (NGS) offering an unprecedented window into the biology of living systems. Competing technologies now abound for long-read sequencing, single-cell sequencing, and spatial sequencing, among others, each regularly finding innovative applications far beyond their original uses. NGS has made possible the science of genomics and has become arguably our primary window into genetics, transcriptomics, epigenetics, translation, genome structure, and RNA structure, with new uses being invented every year. The numerous uses of sequencing today have made it one of the most important tools for modern biomedical research, central to high-profile work on nearly every question in the biological sciences, from fundamental research into evolution, the origins of life, and the regulation of cellular function to applied topics in human health spanning inherited diseases, cancers, neurodegenerative diseases, infectious diseases, autoimmune diseases, and beyond.

The rise of NGS has proven more than just a powerful suite of biotechnologies. It has done more than any single advance to usher in an era of data-driven biological discovery, as dramatic improvements in sequencing technologies have gone hand in hand with similarly dramatic advances in the computational biology and data science needed to make sense of the data they produce. Fundamental computational problems such as sequence searching, alignment, variant calling, and phylogenetics continue to be active areas of research as they drive and are driven by new advances in sequencing biotechnologies and the vast amounts of data they are producing. Today, we can search petabase sequence archives, build phylogenies of tens of thousands of whole genomes, or conduct genome-wide association studies of millions of individuals. New applications and whole fields of study continue to emerge from this fusion of advanced biotechnology and computational biology.

This book fills a need for an up-to-date review of major technologies and problems in the rapidly evolving field of NGS. It provides a review of the fundamentals of the sequencing technologies now available, diverse applications of these technologies in biomedical discovery, important classes of computational and statistical analyses used to interpret them, and a survey of some of the popular tools and pipelines available to put these analyses into practice. This range of topics can serve as an excellent foundation for newcomers to NGS or a timely refresher for experienced practitioners looking to catch up on current directions for this powerful suite of technologies.

Russell Schwartz, PhD
Professor and Head, Ray & Stephanie Lane Computational Biology Department and Professor of Biological Sciences
Carnegie Mellon University
Pittsburgh, PA, USA

Preface

Like the careful unveiling of a story's elements that are essential to writing, sequencing methodically reveals the molecular narrative encoded in nucleic acids.

Next-generation sequencing (NGS), or what could aptly be called '*now*-generation sequencing', has become an established discipline that relies on two fundamental pillars of standardisation: rigorous standard operating procedures (SOPs) and well-validated gold standard datasets (also called truth sets). Yet, its technologies evolve so rapidly—rivalling Moore's Law—that end-users must constantly adapt both their methods and their benchmarking approaches.

The contributors to this book are genomics researchers who use bioinformatics, systems genomics, and NGS tools daily, grappling with this relentless pace of innovation. Through our research and publications, we have observed that advancing technologies require robust validation against reference datasets to assess the accuracy and reliability of new methods and tools. While the field has made remarkable progress, comprehensive resources addressing both procedural standards and these gold standard datasets for NGS remain scarce, highlighting the urgent need for up-to-date guidance in both areas.

After an intensive review process, the editors have assembled 17 chapters that provide comprehensive SOPs for various aspects of NGS, while also addressing the critical role of reference datasets in method validation and quality assessment. A brief overview of the chapter contents is provided in the following sections.

Raja et al., in the chapter titled *Technologies, Computations, and Data Analysis for Next-Generation Sequencing*, address the pressing challenges and limitations of NGS, along with commonly used algorithmic tools and machine learning techniques. While heuristics are mentioned among the methods discussed, the chapter delves into the broader potential of NGS in transforming healthcare and personalised medicine. It concludes by outlining future directions for NGS data analysis and their implications for advancing research and clinical practice.

Selvamani et al., in the chapter titled *Epigenetics: RNA-seq, ChIP-seq, ATAC-seq, and MeDIP-seq*, describe epigenetic regulation that broadly encompasses chromatin accessibility, DNA methylation, DNA-protein interaction, chromosome conformations, histone modifications, and replication dynamics. They explore how non-histone macromolecules' access to chromatin is a major determinant of gene expression regulation and articulate how these genomic tools are vital to understanding biological processes and deciphering epigenetic markers. The authors provide detailed workflows and applications of these techniques, discussing their integration in constructing genome-wide transcriptional regulatory networks, while also examining the advantages and limitations of bioinformatics tools used in NGS-based epigenetics analyses.

Mallya et al., in their chapter titled *Streamlining Next-Generation Sequencing Data Analysis with Nextflow and nf-core Pipelines*, introduce nf-core while describing key pipeline management tools, namely Snakemake, Cromwell, Docker, Galaxy, and Nextflow. The authors highlight how Nextflow and nf-core provide standardised pipeline development, open-source accessibility, flexible deployment, and reproducibility.

Snijesh et al., in their chapter titled *Best Practices for Variant Calling Using the Genome Analysis Toolkit*, provide an overview of the Genome Analysis Toolkit (GATK), a widely used software suite that provides tools for variant calling from NGS data. The authors describe the core variant-calling algorithm, namely HaplotypeCaller, and explain how it replaced UnifiedGenotyper, a previous GATK variant-calling tool that had limitations in detecting indels, handling complex variants, and managing computational resources and scalability. The chapter concludes by describing the effective use of GATK in quality control and filtering to minimise false positives and ensure accurate variant calls.

Thakur et al., in their chapter titled *Implementation of WGCNA for Identifying Regulatory Modules in Biological Networks*, describe weighted gene co-expression network analysis (WGCNA) as a powerful computational method for analysing gene expression data and constructing gene co-expression networks. This approach provides insights into the underlying biological processes and regulatory mechanisms of biological systems. The authors introduce WGCNA's application for identifying regulatory modules in biological networks and demonstrate its utility in revealing novel insights into gene regulation. Using plant stress response as a case study, they detail the computational steps involved in applying WGCNA, including data preprocessing, network construction, module identification, and hub gene identification.

Madamarandawala, in her chapter titled *Meta-Analysis of RNA-seq and Microarray Data*, describes gene expression data from RNA sequencing (RNA-seq) and microarray platforms and explores current tools and protocols for meta-analysis of RNA-seq and microarray data, as well as integrated approaches combining both technologies.

Galati et al., in their chapter titled *Best Practices in Single-Cell RNA-seq Data Analysis*, provide a holistic view of advanced scRNA-seq analyses, such as cell-to-cell communication analysis, gene regulatory network inference, genotyping analysis, and the RNA velocity approach. The authors discuss the integrative potential of time-series, spatial, and multi-omics dimensions of single-cell analysis, emphasising their synergistic role in revealing biological complexity. They conclude by covering major unified resources made available through the proliferation of scRNA-seq technologies, which enhance our global understanding of cellular diversity across scales and species.

Srinivasan, in her chapter titled *Integration of Spatial Transcriptomics and Single-Cell RNA-seq*, describes the use of single-cell technologies for biological analysis while highlighting how spatial transcriptomics has emerged as a valuable approach for understanding complex biological systems. This technology provides three-dimensional context to gene expression analysis within tissue environments. The author provides an overview of the integration between spatial transcriptomics and single-cell analysis, discussing both the capabilities and challenges of these emerging technologies.

Pandey et al., in their chapter titled *Metagenomics Analysis Pipelines for Microbiome Studies: QIIME and mothur*, delve into metagenomics of microbial samples using quantitative insights into microbial ecology (QIIME) and mothur. These popular tools characterise the composition and function of microbial communities in diverse habitats (microbiota) such as soil, water, plants, fermented food, and the human gut. The authors discuss key steps, including quality control of sequence reads, extraction of taxonomic features, taxonomic classification, functional annotation, statistical analysis, and data visualisation. They provide step-by-step guidelines and SOPs for using the software in quality control, taxonomic classification, phylogeny annotation, microbial community statistical analysis, and data visualisation. The chapter concludes with a discussion of how advances in artificial intelligence (AI) and machine learning (ML) address challenges in understanding microbial communities.

Bedse and Rao, in their chapter titled *Standard Operating Procedure and Applications in Single-Cell Transcriptomics*, discuss single-cell transcriptomics (SCT), providing detailed guidelines and recommendations for each step of the experimental workflow. These SOPs have been applied in various SCT studies, including the identification of rare cell populations, characterisation of cell states, and discovery of novel cell types. The authors describe how the adoption of SOPs has enhanced the sharing and integration of SCT datasets across different studies and platforms, enabling meta-analyses and the generation of data-driven insights. They conclude by examining recent advancements in SCT protocols and their applications in diverse fields of biology and medicine.

Kumar and Babu, in their chapter titled *Benchmarking and Evaluation of de novo Assembly Tools for Prokaryotic Long Reads from Oxford Nanopore Technologies*, discuss the emergence of long-read sequencing technologies, such as Oxford Nanopore Technologies (ONT), which have revolutionised the field of genomics. These technologies provide the ability to generate long sequencing reads in real-time, spanning thousands of nucleotides. The authors describe the benchmarking of *de novo* assembly tools that employ these technologies.

Satagopam and Narayanasamy, in their chapter titled *Best Practices for Reproducible Microbial Genomics Analysis*, discuss how automation and standardisation are critical for ensuring reproducibility and replicability in *in silico* protocols. The authors describe core tools required for microbial genomics, specifically for preprocessing, *de novo* assembly, and gene annotation, while also exploring metabolic network reconstruction. They conclude by addressing general principles that can help users evaluate state-of-the-art open-source tools.

Baker et al., in their chapter titled *Single-Cell RNA-seq Data Analysis in the Era of Artificial Intelligence*, introduce RNA-seq, detailing bulk and single-cell methods, and describe the steps involved in each along with postprocessing steps after scRNA-seq experiments. The authors provide a general overview of AI, discussing its main tasks, classical methods, and advanced techniques. They illustrate the role of machine learning, deep learning, and generative AI in scRNA-seq and demonstrate how these techniques can improve the efficiency and precision of the sequencing process. The chapter concludes by describing the impact of AI-powered scRNA-seq on the medical field, emphasising how these innovations empower clinicians to make more informed and effective decisions in healthcare.

Prasad and Prasannakumar, in their chapter titled *Towards Single-Molecule Protein Sequencing*, describe how proteins play a crucial role in maintaining metabolism and regulating physiological functions in organisms, making their analysis essential for understanding biological processes. The authors explain that despite the importance of protein analysis, the limited number of available methods for determining protein sequences has constrained our ability to fully explore the proteome.

Charles et al., in their chapter titled *Standard Operating Procedures for Effective Galaxy Workflows*, highlight how the proliferation of computational tools and complex data analysis requirements has created a divide between experimental and computational biologists. The authors explain how Galaxy, bolstered by the community-driven ToolShed, bridges this gap. This chapter focuses on the Galaxy platform's utility in deciphering complex sequencing datasets, the applicability of existing NGS workflows, and the opportunities for training and teaching using preset workflows. The authors also discuss the platform's adaptability to evolving bioinformatics tool design and its prospects within the global Galaxy community.

Nikhade et al., in their chapter titled *Motif Prediction Using ChIP-seq Data Analysis with Galaxy*, discuss the applications of ChIP-seq for multiple myeloma (MM) treatment using accessible bioinformatics workflows. The authors provide a foundation for investigating bortezomib's mechanism of action and potential resistance mechanisms, contributing to the development of precision medicine approaches in MM treatment.

Singh et al., together with the editors, in their chapter titled *Developing a Whole Exome Consensus Variant-Calling Pipeline to Infer Causal Pathogenic Variants*, discuss detecting germline variants from short-read sequencing. They aim to automate key steps of variant prioritisation by developing a consensus variant exome pipeline using bash. They anticipate that this pipeline will serve as a valuable tool for researchers in the field of variant detection and interpretation.

We hope that these contributions will not only offer practical guidance but also inspire further exploration of emerging questions in this fast-paced field.

Yours for Science,

Prashanth N. Suravajhala
Bioclues.org, Hyderbad, India

Jeffrey W. Bizzaro
Bioinformatics.org, Pennsylvania, USA

About the Editors

Prashanth N. Suravajhala (Prash) is known for his work in bioinformatics, computational biology, and systems biology with his research primarily focusing on understanding complex biological systems, including the study of various biological networks, protein function prediction, and the development of computational tools for studying molecular biology. He is interested in exploring the known unknown regions in the human genomes, primarily working on the top-down systems biology approaches of hypothetical proteins besides long noncoding RNAs in humans, elucidating the mechanisms underpinning small molecular interactions through clinical exomes. His group has benchmarked pipelines and developed methods for systems genomic integration. He has over 100+ publications in peer-reviewed journals and has edited four books. Having traveled to over 90 countries, he loves mentoring students, and he founded Bioclues.org, India's largest bioinformatics society, in 2005.

Jeffrey W. Bizzaro is a bioinformatician with over 25 years of experience in the field. He co-founded Bioinformatics.org, a professional network that advocates for open science and provides resources to students and professionals. He holds a bachelor's degree from the University of Massachusetts Lowell and a master's degree from Boston College. His research has focused on the quantitative analysis of DNA sequences and on comparative genomics.

Contributors

Ankit Agrawal
Institute for Systems Immunology
Universität Würzburg
Germany

Vrushabh Anil Nikhade
GenePowerx
Hyderabad, India

Satheesh Babu T G
Biosensor Lab
Amrita Vishwa Vidyapeetham
Coimbatore, India

Qanita Bani Baker
Department of Computer Science
Jordan University of Science and Technology, Jordan

I Balaur
Luxembourg Centre for Systems Biomedicine (LCSB)
University of Luxembourg
Esch-sur-Alzette, Luxembourg

Mandar Bedse
BIOinformatics CLUb for Experimenting Scientists (Bioclues.org)
Hyderabad, India

Jeffrey W. Bizzaro
Bioinformatics.org
Pennsylvania, USA

Sona Charles
Bioclues.org
Hyderabad, India
and
ICAR-Indian Institute of Spices Research
Kozhikode, India

Maheswari Dasgupta
Department of Bioinformatics
University of North Bengal
Siliguri, West Bengal, India

M Galati
Luxembourg Centre for Systems Biomedicine (LCSB)
University of Luxembourg
Esch-sur-Alzette, Luxembourg

Subhajit Ghosh
Department of Bioinformatics
University of North Bengal
Siliguri, West Bengal, India

Nikhil Gunala
Department of Bioinformatics
Manipal School of Life Sciences, Manipal Academy of Higher Education
Manipal, India

Darshana Joshi
Government Polytechnic Ahmedabad
Gujarat, India

Hima Jyothi Challa
GenePowerx
Hyderabad, India

Anil Kant
Department of Biotechnology and Bioinformatics
Jaypee University of Information Technology (JUIT)
Solan, Himachal Pradesh, India

Shivakumar K
Electronics & Communication Engineering
Excel College of Technology
Namakkal, Tamil Nadu, India

Srinivas Kethavath
GenePowerx
Hyderabad, India

Punya KM
Department of Psychiatry
National Institute of Mental Health and Neuro Sciences
Bangalore, Karnataka, India

Sai Kiran KS
Department of Bioinformatics
Manipal School of Life Sciences, Manipal Academy of Higher Education
Manipal, India

Sachin Kumar
Animal Nutrition Division
ICAR-National Dairy Research Institute
Karnal, Haryana, India

Megha S Kumar
Biosensor Lab
Amrita Vishwa Vidyapeetham
Coimbatore, India

Pavithra Madamarandawala
University of Missouri-St. Louis
St. Louis, MO, USA

Sandeep Mallya
Department of Bioinformatics
Manipal School of Life Sciences, Manipal Academy of Higher Education
Manipal, India

Ahmed Moussa
Ecole Nationale Des Sciences Appliquees de Tanger
Tangier, Morocco

Sneha Murmu
Division of Agricultural Bioinformatics
ICAR-IASRI Library Avenue Road, Pusa
New Delhi, India

Ashitha S Niranjana Murthy
Genetics and Genomics Laboratory
Department of Studies in Genetics and Genomics
University of Mysore, Manasagangotri, Karnataka, India.

Kaviyapriya Muthusamy
Department of Plant Biotechnology
Centre for Plant Molecular Biology and Biotechnology
Tamil Nadu Agricultural University
Coimbatore

Saranya Nallusamy
Department of Plant Molecular Biology and Bioinformatics, Centre for Plant Molecular Biology and Biotechnology
Tamil Nadu Agricultural University
Coimbatore

Amartya Nandi
Amrita School of Biotechnology
Amrita Vishwa Vidyapeetham
Kollam, Kerala, India

S Narayanasamy
Luxembourg Centre for Systems Biomedicine (LCSB)
University of Luxembourg
Esch-sur-Alzette, Luxembourg

Kalyani Palasamudram
GenePowerx
Hyderabad, India

Bharati Pandey
Animal Biotechnology Division
ICAR-National Dairy Research Institute
Karnal, Haryana, India

Bobby Paul
Department of Bioinformatics
Manipal School of Life Sciences, Manipal Academy of Higher Education
Manipal, India

Sateesh Peri
ClearNote Health
San Mateo, California, USA

Shri Hari Prasad
Centre for Crop and Food Innovation
Murdoch University
West Australia
and
Centre for Plant Biotechnology and Molecular Biology
Kerala Agricultural University
Thrissur, Kerala, India

Adarsh Pooradan Prasannakumar
Kasturba Medical College
MAHE, Manipal, Karnataka

Adnan Raja
Department of Biotechnology and Bioinformatics
Jaypee University of Information Technology (JUIT)
Solan, Himachal Pradesh, India
and
Government Teacher at Zonal Education Office
Kishtwar, J&K, India

Shalini Rajagopal
Bioclues.org
Hyderabad, India

Kalyan Ram Uppaluri
GenePowerx
Hyderabad, India

Ranjana M Raju
Amrita School of Biotechnology
Amrita Vishwa Vidyapeetham
Clappana, Kerala, India

R. Rawal
Luxembourg Centre for Systems Biomedicine (LCSB)
University of Luxembourg
Esch-sur-Alzette, Luxembourg

V Satagopam
Luxembourg Centre for Systems Biomedicine (LCSB)
University of Luxembourg
Esch-sur-Alzette, Luxembourg

Selva Babu Selvamani
Department of Plant Molecular Biology and Bioinformatics
Centre for Plant Molecular Biology and Biotechnology
Tamil Nadu Agricultural University
Coimbatore

Upasna Selvapandian
Department of Plant Biotechnology
Centre for Plant Molecular Biology and Biotechnology
Tamil Nadu Agricultural University
Coimbatore

Madiha Natchi Samu Shihabdeen
Department of Plant Biotechnology, Centre for Plant Molecular Biology and Biotechnology
Tamil Nadu Agricultural University
Coimbatore

M Shoaib
Luxembourg Centre for Systems Biomedicine (LCSB)
University of Luxembourg
Esch-sur-Alzette, Luxembourg

Rohit Shukla
Department of Biotechnology and Bioinformatics
Jaypee University of Information Technology (JUIT)
Solan, Himachal Pradesh, India

Harpreet Singh
Bioclues.org
Hyderabad, India
Hans Raj Mahila Maha Vidyalaya
Jalandhar, Punjab, India
APBioNET.org, Singapore

Moirangthem Goutam Singh
Centre for Infectious Diseases, Biological Sciences and Technology Division
CSIR-North East Institute of Science and Technology
Jorhat, Assam, India
and
Faculty of Biological Science, Academy of Scientific and Innovative Research (AcSIR)
Ghaziabad, Uttar Pradesh, India

Sakshi Singh
Core Unit for Molecular Tumor Diagnostic (CMTD), National Center for Tumor Diseases(NCT/UCC)
Dresden, Germany
and
German Cancer Research Center (DKFZ)
Heidelberg, Germany

Tiratha Raj Singh
Department of Biotechnology and Bioinformatics
Jaypee University of Information Technology (JUIT)
Solan, Himachal Pradesh, India
and
Centre of Excellence in Healthcare Technologies and Informatics (CHETI), Department of Biotechnology and Bioinformatics
Jaypee University of Information Technology (JUIT)
Solan, Himachal Pradesh, India

Ujjwal Prathap Singh
Amrita School of Biotechnology
Amrita Vishwa Vidyapeetham
Clappana, Kerala, India

V Srinivasan
Stoke Therapeutics
Boston, Massachusetts, United States

K. Sri Manjari
GenePowerx
Hyderabad, India

Prashanth N. Suravajhala
Bioclues.org
Hyderabad, India
and
Amrita School of Biotechnology
Amrita Vishwa Vidyapeetham
Clappana, Kerala, India

Kiran K Telukunta
Bioclues.org
Hyderabad, India
and
Tarunavadaanenasaha Muktbharatonnayana Samstha (TMS) Foundation

Subarna Thakur
Department of Bioinformatics
University of North Bengal
Siliguri, West Bengal, India

Brigitte Vannier
University of Poitiers
Poitiers, France

Snijesh VP
Division of Molecular Medicine
St. John's Research Institute,
St. John's National Academy of Health Sciences
Bangalore, Karnataka, India

Romi Wahengbam
Centre for Infectious Diseases, Biological Sciences and Technology Division, CSIR-North East Institute of Science and Technology
Jorhat, Assam, India
and
Faculty of Biological Science, Academy of Scientific and Innovative Research (AcSIR)
Ghaziabad, Uttar Pradesh, India

X Wang
Luxembourg Centre for Systems Biomedicine (LCSB)
University of Luxembourg
Esch-sur-Alzette, Luxembourg

Rina Yumnam
Centre for Infectious Diseases, Biological Sciences and Technology Division, CSIR-North East Institute of Science and Technology
Jorhat, Assam, India

1 Technologies, Computations, and Data Analysis for Next-Generation Sequencing

*Adnan Raja**, *Rohit Shukla**, *Brigitte Vannier, Ahmed Moussa, Anil Kant, and Tiratha Raj Singh*

NEXT-GENERATION SEQUENCING (NGS) IN DETAIL

The first commercial NGS platform was introduced in 2005. Since then, the field has rapidly evolved, with new technologies and applications emerging regularly [1]. It has enabled researchers to study genetic variation, gene expression, epigenetic modifications, and microbiome diversity on an unprecedented scale [2,3]. Today, NGS is being used in a wide range of fields, including basic research, clinical diagnostics, personalised medicine, and environmental monitoring [4,5].

ADVANTAGES OF NGS OVER TRADITIONAL SANGER SEQUENCING

NGS has several advantages over traditional Sanger sequencing [6]:

High throughput: NGS can sequence millions to billions of DNA fragments simultaneously, generating large amounts of data in a single experiment. This makes it possible to sequence entire genomes, exomes, transcriptomes, and epigenomes in a single run.

Speed: NGS can generate results in a matter of hours to days, whereas Sanger sequencing can take weeks to months.

Cost-effectiveness: The cost per base of NGS has decreased dramatically over the past decade, making it more affordable than Sanger sequencing for many applications.

Accuracy: NGS technologies have improved accuracy compared to Sanger sequencing, reducing errors due to sample contamination or amplification bias.

Flexibility: NGS can be used for a wide range of applications, including whole-genome sequencing (WGS), targeted sequencing, transcriptome analysis, epigenetic analysis, and metagenomics.

Scalability: NGS is highly scalable, allowing for the analysis of large and complex datasets.

BRIEF HISTORY OF NGS DEVELOPMENT

The development of NGS can be traced back to the early 1990s when automated fluorescent DNA sequencing was introduced. This technology was a significant improvement over the manual Sanger sequencing method, but it was still limited in terms of speed and scalability. In 2005, the first commercial NGS platform was introduced by 454 Life Sciences. This platform used pyrosequencing technology to sequence millions of DNA fragments simultaneously, generating large amounts of data in a short amount of time. Shortly after, Illumina introduced their own platform, which used reversible terminator sequencing to generate highly accurate data. Over the next decade, several other

* These authors contributed equally to this work.

DOI: 10.1201/9781003354062-1

platforms were developed, including Ion Torrent, Pacific Biosciences (PacBio), and Oxford Nanopore Technologies (ONT). Each of these has its strengths and weaknesses, making them suitable for different applications. The continued evolution of NGS technology is expected to drive further advances in genomics research and facilitate the translation of genomics into clinical practice [6].

NGS PLATFORMS AND TECHNOLOGIES

Several types of NGS technologies are available, each possessing unique strengths and limitations. Presented in this section are some of the most frequently utilised platforms. Although more comprehensive details on these technologies may be given in subsequent chapters, we will briefly examine them in the following sections:

Illumina/Solexa Sequencing

Illumina sequencing, also known as Solexa sequencing, is a massively parallel DNA sequencing technology that uses reversible terminator chemistry to generate highly accurate and high-throughput sequencing data. With this technology, DNA fragments are first randomly fragmented into smaller pieces and then ligated to adapters, which are sequences that allow the fragments to be amplified and sequenced. The fragments are then attached to a solid support, such as a glass slide or flow cell, where they undergo amplification to create clusters of identical fragments [7].

The sequencing process involves the use of fluorescently labelled reversible terminators that are complementary to the nucleotides present in the fragments. As the nucleotides are incorporated into the growing DNA strands, the fluorescent label emits a specific colour of light. Both the colour and the intensity of this emitted light are detected and used to determine the identity of the incorporated nucleotide. Illumina sequencing generates short read lengths, typically ranging from 50 to 300 base pairs but can generate billions of reads in a single run. The high throughput and accuracy of the technology make it well suited for a wide range of applications, including whole-genome sequencing, targeted sequencing, transcriptome analysis, and epigenetic analysis. Illumina sequencing has enabled researchers to generate large amounts of high-quality data in a cost-effective and efficient manner. Its widespread adoption has facilitated many groundbreaking discoveries in fields such as clinical diagnostics, personalised medicine, metagenomics, and environmental monitoring [8].

Ion Torrent Sequencing

Ion Torrent sequencing is a type of NGS technology that uses semiconductor sequencing to detect hydrogen ions that are released during DNA synthesis. In Ion Torrent sequencing, DNA fragments are first ligated to adapters and then attached to beads, which are contained in wells on a semiconductor chip. Each well contains a single bead, and the DNA fragments on the bead undergo polymerase chain reaction (PCR) amplification to create a clonal population of identical fragments. During sequencing, nucleotides are added to the DNA strand, and if the nucleotide being added is complementary to the next base in the DNA strand, a hydrogen ion is released. This ion release is detected by a sensor in the semiconductor chip, and the data are recorded. Ion Torrent sequencing generates shorter read lengths than other NGS technologies, typically ranging from 100 to 400 base pairs but has a simpler library preparation process and can be used for targeted sequencing applications [9].

The simplicity and cost-effectiveness of Ion Torrent sequencing make it well suited for applications such as clinical sequencing, microbial genomics, and targeted sequencing [10]. The technology has been used to sequence the human genome, as well as to identify mutations in cancer cells and infectious agents. Ion Torrent sequencing has made high-quality sequencing data more accessible to researchers and clinicians, enabling advances in personalised medicine and the development of new therapies for a wide range of diseases [11].

PacBio Sequencing

PacBio sequencing is a type of NGS technology that uses single DNA molecules to generate long sequencing reads. In PacBio sequencing, DNA fragments are first randomly sheared into small pieces and then ligated to adapters for sequencing. The fragments are then loaded onto a zero-mode waveguide (ZMW) chip, which contains tiny wells that are designed to hold a single DNA molecule. The sequencing process involves the use of polymerase enzymes that incorporate fluorescently labelled nucleotides into the growing DNA strands. As the nucleotides are incorporated, the fluorescence is detected and recorded, allowing the sequence to be determined. PacBio sequencing can generate sequencing reads that are tens of kilobases in length, making it well-suited for applications such as *de novo* sequencing, genome assembly, and detecting structural variants. However, the technology has a relatively high error rate, which can be reduced by using multiple passes of the same DNA molecule to generate a consensus sequence [12,13].

PacBio sequencing has been used to sequence the genomes of many organisms, including humans, plants, and animals, and has been instrumental in uncovering the complexity of the genome, such as detecting previously hidden structural variants and identifying epigenetic modifications. PacBio sequencing offers a powerful tool for generating long sequencing reads, which can help researchers to better understand the complex biology of living organisms and improve our ability to diagnose and treat diseases [14].

Oxford Nanopore Technologies

Oxford Nanopore Technologies (ONT) sequencing is a type of NGS technology that uses nanopores to directly sequence DNA strands. In ONT, DNA is first fragmented and ligated to adapters for sequencing. The DNA fragments are then loaded onto a flow cell that contains thousands of nanopores, which are tiny channels that allow a single strand of DNA to pass through. As the DNA strand passes through the nanopore, changes in the electrical current are measured and recorded, which are caused by different nucleotides in the DNA strand. These changes in current are then used to determine the sequence of the DNA strand. ONT sequencing can generate sequencing reads that are hundreds of kilobases in length, making it well suited for applications such as *de novo* sequencing, genome assembly, and detecting structural variants. The technology can also be used for real-time analysis of RNA transcripts, DNA modifications, and protein interactions [15,16].

The portability and ease of use of ONT sequencing have enabled it to be used in many different settings, including fieldwork and point-of-care (POC) diagnostics. The technology has been used to sequence the genomes of many organisms, including humans, plants and animals, and has been instrumental in identifying new species and detecting disease-causing pathogens in the environment. ONT offers a powerful tool for generating long sequencing reads and real-time analysis of biological molecules, which can help researchers to better understand the complex biology of living organisms and improve our ability to diagnose and treat diseases [17,18].

Sequencing by Oligonucleotide Ligation and Detection (SOLiD)

The use of SOLiD sequencing has declined significantly over the years. It was based on ligation-based chemistry, which allowed for accurate base calling. The technology generated shorter read lengths than Illumina, typically ranging from 50 to 75 base pairs, but had high accuracy and could be used for targeted sequencing applications. This DNA sequencing technology was developed by Applied Biosystems (now it is part of Thermo Fisher Scientific). It was first introduced in 2006 as a promising alternative to traditional Sanger sequencing.

In the SOLiD workflow DNA was fragmented into shorter segments, and adapters were ligated to their ends. These adapters contained sequences that were complementary to a specific set of oligonucleotides. Each oligonucleotide represented a specific base or colour, resulting in a unique two-base code for each possible DNA combination. In the next step, the ligated fragments were immobilised on a glass slide or bead surface, and PCR amplification was performed to create clonal copies [9].

After amplification, the fragments were subjected to a series of ligation reactions. In each reaction, a specific fluorescently labelled oligonucleotide probe was hybridised to the DNA template. After that, fluorescence detection was performed to determine the identity of the bases. The colour signals emitted by the fluorophores were captured and analysed by specialised software to reconstruct the sequence.

SOLiD sequencing offered high accuracy and relatively long read lengths compared to other NGS technologies available at the time. However, it also had some limitations. It was relatively expensive, had higher error rates than other sequencing methods, and had a complex data analysis process. Since its introduction, SOLiD has been largely superseded by newer technologies, such as Illumina sequencing, which have become more popular due to their lower cost, higher throughput, and improved accuracy [19].

Each NGS platform has its advantages and limitations, and the choice of platform depends on the specific research question and the type of data required. Users can select the method as per the requirements of their budget, research topic, and other important and necessary accomplishments.

NGS WORKFLOWS

In general, the NGS workflow generally involves the following steps, which are described in detail as follows [20–22]:

Sample preparation: The first step in the NGS workflow is to extract DNA or RNA from the biological sample of interest. The DNA or RNA is then fragmented, and adapters are added to the ends of the fragments to prepare them for sequencing.

Library preparation: The prepared fragments are then amplified and sequenced using one of the NGS platforms. During library preparation, PCR amplification is performed to create millions of copies of the fragments.

Sequencing: The amplified fragments are loaded onto the NGS instrument, which performs the sequencing reaction. Different NGS platforms use different technologies, but they all work by detecting the sequence of nucleotides in the DNA or RNA fragments.

Data analysis: Once the sequencing is complete, the raw data are processed and analysed to generate usable results. This involves several steps, including quality control, read alignment, variant calling, and annotation. There are a variety of software tools available for each of these steps.

Interpretation: The final step in the NGS workflow is to interpret the results of the analysis in the context of the research question. This often involves comparing the results to known reference sequences or databases, as well as integrating other relevant data, such as functional annotation and pathway analysis.

Quality Control and Preprocessing of Raw Sequencing Data

Quality control and preprocessing are important steps in the analysis of NGS data. These steps help to identify and remove low-quality reads, adapter sequences, and contaminants, which can negatively impact downstream analysis [22,23]. Here are the common quality control and preprocessing steps for NGS data:

Raw data quality assessment: The first step is to check the quality of the raw data using software tools such as FastQC (https://www.bioinformatics.babraham.ac.uk/projects/fastqc/). This step involves checking the quality of the reads, the length of the reads, the presence of adapter sequences, and the overall distribution of the base quality scores.

Trimming and filtering: Low-quality reads, adapter sequences, and contaminants can be removed using trimming and filtering tools such as Trimmomatic [24] and Cutadapt (https://cutadapt.readthedocs.io/en/stable/). These tools can remove low-quality bases from the ends of reads, remove adapter sequences, and filter out reads that match known contaminants.

Removal of duplicate reads: Duplicate reads can be generated during the library preparation step, and they can lead to biased results in downstream analysis. Tools such as Picard can be used to remove duplicate reads.

Alignment to reference genome: The reads are aligned to a reference genome using alignment software such as Bowtie2 [25], BWA [26], or HISAT2 [27]. This step involves mapping the reads to the reference genome and identifying any reads that do not align or align ambiguously.

Quality control after alignment: After alignment, quality control measures can be taken to assess the quality of the alignment. These measures include the percentage of mapped reads, the distribution of reads across the genome, and the coverage depth of the reads.

Variant calling: Variant calling involves identifying single nucleotide polymorphisms (SNPs) and small insertions or deletions (indels) in the aligned reads [28]. Software tools such as GATK [29], SAMtools [30], and FreeBayes (https://github.com/freebayes/freebayes) can be used for variant calling.

Annotation and interpretation: After variant calling, the identified variants can be annotated using databases such as dbSNP (https://www.ncbi.nlm.nih.gov/snp/) and ClinVar (https://www.ncbi.nlm.nih.gov/clinvar/). The annotated variants can then be interpreted in the context of the research question.

As far as NGS data analysis is concerned, quality control and preprocessing are critical steps, as they help to ensure that the data are of high quality and suitable for downstream analysis. Once this step is completed with good accuracy, it is possible that further analyses will be of a high standard and produce relevant information.

READ MAPPING TO A REFERENCE GENOME OR *DE NOVO* ASSEMBLY: PROS AND CONS

After sequencing, the next step in NGS data analysis is to process the raw reads to obtain useful information. One key decision that researchers need to make is whether to map the reads to a reference genome or to perform *de novo* assembly, the latter of which is employed when there is no reference genome [31]. Read mapping to a reference genome involves aligning the raw reads to a known reference genome. This method is faster and sometimes more straightforward than *de novo* assembly and can be used to identify mutations, structural variations, and other genomic features. However, it relies on the quality and completeness of the reference genome, which can be a challenge for non-model organisms or highly variable genomes [32].

De novo assembly involves assembling the raw reads into contiguous sequences, called contigs, without the use of a reference genome. This method is more complex and computationally intensive than read mapping and is generally used when a reference genome is not available or is of poor quality. *De novo* assembly can be used to discover novel genes, gene structures, and regulatory elements, and can be useful for comparative genomics studies. Both read mapping and *de novo* assembly have advantages and disadvantages, and the choice of method will depend on the specific research question and the quality of the available data [33]. Here are some of the pros and cons of each method:

Read mapping:

Pros:
Faster and more straightforward than *de novo* assembly.
Can be used to identify mutations and structural variations.
Can be used for comparative genomics studies.

Cons:
Relies on the quality and completeness of the reference genome.
May miss novel genomic features or regions with high variation and may introduce bias due to mapping errors or differences between the reference genome and the sample being analysed.

De novo *assembly:*

Pros:
Can be used to discover novel genomic features and regulatory elements.
Not limited by the quality or completeness of a reference genome.
Can be used for non-model organisms or highly variable genomes.
Cons:
More complex and computationally intensive than read mapping.
Can be difficult to resolve complex genomic regions or repetitive sequences.
May result in fragmented or incomplete assemblies.

Transcriptome Analysis and Differential Gene Expression

Transcriptome analysis is the study of all RNA transcripts present in a cell or tissue at a given time. NGS technologies have allowed researchers to sequence and quantify RNA transcripts with high accuracy and sensitivity. One important application of transcriptome analysis is the identification of differentially expressed genes (DEGs), which are genes that are expressed at significantly different levels between two or more biological conditions. DEGs can provide important insights into the biological processes underlying different phenotypes, such as disease states or developmental stages [22]. Here are the general steps involved in transcriptome analysis and DEG identification:

RNA extraction: Total RNA is extracted from the cells or tissues of interest and purified to remove any contaminants before converting to a complementary DNA (cDNA).
RNA sequencing: The cDNA is quantified, and the resulting reads are preprocessed to remove any low-quality reads or adaptor sequences.
Read mapping: The preprocessed reads are mapped to a reference genome or transcriptome to identify the transcripts expressed in the sample.
Transcript quantification: The abundance of each transcript is quantified using methods such as fragments per kilobase of transcript per million mapped reads (FPKM) or transcripts per million (TPM).
Differential gene expression analysis: The abundance of each transcript is compared between different biological conditions, such as diseased versus healthy tissue or treated versus untreated cells. Various methods, such as DESeq2, EdgeR or the (now deprecated) Cufflinks-Cuffdiff, are used to identify DEGs, which are genes that are expressed at significantly different levels between the conditions.
Functional analysis: The identified DEGs are then analysed to identify the biological processes and pathways that are affected by the differential gene expression.

Transcriptome analysis and DEG identification are powerful tools for understanding the molecular basis of different biological processes, such as disease states or developmental stages. These methods can provide insights into the underlying molecular mechanisms and can identify potential targets for therapeutic intervention.

Epigenetic Analysis

Epigenetics refers to the study of changes in gene expression that do not involve changes in the DNA sequence. Such modifications can be inherited and can affect the phenotype of an organism. NGS technologies have now allowed researchers to study these modifications on a genome-wide scale, and the analyses can provide insights into the molecular mechanisms underlying different biological processes, such as disease states or developmental stages [34]. NGS can identify potential targets for therapeutic intervention and can be used to develop new epigenetic therapies [35]. The following are the general steps involved:

Chromatin immunoprecipitation (ChIP): The first step in epigenetic analysis is to isolate the chromatin from the cells of interest. ChIP is a technique used to isolate specific chromatin fragments that are bound by a particular protein or modified by a specific epigenetic mark [36].

DNA sequencing: The isolated chromatin fragments are then sequenced using NGS technologies, such as Illumina or PacBio. The resulting reads are then preprocessed to remove any low-quality reads or adapter sequences.

Read mapping: The preprocessed reads are mapped to a reference genome to identify the genomic regions that are enriched in the specific protein or epigenetic mark of interest.

Peak calling: The enriched regions are then identified using peak calling algorithms, which identify regions of the genome that are significantly enriched for the specific protein or epigenetic mark.

Differential analysis: The enrichment of the protein or epigenetic mark is compared between different biological conditions, such as diseased versus healthy tissue or treated versus untreated cells. Statistical methods such as MACS2 or SICER are used to identify differentially enriched regions.

Functional analysis: The differentially enriched regions that have been identified are then analysed to identify the biological processes and pathways that are affected.

Metagenomic Analysis

Metagenomic analysis is the study of genetic material recovered directly from environmental samples, such as soil, water, or faeces, to understand the microbial community present in the sample [37]. NGS technologies have allowed researchers to sequence and quantify microbial DNA with high accuracy and sensitivity. Metagenomic analysis can provide insights into the diversity and function of microbial communities [38]. The NGS technology comprised whole genome sequencing and amplicon sequencing (16srRNA sequencing) for the metagenomics sequencing analysis. These methods can be used to identify potential targets for biotechnological applications, such as the development of new enzymes or biofuels, and can also be used to understand the role of microbial communities in human health and disease. The following are the steps involved in metagenomic analysis:

Sample collection: The first step in metagenomic analysis is to collect the environmental sample of interest. The sample is then purified to remove any contaminants.

DNA extraction: The purified sample is then processed to extract the DNA from the microbial community present in the sample.

DNA sequencing: The extracted DNA is then sequenced using NGS technologies, such as Illumina or PacBio sequencing. The resulting reads are then preprocessed to remove any low-quality reads or adapter sequences.

Read quality control: The preprocessed reads are then subjected to quality control to identify and remove any reads that do not meet certain quality standards, such as those that contain sequencing errors or contaminants.

Assembly or read mapping: The high-quality reads are then assembled into contigs or mapped to a reference database to identify the microbial species present in the sample.

Taxonomic classification: The identified microbial species are then taxonomically classified using databases such as NCBI nr or SILVA (SSU and LSU Ref NR datasets).

Functional analysis: The genes and metabolic pathways present in the microbial community are then identified and analysed to understand the biological processes and functions present in the community.

PATHWAY AND FUNCTIONAL ENRICHMENT ANALYSIS

Pathway and functional enrichment analysis is a bioinformatics approach that aims to identify the biological pathways and functions that are overrepresented in a list of genes or proteins of interest. Pathway and functional enrichment analysis can provide valuable insights into the biological processes and molecular mechanisms involved in different biological conditions, such as disease states or drug treatments. There are some basic steps required for the annotation of biological pathways. The first step in pathway and functional enrichment analysis is to identify the genes or proteins that are expressed or enriched differentially in a specific biological condition of interest. This can be achieved through differential gene expression analysis, epigenetic analysis, or other NGS-based analyses [39]. The selected genes or proteins are then compared to those present in the pathway database to identify the pathways and functions that are overrepresented in the list [40]. Statistical methods such as Fisher's exact test or hypergeometric test are commonly used for this analysis [41].

BIOINFORMATICS TOOLS AND RESOURCES FOR NGS DATA MANAGEMENT AND ANALYSIS

Various bioinformatics tools and resources are essential for the analysis and interpretation of NGS data. The following are some commonly used tools and resources for NGS data analysis through the bioinformatics approach:

Sequence alignment tools: These tools are used to align sequencing reads to a reference genome or transcriptome. Some popular examples include Bowtie2 [25], BWA [26], Hisat2, and STAR [42].

Assembly tools: These tools are used to assemble sequencing reads into contigs or transcripts. Some popular examples include Trinity [43], SPAdes [44], and SOAPdenovo [45].

Quality control tools: These tools are used to assess the quality of sequencing reads and filter out those that are low quality. Some popular examples include FastQC (https://www.bioinformatics.babraham.ac.uk/projects/fastqc/, last accessed in Jan 2025), Trimmomatic [24], and Cutadapt (https://cutadapt.readthedocs.io/en/stable/, last accessed in Jan 2025).

Variant calling tools: These tools are used to identify genetic variants in sequencing data, such as SNPs and indels. Some popular examples include GATK [29], SAMtools [30], and FreeBayes (https://github.com/freebayes/freebayes, last accessed in Jan 2025).

Annotation databases: These databases contain information on gene function, pathway and functional enrichment analysis, and other biological features. Some popular examples include Ensembl (https://useast.ensembl.org/index.html, last accessed in Jan 2025), NCBI (https://www.ncbi.nlm.nih.gov/, last accessed in Jan 2025), and UniProt (https://www.uniprot.org/, last accessed in Jan 2025).

Gene expression analysis tools: These tools are used to identify DEGs between different biological conditions. Some popular examples include DESeq2 [46], edgeR [47], and limma [48].

Data visualisation tools: These tools are used to visualise and interpret NGS data. Some popular examples include Integrative Genomics Viewer (IGV) [49] and Circos [50].

In addition to these tools, there are also many online resources and databases available for bioinformatics analysis, such as the NCBI's GenBank database (https://www.ncbi.nlm.nih.gov/genbank/, last accessed in Jan 2025), the European Bioinformatics Institute (EBI) (https://www.ebi.ac.uk/, last accessed in Jan 2025), and the Bioconductor project (https://www.bioconductor.org/, last accessed in Jan 2025).

Popular Tools for NGS Data Analysis

Table 1.1 lists some popular tools used for NGS data analysis, categorised by domain.

TABLE 1.1
Popular Tools for NGS Data Analysis

Domain	Tool	Use	URL
Quality control and preprocessing	FastQC	Quality control tool that generates reports on the quality of sequencing reads	bioinformatics.babraham.ac.uk/projects/fastqc/
	Trimmomatic	Filtering out low-quality reads and adapter sequences	github.com/timflutre/trimmomatic
	Cutadapt	Trimming adapter sequences from sequencing reads	cutadapt.readthedocs.io/en/stable/
Read mapping and alignment	Bowtie2	Alignment tool that maps reads to a reference genome	github.com/BenLangmead/bowtie2
	BWA	Alignment tool that maps reads to a reference genome or transcriptome	github.com/lh3/bwa
	STAR	Alignment tool specifically designed for RNA sequencing data	github.com/alexdobin/STAR
Variant calling	GATK	SNP and indel calling and for genotyping	gatk.broadinstitute.org/hc/en-us
	FreeBayes	SNP and indel calling	github.com/freebayes/freebayes
	SAMtools	Variant calling and alignment manipulation	github.com/samtools/samtools
Transcriptome analysis	Cufflinks	Transcriptome assembly and differential expression analysis	github.com/cole-trapnell-lab/cufflinks
	StringTie	Transcriptome assembly and differential expression analysis	ccb.jhu.edu/software/stringtie/
	Salmon	Quantification of transcript expression levels	combine-lab.github.io/salmon/
Epigenetic analysis	Bismark	Bisulphite mapping and methylation analysis	github.com/FelixKrueger/Bismark
	MACS	Identification of ChIP-seq peaks and regions of transcription factor binding	macs3-project.github.io/MACS/
	HOMER	Motif discovery and ChIP-seq analysis	github.com/bastienwirtz/homer
Metagenomic analysis	QIIME2	16S rRNA gene analysis and microbiome profiling	qiime2.org
	Kraken2	Taxonomic classification of metagenomic reads based on sequence alignment	github.com/DerrickWood/kraken2
	Metaphlan4	Taxonomic classification of metagenomic reads based on marker clades	github.com/biobakery/MetaPhlAn
	HUMAnN3	Taxonomic and functional analysis of metagenomic data	github.com/biobakery/humann
Pathway and functional enrichment analysis	DAVID	Functional annotation and pathway analysis of gene lists	david.ncifcrf.gov
	KOBAS	Pathway analysis and identification of significantly enriched pathways	kobas.cbi.pku.edu.cn
	KEGGMapper	Web server used to find the potential pathways	genome.jp/kegg/mapper/
	MetaboAnalyst	Metabolomics data analysis and pathway analysis	metaboanalyst.ca

HEURISTIC-BASED APPROACHES FOR NGS DATA ANALYSIS

'NGS data' refers to the high-throughput DNA sequencing data generated by modern sequencing technologies. It provides a wealth of genetic information and has greatly impacted various fields, including genomics, genetics, and personalised medicine. Heuristics, in the context of NGS data analysis, are computational techniques or rules of thumb used to guide the analysis and interpretation of the data. They help researchers and analysts make informed decisions and draw meaningful conclusions from the vast amount of sequencing data. Some of the commonly employed heuristics in NGS data analysis can be described in terms of quality control, read alignment, variant calling, preprocessing of the data, and its statistical or computational inferences [51].

NGS data typically undergo quality control to ensure the reliability of the results. Heuristics, such as examining the distribution of quality scores, assessing read length, and evaluating the proportion of low-quality bases, can be employed to filter out poor-quality reads. In NGS, reads are aligned to a reference genome to identify their origin and location. Heuristics are then used to optimise this alignment process, such as by filtering out low-quality alignments, setting appropriate alignment parameters, and handling repetitive regions. Heuristics play a crucial role in identifying genetic variations or variants from NGS data. Techniques like base quality score recalibration, variant quality score recalibration, and variant filtering are applied to reduce false-positive and false-negative variant calls [52].

NGS data may require preprocessing steps like adapter trimming, removing PCR duplicates, and correcting for sequencing biases before the final analysis begins. Heuristics guide these preprocessing steps by defining the criteria for identifying and handling artefacts and biases. They are often used in the statistical analysis of NGS data, including differential expression analysis, variant association studies, and genomic region enrichment analysis. They define thresholds, test statistics, multiple testing corrections, and other parameters to make robust and reliable inferences about the data. However, as valuable as they are to NGS data analysis, heuristics are not absolute rules but can vary depending on the specific experiment, technology, and research question being asked [51]. Their application requires careful consideration, validation, and adjustment based on the specific context and data characteristics.

MACHINE LEARNING (ML) AND DEEP LEARNING (DL) IN NGS DATA ANALYSIS

ML AND NGS

ML techniques have been widely used in the context of NGS data analysis to extract valuable insights, classify samples, identify patterns, and make predictions from high-dimensional genomic data [53,54]. ML models can be trained to classify NGS samples based on their genomic features. For example, supervised learning algorithms like support vector machines (SVM), random forests (RF) [55], and artificial neural networks (ANNs) can be used to predict disease subtypes, sample types, or clinical outcomes based on genomic profiles [56].

NGS data are complex and contain many features (e.g., gene expression levels and genomic variants) compared to the number of samples. ML techniques such as feature selection and dimensionality reduction (e.g., principal component analysis [PCA] and t-Distributed Stochastic Neighbour Embedding [t-SNE]) can help in identifying the most informative features and reduce the dimensionality of the data for downstream analysis. Unsupervised learning algorithms, such as *k*-means clustering or hierarchical clustering, can group NGS samples based on their genomic similarities or patterns, allowing for the identification of subgroups or novel classes within the data [57,58].

Other ML algorithms, including hidden Markov models (HMMs) and convolutional neural networks (CNNs), can learn patterns and motifs in DNA sequences from NGS data [53], enabling the discovery of regulatory elements like transcription factor binding sites or enhancer regions [59]. ML models can also aid in variant calling and prioritisation by learning the patterns associated with true variants and distinguishing them from sequencing errors or artefacts, if any. The accuracy and efficiency can also be increased for identifying genetic variations from NGS data. ML models trained on NGS data can be used in combination with other omics data (e.g., transcriptomics, proteomics, metabolomics, etc.) to gain a more comprehensive understanding of biological systems [60]. The success of ML in NGS data analysis depends on several factors, including the quality and representativeness of the data, appropriate model selection and validation, careful feature engineering, and domain knowledge integration. However, ML models should be interpreted and validated in collaboration with biological experts to ensure meaningful and biologically relevant findings.

DL and NGS

DL is a powerful ML technique that has gained significant popularity recently in various domains, including NGS data analysis [61]. DL algorithms, specifically deep neural networks, have shown their capabilities in extracting meaningful patterns and making accurate predictions from complex and high-dimensional NGS data. DL models can be trained to directly predict genetic variants from raw NGS data, bypassing the traditional pipeline of alignment and variant calling, even with high accuracy. DL can also be used to analyse gene expression data obtained from NGS experiments, such as RNA-seq. Deep neural networks can also capture intricate relationships between gene expression profiles and various biological conditions [62].

DL models have been applied to identify regulatory elements in the genome, such as transcription factor binding sites (TFBS) and enhancer regions. These models can also learn sequence motifs and patterns associated with regulatory elements easily and enable the prediction of novel regulatory regions through NGS data analysis [63].

DL algorithms excel in analysing genomic sequences, such as DNA and RNA sequences. Models like CNNs and recurrent neural networks (RNNs) can learn sequence patterns, motifs, and higher-level representations that are informative for tasks like sequence classification, variant prioritisation, and functional annotation. DL can also be utilised to develop predictive models for clinical outcomes, treatment response, drug responses, and disease classification using NGS data. By integrating genomic, clinical, drug, and other relevant data, deep learning models can assist in personalised medicine by providing customised treatment recommendations and prognostic insights. The interpretability of DL models in the context of genomics can be challenging. However, with careful model design, training, and validation, DL has the potential to enhance the analysis and interpretation of NGS data and facilitate advancements in genomic research and precision medicine [64].

Application of DL on Different NGS Methods

Nowadays researchers are using deep learning techniques in different NGS domains to analyse and infer the meaningful pattern from the large datasets.

Variant Calling and Genomic Variant Interpretation

Variant calling, the process of identifying genetic variations from NGS data, is a fundamental step in genomic analysis. DL algorithms can be trained to predict and classify different types of genetic variants accurately. These algorithms can learn complex patterns and features directly from the raw sequencing data, potentially improving variant calling accuracy and reducing false-positive and false-negative rates. Moreover, DL models can be used for variant interpretation by integrating functional genomics data, evolutionary conservation, and disease-specific information to predict the pathogenicity of genetic variants [65].

Gene Expression Analysis

DL techniques have been employed to analyse gene expression patterns from NGS data. CNNs and RNNs have been used to predict gene expression levels, identify gene regulatory networks, and discover novel gene expression patterns. By integrating genomic data with DL models, researchers can uncover complex relationships between genetic variations and gene expression, leading to a better understanding of disease mechanisms and the development of targeted therapies [66,67].

Epigenomics

Epigenomic modifications play a crucial role in gene regulation and cellular differentiation. DL algorithms have been applied to analyse epigenomic data, including DNA methylation and histone modification patterns. These models can identify epigenetic signatures associated with specific diseases or developmental stages and provide insights into the functional impact of epigenetic changes on gene expression [68,69].

Enhancing Clinical Diagnosis and Prognosis

DL models trained on large-scale genomic and clinical data have the potential to improve diagnostic accuracy and prognostic predictions. By integrating patient-specific genomic data, clinical records, and outcomes, DL algorithms can assist in personalised medicine by identifying genetic markers associated with disease susceptibility, drug response, and treatment outcomes. These models can also predict disease progression and patient survival, aiding clinicians in making informed decisions [70].

Drug Discovery and Precision Medicine

DL has the potential to revolutionise drug discovery and precision medicine by leveraging genomic and molecular data. Such models can analyse large-scale genomic and chemical datasets to predict interactions between drug and target, identify potential drug candidates, and optimise drug design. They can accelerate the drug discovery process and enable the development of personalised therapies based on an individual's genomic profile [71,72].

Challenges and Future Directions

Despite the remarkable progress, several challenges need to be addressed for the wider adoption of DL in NGS data analysis. These include the need for large and diverse datasets, interpretability of DL models, generalisation across different populations, and ethical considerations surrounding privacy and data sharing.

Challenges and Limitations of NGS Data Analysis

Despite its impact on genomics and other fields of research, some challenges and limitations are still associated with NGS data analysis. The first major challenge is the sheer volume of data generated, which can make processing and analysis time-consuming and computationally intensive [73,74]. This can be a major obstacle for researchers who may not have access to high-performance computing resources or expertise in bioinformatics. The next challenges are sequencing biases, PCR duplicates, and base-calling errors. Therefore, it is important to carefully evaluate and filter out low-quality reads and other sources of noise before proceeding with further analyses.

NGS data analysis often relies on a reference genome for read mapping and variant calling. However, the reference genome may not accurately represent the genetic variation of the sample being analysed, which can introduce bias and lead to false positives or false negatives. As the size of NGS datasets continues to grow, the computational resources required for data analysis are becoming increasingly expensive and complex. NGS data analysis can provide a wealth of information about the genome, transcriptome, epigenome, and microbiome, but the interpretation of these data can be challenging [28]. The challenge for the researcher is to accurately identify and prioritise biologically relevant information in order to draw meaningful conclusions from their data [75].

Data integration and ethical constraints are two other considerations. Integrating data from different platforms and technologies can be difficult due to differences in data formats, sample preparation methods, and experimental conditions. NGS data analysis can raise ethical concerns related to privacy, data ownership, and informed consent. Taken as a whole, this technology presents challenges, as well as opportunities, for researchers. By addressing this, researchers can continue to harness the power of NGS to advance our understanding of the genome and other biological domains and systems [73].

FUTURE DIRECTIONS IN NGS DATA ANALYSIS AND ITS POTENTIAL IMPACT ON HEALTHCARE AND PERSONALISED MEDICINE

The field of NGS data analysis is rapidly evolving, and there are many exciting future directions to explore. Single-cell sequencing technology has the potential to shake up our understanding of cell biology by enabling the analysis of individual cells. It, too, is rapidly developing, and there is significant interest in developing computational tools to analyse the large datasets generated by these experiments. NGS technologies have traditionally generated short reads, which can make it difficult to assemble complex genomes or analyse complex transcriptomes. However, recent advances in long-read sequencing technologies, such as PacBio and ONT, have the potential to overcome this. Integration of NGS data with other types of omics data, such as proteomics and metabolomics, has the potential to provide a more comprehensive understanding of biological systems [76]. Spatial transcriptomics (ST), combining imaging and sequencing, will also provide new insights in biology [77]. The measurement of transcriptional activity with spatial position of cells is used in embryology, neurobiology, and cancer studies. Several technologies (LCM, Visium, GeoMx DSP, and CosMx, for instance) exist today for ST, and integration with proteomics is the next step [78]. As with the aforementioned technologies, storage is an issue, as single-cell resolution produces an enormous amount of data. AI based solutions will be the future of NGS and its applications will be pursued in error correction, variant calling, read alignment and expression analysis [79].

CONCLUSIONS AND FUTURE PERSPECTIVES

NGS data analysis is a complex process that requires specialised bioinformatics tools and resources to handle large amounts of data. Despite the challenges and limitations, such as the need for high computational and storage resources, NGS data analysis has led to important scientific discoveries and has already begun to transform healthcare with personalised medicine.

Future directions in NGS data analysis include the development of more efficient and accurate bioinformatics tools, integration of NGS data with other types of data, and application of the technology to other areas of research, such as environmental microbiology and metagenomics. As NGS technology will continue to improve and the cost of sequencing continue to decrease, we can expect to see even more significant impacts in the years to come.

REFERENCES

1. Qin D. Next-generation sequencing and its clinical application. Cancer Biol Med. 2019 Feb;16(1):4–10. doi: 10.20892/j.issn.2095-3941.2018.0055.
2. Gupta N, Verma VK. Next-generation sequencing and its application: empowering in public health beyond reality. Microb Technol Welf Soc. 2019 Sep;17:313–41. doi: 10.1007/978-981-13-8844-6_15.
3. Cao Y, Fanning S, Proos S, Jordan K, Srikumar S. A review on the applications of next generation sequencing technologies as applied to food-related microbiome studies. Front Microbiol. 2017;8. Accessed: Jul. 04, 2023. [Online]. Available: https://www.frontiersin.org/articles/10.3389/fmicb.2017.01829
4. Buermans HPJ, den Dunnen JT. Next generation sequencing technology: advances and applications. Biochim Biophys Acta BBA—Mol Basis Dis. 2014 Oct;1842(10):1932–41. doi: 10.1016/j.bbadis.2014.06.015.
5. Besser J, Carleton HA, Gerner-Smidt P, Lindsey RL, Trees E. Next-generation sequencing technologies and their application to the study and control of bacterial infections. Clin Microbiol Infect. 2018 Apr;24(4):335–41. doi: 10.1016/j.cmi.2017.10.013.
6. Pervez MT, Hasnain MJU, Abbas SH, Moustafa MF, Aslam N, Shah SM. A comprehensive review of performance of next-generation sequencing platforms. BioMed Res Int. 2022;2022:3457806. doi: 10.1155/2022/3457806.
7. Slatko BE, Gardner AF, Ausubel FM. Overview of next generation sequencing technologies. Curr Protoc Mol Biol. 2018 Apr;122(1):e59. doi: 10.1002/cpmb.59.
8. Hu Z, Cheng L, Wang H. The Illumina-Solexa sequencing protocol for bacterial genomes. Methods Mol Biol. 2015;1231:91–97. doi: 10.1007/978-1-4939-1720-4_6.
9. Heather JM, Chain B. The sequence of sequencers: the history of sequencing DNA. Genomics. 2016 Jan;107(1, pp):1–8. doi: 10.1016/j.ygeno.2015.11.003.
10. Merriman B, Rothberg JM, I. T. R&D Team. Progress in ion torrent semiconductor chip based sequencing. Electrophoresis. 2012;33(23):3397–417. doi: 10.1002/elps.201200424.
11. Pereira R, Oliveira J, Sousa M. Bioinformatics and computational tools for next-generation sequencing analysis in clinical genetics. J Clin Med. 2020 Jan;9(1):132. doi: 10.3390/jcm9010132.

12. Rhoads A, Au KF. PacBio sequencing and its applications. Genomics Proteomics Bioinformatics. 2015 Oct;13(5):278–89. doi: 10.1016/j.gpb.2015.08.002.
13. Athanasopoulou K, Boti MA, Adamopoulos PG, Skourou PC, Scorilas A. Third-generation sequencing: the spearhead towards the radical transformation of modern genomics. Life. 2022 Jan;12(1):30. doi: 10.3390/life12010030.
14. Marx V. Method of the year: long-read sequencing. Nat Methods. 2023 Jan;20(1):6–11. doi: 10.1038/s41592-022-01730-w.
15. MacKenzie M, Argyropoulos C. An introduction to nanopore sequencing: past, present, and future considerations. Micromachines. 2023 Feb;14(2):459. doi: 10.3390/mi14020459.
16. Lu H, Giordano F, Ning Z. Oxford Nanopore MinION sequencing and genome assembly. Genomics Proteomics Bioinformatics. 2016 Oct;14(5):265–79. doi: 10.1016/j.gpb.2016.05.004.
17. Wang Y, Yang Q, Wang Z. The evolution of nanopore sequencing. Front Genet. 2015 Jan;5:449. doi: 10.3389/fgene.2014.00449.
18. Wang Y, Zhao Y, Bollas A, Wang Y, Au KF. Nanopore sequencing technology, bioinformatics and applications. Nat Biotechnol. 2021 Nov;39(11):1348–65. doi: 10.1038/s41587-021-01108-x.
19. Anderson MW, Schrijver I. Next generation DNA sequencing and the future of genomic medicine. Genes. 2010 Jun;1(1), 38–69. doi: 10.3390/genes1010038.
20. Hess JF *et al.* Library preparation for next generation sequencing: a review of automation strategies. Biotechnol Adv. 2020 Jul;41:107537. doi: 10.1016/j.biotechadv.2020.107537.
21. Wadapurkar RM, Vyas R. Computational analysis of next generation sequencing data and its applications in clinical oncology. Inform Med Unlocked. 2018 Jan;11:75–82. doi: 10.1016/j.imu.2018.05.003.
22. Conesa A *et al.* A survey of best practices for RNA-seq data analysis. Genome Biol. 2016 Jan;17(1):13. doi: 10.1186/s13059-016-0881-8.
23. Yang IS, Kim S. Analysis of whole transcriptome sequencing data: workflow and software. Genomics Inform. 2015 Dec;13(4):119–25. doi: 10.5808/GI.2015.13.4.119.
24. Bolger AM, Lohse M, Usadel B. Trimmomatic: a flexible trimmer for Illumina sequence data. Bioinformatics. 2014 Aug;30(15):2114–20. doi: 10.1093/bioinformatics/btu170.
25. Langmead B, Salzberg SL. Fast gapped-read alignment with Bowtie 2. Nat Methods. 2012 Mar;9(4):357–59. doi: 10.1038/nmeth.1923.
26. Li H, Durbin R. Fast and accurate short read alignment with Burrows–Wheeler transform. Bioinformatics. 2009 Jul;25(14):1754–60. doi: 10.1093/bioinformatics/btp324.
27. Kim D, Langmead B, Salzberg SL. HISAT: a fast spliced aligner with low memory requirements. Nat. Methods. 2015 Apr;12(4):357–60. doi: 10.1038/nmeth.3317.
28. Koboldt DC. Best practices for variant calling in clinical sequencing. Genome Med. 2020 Oct;12(1):91. doi: 10.1186/s13073-020-00791-w.
29. McKenna A *et al.* The genome analysis toolkit: a MapReduce framework for analyzing next-generation DNA sequencing data. Genome Res. 2010 Sep;20(9):1297–303. doi: 10.1101/gr.107524.110.
30. Li H *et al.* The Sequence Alignment/Map format and SAMtools. Bioinformatics. 2009 Aug;25(16):2078–79. doi: 10.1093/bioinformatics/btp352.
31. Gondane A, Itkonen HM. Revealing the history and mystery of RNA-seq. Curr Issues Mol Biol. 2023 Feb;45(3):1860–74. doi: 10.3390/cimb45030120.
32. Li B *et al.* Evaluation of de novo transcriptome assemblies from RNA-Seq data. Genome Biol. 2014;15(12):553. doi: 10.1186/s13059-014-0553-5.
33. Raghavan V, Kraft L, Mesny F, Rigerte L. A simple guide to de novo transcriptome assembly and annotation. Brief Bioinform. 2022 Mar;23(2):bbab563. doi: 10.1093/bib/bbab563.
34. Samantara K, Shiv A, de Sousa LL, Sandhu KS, Priyadarshini P, Mohapatra SR. A comprehensive review on epigenetic mechanisms and application of epigenetic modifications for crop improvement. Environ Exp Bot. 2021 Aug;188:104479. doi: 10.1016/j.envexpbot.2021.104479.
35. Mehrmohamadi M, Sepehri MH, Nazer N, Norouzi MR. A comparative overview of epigenomic profiling methods. Front Cell Dev Biol. 2021;9. Accessed: Jul. 04, 2023. [Online]. Available: https://www.frontiersin.org/articles/10.3389/fcell.2021.714687
36. Bailey T *et al.* Practical guidelines for the comprehensive analysis of ChIP-seq data. PLOS Comput Biol. 2013 Nov;9(11):e1003326. doi: 10.1371/journal.pcbi.1003326.
37. Bharti R, Grimm DG. Current challenges and best-practice protocols for microbiome analysis. Brief Bioinform. 2021 Jan;22(1):178–93. doi: 10.1093/bib/bbz155.
38. Yang C *et al.* A review of computational tools for generating metagenome-assembled genomes from metagenomic sequencing data. Comput Struct Biotechnol J. 2021 Jan;19:6301–14. doi: 10.1016/j.csbj.2021.11.028.

39. Zhao K, Rhee SY. Interpreting omics data with pathway enrichment analysis. Trends Genet. 2023 Apr;39(4):308–19. doi: 10.1016/j.tig.2023.01.003.
40. Nguyen T-M, Shafi A, Nguyen T, Draghici S. Identifying significantly impacted pathways: a comprehensive review and assessment. Genome Biol. 2019 Oct;20:203. doi: 10.1186/s13059-019-1790-4.
41. Chicco D, Agapito G. Nine quick tips for pathway enrichment analysis. PLoS Comput Biol. 2022 Aug;18(8):e1010348. doi: 10.1371/journal.pcbi.1010348.
42. Dobin A *et al.* STAR: ultrafast universal RNA-seq aligner. Bioinformatics. 2013 Jan;29(1):15–21. doi: 10.1093/bioinformatics/bts635.
43. Grabherr MG *et al.* Trinity: reconstructing a full-length transcriptome without a genome from RNA-Seq data. Nat Biotechnol. 2011 May;29(7):644–52. doi: 10.1038/nbt.1883.
44. Bankevich A *et al.* SPAdes: a new genome assembly algorithm and its applications to single-cell sequencing. J Comput Biol. 2012 May;19(5):455–77. doi: 10.1089/cmb.2012.0021.
45. Luo R *et al.* SOAPdenovo2: an empirically improved memory-efficient short-read de novo assembler. GigaScience. 2012 Dec;1:18. doi: 10.1186/2047-217X-1-18.
46. Love MI, Huber W, Anders S. Moderated estimation of fold change and dispersion for RNA-Seq data with DESeq2. Genome Biol. 2014 Dec;15(12):550. doi: 10.1186/s13059-014-0550-8.
47. Robinson MD, McCarthy DJ, Smyth GK. edgeR: a Bioconductor package for differential expression analysis of digital gene expression data. Bioinformatics. 2010 Jan;26(1):139–40. doi: 10.1093/bioinformatics/btp616.
48. Ritchie ME *et al.* *limma* powers differential expression analyses for RNA-sequencing and microarray studies. Nucleic Acids Res. 2015 Apr;43(7):e47. doi: 10.1093/nar/gkv007.
49. Robinson JT *et al.* Integrative genomics viewer. Nat Biotechnol. 2011 Jan;29(1):24–26. doi: 10.1038/nbt.1754.
50. Krzywinski M *et al.* Circos: an information aesthetic for comparative genomics. Genome Res. 2009 Sep;19(9)1639–45. doi: 10.1101/gr.092759.109.
51. Hedtke I, Lemnian I, Grosse I, Muller-Hannemann M. Optimal block-based trimming for next generation sequencing. IEEE/ACM Trans Comput Biol Bioinform. 2018;15(2):364–76. doi: 10.1109/TCBB.2017.2696525.
52. Ohanian M, Otway R, Fatkin D. Heuristic methods for finding pathogenic variants in Gene coding sequences. J Am Heart Assoc. 2012 Oct;1(5):e002642. doi: 10.1161/JAHA.112.002642.
53. Yang A, Zhang W, Wang J, Yang K, Han Y, Zhang L. Review on the application of machine learning algorithms in the sequence data mining of DNA. Front Bioeng Biotechnol. 2020;8. Accessed: Jul. 04, 2023. [Online]. Available: https://www.frontiersin.org/articles/10.3389/fbioe.2020.01032
54. Yadav AK, Shukla R, Singh TR. Machine learning in expert systems for disease diagnostics in human healthcare. in *Machine learning, big data, and IoT for medical informatics*, P. Kumar, Y. Kumar, and M. A. Tawhid, Eds., Intelligent Data-Centric Systems. Academic Press, 2021, pp. 179–200. doi: 10.1016/B978-0-12-821777-1.00022-7.
55. Pellegrino E *et al.* Machine learning random forest for predicting oncosomatic variant NGS analysis. Sci Rep. 2021 Nov;11(1):21820. doi: 10.1038/s41598-021-01253-y.
56. Jovcevska I. Next generation sequencing and machine learning technologies are painting the epigenetic portrait of glioblastoma. Front Oncol. 2020;10. Accessed: Jul. 04, 2023. [Online]. Available: https://www.frontiersin.org/articles/10.3389/fonc.2020.00798
57. Oliveira FHM, Machado ARP, Andrade AO. On the use of distributed stochastic neighbor embedding for data visualization and classification of individuals with Parkinson's disease. Comput Math Methods Med. 2018 Nov;2018: e8019232. doi: 10.1155/2018/8019232.
58. Cieslak MC, Castelfranco AM, Roncalli V, Lenz PH, Hartline DK. t-Distributed Stochastic Neighbor Embedding (t-SNE): a tool for eco-physiological transcriptomic analysis. Mar Genomics. 2020 Jun;51: 100723. doi: 10.1016/j.margen.2019.100723.
59. Gunasekaran H, Ramalakshmi K, Rex Macedo Arokiaraj A, Deepa Kanmani S, Venkatesan C, Suresh Gnana Dhas C. Analysis of DNA sequence classification using CNN and hybrid models. Comput Math Methods Med. 2021 Jul;2021:e1835056. doi: 10.1155/2021/1835056.
60. Rauschert S, Raubenheimer K, Melton PE, Huang RC. Machine learning and clinical epigenetics: a review of challenges for diagnosis and classification. Clin Epigenetics. 2020 Apr;12(1):51. doi: 10.1186/s13148-020-00842-4.
61. Shukla R, Yadav AK, Singh TR. Application of deep learning in biological big data analysis. in *Large-scale data streaming, processing, and blockchain security*, IGI Global, 2021, pp. 117–48. doi: 10.4018/978-1-7998-3444-1.ch006.

62. Schmidt B, Hildebrandt A. Deep learning in next-generation sequencing. Drug Discov. Today. 2021 Jan;26(1):173–80. doi: 10.1016/j.drudis.2020.10.002.
63. Koo PK, Ploenzke M. Deep learning for inferring transcription factor binding sites. Curr Opin Syst Biol. 2020 Feb;19:16–23. doi: 10.1016/j.coisb.2020.04.001.
64. Zheng A, Lamkin M, Zhao H, Wu C, Su H, Gymrek M. Deep neural networks identify sequence context features predictive of transcription factor binding. Nat Mach Intell. 2021 Feb;3(2):172–80. doi: 10.1038/s42256-020-00282-y.
65. Jo T, Nho K, Bice P, Saykin AJ, For The Alzheimer's Disease Neuroimaging Initiative. Deep learning-based identification of genetic variants: application to Alzheimer's disease classification. Brief Bioinform. 2022 Mar;23(2):bbac022. doi: 10.1093/bib/bbac022.
66. Li Z, Gao E, Zhou J, Han W, Xu X, Gao X. Applications of deep learning in understanding gene regulation. Cell Rep Methods. 2023 Jan;3(1):100384. doi: 10.1016/j.crmeth.2022.100384.
67. Shen X, Jiang C, Wen Y, Li C, Lu Q. A brief review on deep learning applications in genomic studies. Front Syst Biol. 2022;2. Accessed: Jul. 04, 2023. [Online]. Available: https://www.frontiersin.org/articles/10.3389/fsysb.2022.877717
68. Brasil S *et al.* Artificial intelligence in epigenetic studies: shedding light on rare diseases. Front Mol Biosci. 2021;8. Accessed: Jul. 04, 2023. [Online]. Available: https://www.frontiersin.org/articles/10.3389/fmolb.2021.648012
69. Wang Y *et al.* A deep learning approach to automate whole-genome prediction of diverse epigenomic modifications in plants. New Phytol., vol. 232, no. 2, pp. 880–97, 2021, doi: 10.1111/nph.17630.
70. Yang S, Zhu F, Ling X, Liu Q, Zhao P. Intelligent health care: applications of deep learning in computational medicine. Front Genet. 2021;12. Accessed: Jul. 04, 2023. [Online]. Available: https://www.frontiersin.org/articles/10.3389/fgene.2021.607471
71. Jiménez-Luna J, Grisoni F, Schneider G. Drug discovery with explainable artificial intelligence. Nat Mach Intell. 2020 Oct;2(10):573–84. doi: 10.1038/s42256-020-00236-4.
72. Askr H, Elgeldawi E, Aboul Ella H, Elshaier YAMM, Gomaa MM, Hassanien AE. Deep learning in drug discovery: an integrative review and future challenges. Artif Intell Rev. 2023 Jul;56(7):5975–6037. doi: 10.1007/s10462-022-10306-1.
73. Martinez-Martin N, Magnus D. Privacy and ethical challenges in next-generation sequencing. Expert Rev Precis Med Drug Dev. 2019;4(2):95–104. doi: 10.1080/23808993.2019.1599685.
74. Luthra R, Chen H, Roy-Chowdhuri S, Singh RR. Next-generation sequencing in clinical molecular diagnostics of cancer: advantages and challenges. Cancers. 2015 Oct;7(4):2023–36. doi: 10.3390/cancers7040874.
75. McInnes G *et al.* Opportunities and challenges for the computational interpretation of rare variation in clinically important genes. Am J Hum Genet. 2021 Apr;108(4):535–48. doi: 10.1016/j.ajhg.2021.03.003.
76. Haynes E, Jimenez E, Pardo MA, Helyar SJ. The future of NGS (Next Generation Sequencing) analysis in testing food authenticity. Food Control. 2019 Jul;101:134–43. doi: 10.1016/j.foodcont.2019.02.010.
77. Walker BL, Cang Z, Ren H, Bourgain-Chang E, Nie Q. Deciphering tissue structure and function using spatial transcriptomics. Commun Biol. 2022 Mar;5(1):220. doi: 10.1038/s42003-022-03175-5.
78. Moses L, Pachter L. Museum of spatial transcriptomics. Nat Methods. 2022 May;19(5):534–46. doi: 10.1038/s41592-022-01409-2.
79. Dias R, Torkamani A. Artificial intelligence in clinical and genomic diagnostics. Genome Med. 2019 Nov;11(1):70. doi: 10.1186/s13073-019-0689-8.

2 Epigenetics

RNA-seq, ChIP-seq, ATAC-seq, and MeDIP-seq

Selva Babu Selvamani, Upasna Selvapandian, Madiha Natchi Samu Shihabdeen, Kaviyapriya Muthusamy, and Saranya Nallusamy

INTRODUCTION

A complex biological system in an organism is interconnected through the interacting macromolecules (DNA, RNA, and proteins) which plays a role in phenotypic changes and brings about biochemical changes (Bersanelli et al. 2016). Regulators like histone modification, chromosomal accessibility, non-coding RNAs, and nucleic acid-protein interactions are known factors for controlling gene expressions, which directly curbs the disease phenotypes. Understanding these regulators along with gene expression may help unravel the underlying mechanism and decipher the biomarkers associated with the disease and traits. Rapid evolution in the field of bioinformatics has strengthened the analysis of next generation data generated from different omics experiments, viz., genomics (epigenomics), transcriptomics (mRNA, miRNA, and lncRNA), proteomics, and metabolomics (O'Connor et al. 2023). Parallelisation and advancement in computational resources have paved the way for the development of high-end tools and software required for omics data analysis. The regulation of genes by environmental and behavioural factors is known as 'epigenetic effect', i.e., the methylation of DNA without mutation (Handy, Castro, and Loscalzo 2011). In addition, epigenetic regulation also controls gene expression by the adjacent chromosomal organisation (Jaenisch and Bird 2003).

The transcriptome analysis era began with the first published cDNA library of silkworm moths and then progressed to two major techniques: microarray (hybridisation-based technique) analysis and RNA-sequencing (RNA-seq) (a high-throughput sequencing-based technique; Lowe et al. 2017). Transcriptome analysis has evolved from bulk RNA sequencing to single-cell techniques and, more recently, to spatial transcriptomics. The primary goal of transcriptome sequencing is the quantification of gene expression under different conditions. Following the development of high-throughput RNA-seq, the applications of transcriptomics have expanded to identify splice variants, novel variants (insertion/deletions and single nucleotide polymorphisms [SNPs]), gene fusion, novel genes, and isoforms. The use of pipelines and tools can vary based on the experimental design and objective of the study.

Chromatin immunoprecipitation sequencing (ChIP-seq) analysis is an important epigenomic method that reveals the correlation between the epigenetic landscape and various biological processes, including disease, development, and cell identity. It identifies DNA-protein interactions, revealing the genome-wide binding sites for a specific protein of interest (Nakato and Sakata 2021). Chromatin accessibility is a factor that regulates gene expression. Accessible regions account for about 3% of the total genome, and these regions contain approximately 90% of known transcription factor binding sites (TFBSes) (Sahinyan et al. 2022). The variation in accessibility is an essential component of disease and phenotypical changes. Advancements in next-generation sequencing (NGS) have led to the development of a novel approach called assay for transposase-accessible chromatin with sequencing (ATAC-seq). This technique helps document chromatin accessibility

DOI: 10.1201/9781003354062-2

regions across different conditions or cell types. ATAC-seq relies on simultaneously fragmenting Tn5-accessible regions of chromatin and ligating sequencing adapters to these fragments.

Methylation at the fifth position of the pyrimidine ring of cytosine (5mC) and 5-hydroxymethyl-cytosine (5hmC) are epigenetic modifications that can modulate gene expression (Shi et al. 2017). Methylated DNA immunoprecipitation sequencing (MeDIP-seq) (for 5mC) and hMeDIP-seq (for 5hmC) are methods used to study these modifications. These techniques use specific antibodies to immunoprecipitate DNA fragments containing the modification of interest, followed by sequencing of the immunoprecipitated DNA.

To date, there is no detailed description of epigenetics analysis using different bioinformatics methods. In this chapter, we have critically reviewed and listed the standard operating procedures, tools, and databases that can be used for this.

RNA-SEQ

Experimental Design and Transcriptome Sequencing

Conventional transcriptome sequencing starts with the planning of experiments (experimental design), extraction of RNA, library preparation, sequencing, and data analysis and ends with validation. Precise estimation of gene expression between the different conditions relies on good experimental design because poor design leads to higher false positives. When designing a transcriptomics experiment, three main components should be considered: biological, technical, and statistical methods (Bruning et al. 2014). The technical aspects include the choice of sequencing platform and sequencing method, such as single-end or paired-end, and long reads or short reads. Another important technical consideration is the sequencing depth, which refers to the number of reads in millions. The selection of these technical parameters depends on various factors, including the availability of a reference genome for the organism being studied. Each of these choices can significantly impact the quality and quantity of data obtained, as well as the subsequent analysis and interpretation of results. For example, 20 million reads in the paired-end sequencing layout is sufficient for gene expression analysis in *Oryza sativa,* but for non-model organism, depth of sequencing will increase the reliability of gene expression results and high-quality reference transcriptome construction helps to estimate the expression of the transcripts (even the transcripts with low expression). Higher genome size is directly correlated with the complexity of transcriptome analysis. To overcome this, a higher depth of sequencing is required.

The experimental design for transcriptome sequencing must account for various biological aspects. These include time series, which involve sampling at different time intervals; experimental conditions, such as comparing treated or infected samples with controls; developmental stages, which may be based on age or growth phase; and the specific tissue types being studied. Alongside these biological considerations, statistical aspects play a crucial role in experimental design. Researchers need to carefully consider methods for error estimation and control, formulate clear hypotheses, and determine appropriate significance testing approaches. Reducing the deviation in biological and technical components will help to minimise the error in the experiment. Further, successful planning of experiments will lay the foundation for a better understanding of key genes and their expression. A complete workflow of transcriptomics is illustrated in Figure 2.1.

RNA Extraction and Library Preparation

After extracting total RNA from samples, the initial quality check involves assessing 28S and 18S rRNA bands using gel electrophoresis, followed by spectrophotometric analysis with a Nanodrop instrument to measure 260/280 and 260/230 absorbance ratios, which indicate protein contamination levels and RNA quantity, respectively. To further assess RNA quality, the RNA Integrity Number (RIN) is determined, e.g., using an Agilent Bioanalyzer. The RIN scale ranges from 1 to 10, where 1 indicates highly degraded RNA and 10 represents RNA with minimal degradation. For

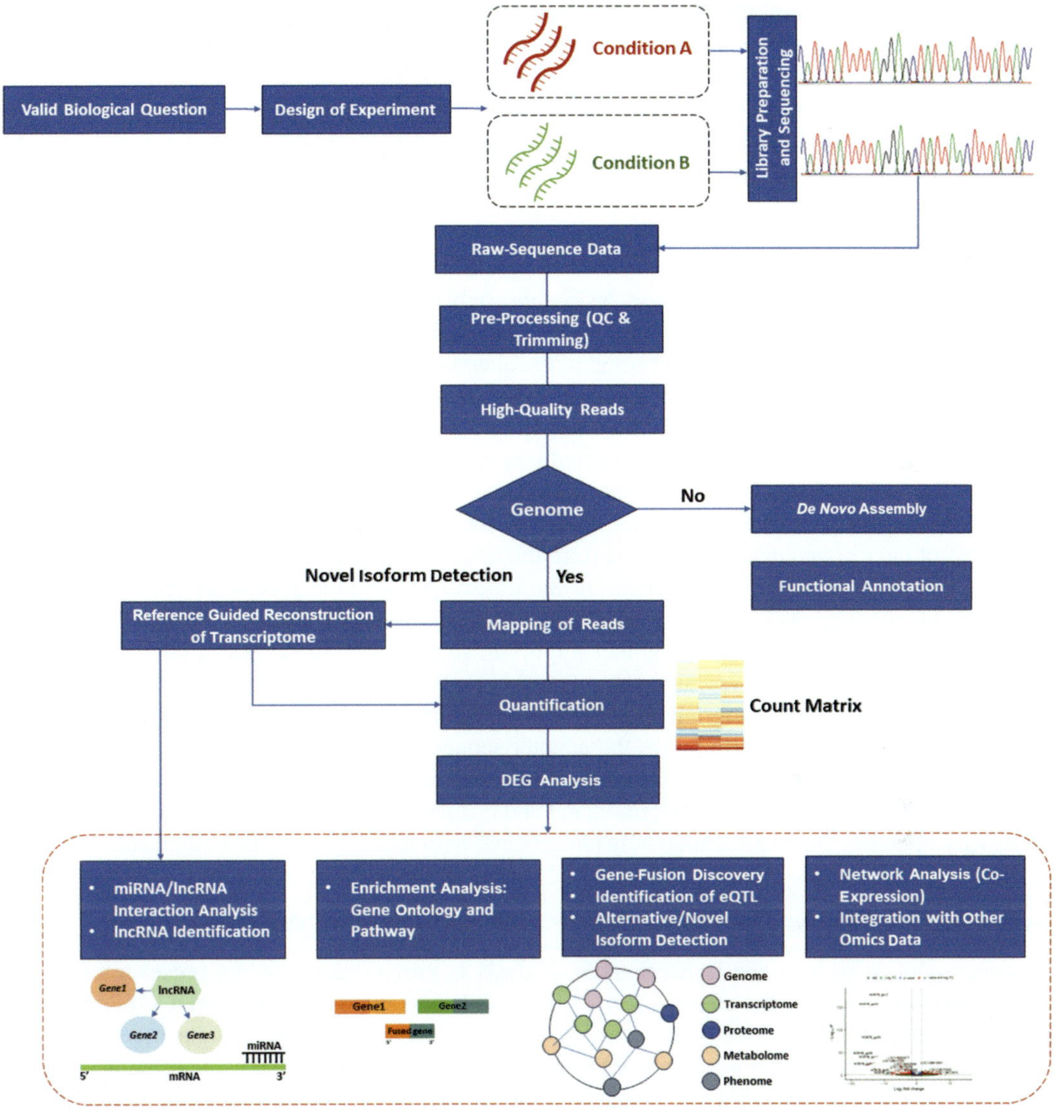

FIGURE 2.1 Illustration of RNA-seq analysis workflow widely used for the identification of DEGS, novel isoforms, lncRNAs, and integration of multi-omics data.

reliable library preparation and sequencing, a RIN value above 7 is typically considered indicative of high-quality RNA with minimal degradation. RNA samples with RIN values below 7 are generally deemed suboptimal, as using such samples can lead to low-confidence sequencing reads and may result in inappropriate conclusions, including false-positive results, during data analysis. Total RNA extracted from the samples consists of more than 90% of ribosomal RNA (rRNA), which must be depleted before library preparation to avoid contamination of other RNAs and to increase the depth of mRNA reads. Many transcriptome sequencing protocols include a step to select mRNAs with polyadenylated (poly-A) tails, which effectively excludes rRNAs from the sequencing process. The primary steps in library preparation are (1) conversion of RNA to cDNA, (2) fragments/size selection, and (3) adapter ligation. The choice of library preparation method depends on the specific aims of the experiment. For studies targeting the whole transcriptome (excluding rRNA) to identify long non-coding RNAs, splice variants, and novel isoforms, a whole transcriptome library should be prepared. In this case, sequencing should be performed using a paired-end layout with longer read lengths, typically ranging from 100 to 300 bp. This approach provides more comprehensive coverage of the transcriptome and allows for better detection and characterisation of complex RNA features. If gene expression studies are the goal of the experiment, then a single-end sequencing layout and mRNA-only library preparation method can be adopted (Li and Wang 2021). A minimum of 100 to 200 bp length reads must be generated for the transcriptome analysis to overcome high repetition regions in mapping; failing to do so will result in false mapping.

Preprocessing

The crucial step in RNA-seq or any other NGS data analysis is preprocessing, which includes assessment of raw data quality, adapter contamination, and trimming of low-quality bases and adapters. RNA-seq data produced from high-throughput sequencing instruments will be in FASTQ format. Base quality, adapter contamination, guanine-cytosine (GC) content, overrepresented sequences, and duplicated reads are the critical parameters to be assessed in quality checking. FastQC (https://www.bioinformatics.babraham.ac.uk/projects/fastqc/; accessed Sep 2024) and HTQC (Yang et al. 2013) are the tools that can be used for the assessment of quality in FASTQ files.

Adapters are short external sequences ligated to the cDNA fragments that do not belong to the organisms of interest and must be removed from the raw data, or else they will be problematic in read mapping to the genome or in construction of the transcriptome. This process of removing external sequences is called adapter trimming, where the specific sequences attached during library preparation are masked and eliminated. Another important step in preprocessing is quality trimming, in which the low-quality bases (typically with a Phred score below 30) are removed from the reads. This reduces mismatches during alignment and improves the efficiency of read alignment. Most of the sequencers currently employed remove the adapter sequences by default. Tools commonly used for quality and adapter trimming are Trimmomatic (Bolger, Lohse, and Usadel 2014), cutadapt (Martin 2011), and fastp (Chen et al. 2018). Trimmomatic and fastp are equipped with built-in Illumina adapter sequences, allowing them to automatically identify and trim these adapters from the data.

Availability of Genome Decides the Strategy of Transcriptome Analysis

Reference-guided or *de novo* transcriptome analysis is purely based on the presence of a high-quality reference genome for the organism of interest. If genome data exist, then it would be advisable to move towards reference-based analysis. The reference-based transcriptome assembly can be employed for the identification of lncRNAs and novel isoforms. However, if the reference genome is absent, the reference transcriptome must be built from RNA-seq reads, which can be further used for mapping and quantification of transcripts. For constructing *de novo* transcriptomes, a higher depth of transcriptome sequencing and a significant number of replications are essential.

Construction of Transcriptome through *De novo* Assembly

Most non-model organisms lack both genome and transcriptome information. In such cases, transcriptome should be constructed from reads that contain all the transcripts (including the splice variants) of sequenced samples. (NB: If transcriptome sequencing is performed between two or more different conditions, a reference transcriptome should be built using reads of all of the samples, or else the transcriptome can lack the novel genes or isoforms present in either of the samples). Trinity (Grabherr et al. 2011) and rnaSPAdes (Bushmanova et al. 2019) are the most used *de novo* transcriptome assemblers for transcriptome construction. Other transcriptome assemblers are described in Table 2.1. A higher percentage of similar transcripts or isoforms are clustered from the assembled transcriptome to reduce redundant transcripts using CD-HIT-EST (Raghavan et al. 2022). The mapping percentage of reads to the developed transcriptome will reflect on the assembly quality; a remapping percentage greater than 85% is considered a good-quality assembly.

TABLE 2.1
Different *de novo* Assemblers Employed for the Construction of Transcriptome

Tool	Description	Source	Reference
Trinity	The *de novo* method comprises three different steps, viz.: (1) Inchworm, (2) Chrysalis, and (3) Butterfly. Assembles each transcript based on a De Bruijn graph. Can perform both genome-guided and *de novo* assembly of the transcriptome.	github.com/trinityrnaseq/trinityrnaseq; accessed Sep 2024	(Grabherr et al. 2011)
rnaSPAdes	Tool developed by modifying the parameters of SPAdes according to the requirement of transcriptome assembly. Relies on the simplified De Bruijn graph for constructing the transcripts graph.	cab.spbu.ru/software/rnaspades/; accessed Sep 2024	(Bushmanova et al. 2019)
Velvet-Oases	Consists of two major steps: Velvet assembles the contigs using the De Bruijn graph method and further assembles transcripts forwarded to Oases, where the tool uses short reads information to construct the isoforms.	github.com/dzerbino/oases; accessed Sep 2024	(Schulz et al. 2012)
Trans-ABySS	Pipeline developed to analyse the transcriptome contigs assembled by ABySS. Reduces the redundancy in the transcriptome and collects information like splicing events, novel and retained introns, skipped and novel exons, and alternative splicing. Also, can be used for detecting gene-fusion events.	github.com/bcgsc/transabyss; accessed Sep 2024	(Birol et al. 2009)
SOAPdenovo-Trans	A tool that adopted DBG construction from SOAPdenovo and low-frequency *k*-mer removal from Trinity. Reads are mapped back to the contigs to develop the scaffold graphs, which are further used to retrieve all possible paths to generate transcripts.	github.com/aquaskyline/SOAPdenovo-Trans; accessed Sep 2024	(Xie et al. 2014)

Removal of Contaminant Sequences

The transcriptome assembly may include off-target sequences, potentially causing errors in downstream analyses. To avoid this problem, contaminant sequences have to be removed. This can be done using KRAKEN (Wood, Lu, and Langmead 2019) and Centrifuge (Kim et al. 2016), which will result in the clean, and high-quality transcriptome. Further, the completeness of the assembled transcriptome can be assessed using Benchmark Universal Single-Copy Orthologs (BUSCO) analysis (Simão et al. 2015). A BUSCO value greater than 80% generally indicates that the assembly is relatively complete. BUSCO analysis gives complete, missing, and fragmented genes based on orthologous analysis; the percentage of fragmentation helps to assess the amount of fragmentation in constructed transcriptomes. In this case, an alternate tool or alternate *k*-mer value can be used to reconstruct the transcriptome.

Indexing and Alignment

The genome or transcriptome should be indexed before proceeding with the alignment. This will make the process faster and allow the aligners to find matches with the genome. Each aligner uses its own method for indexing the genome or transcriptome. In simple terms, the genome index is like the index of a book, which helps for easy and fast access to the content (or match, in the case of aligning).

In both reference-based and *de novo* transcriptome analysis, aligning the reads to the reference genome or reference transcriptome will determine the accuracy in the quantification of gene expression and identification of novel genes or splice variants. Aligning the reads is done using splice-aware

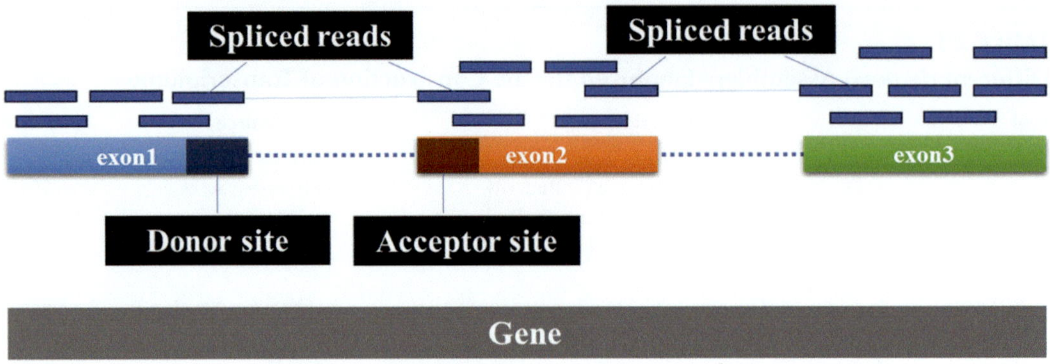

FIGURE 2.2 Illustration of mapping performed by splice-ware aligners.

or ordinary end-to-end aligners (splice-unaware aligners). Splice-aware aligners will be useful if RNA-seq reads are mapped to the reference genome; otherwise, for mapping reads to the transcriptome, end-to-end genome aligners can be utilised. These splice-aware aligners map the reads across the splice junctions which may provide gaps in the alignment (Figure 2.2). Similarly, two methodologies are employed in the alignment of transcriptome reads, viz., pseudo aligners and traditional aligners. Pseudo aligners perform rapid mapping by using *k*-mer-based algorithms, but traditional aligners carry out the seed (short sequence in the read) matching with the target sequence (genome); upon matching aligners, the seed extension procedure is followed. Choosing the aligners relies on the experimental prerequisites. If the genome is available (the reference-based approach), splice-aware aligners may be preferred. Alternatively, if the genome is not available, a reference transcriptome built using RNA-seq reads will be used as the reference for the mapping, where both splice-aware or splice-unaware aligners can be used for mapping reads. For RNA-seq analysis in prokaryotes like bacteria, genome aligners can be used for mapping the RNA-seq reads with the genome, where the chances of splicing are rare. If the experiment aims to identify novel genes, RNA-seq variants, or detection of fusion, traditional aligners are used for mapping. Pseudo aligners are a great choice for the quantification of a known transcript with higher speed and accuracy. Table 2.2 lists software used in the alignment of reads to reference. High-quality mapping is achieved with a minimum mapping

TABLE 2.2
List of Aligners and Their Alignment Methodology

Tool	Method	URL	Reference
HISAT2		daehwankimlab.github.io/hisat2/download/; accessed Sep 2024	(Kim et al. 2019)
Tophat2	Splice-aware and traditional alignment	ccb.jhu.edu/software/tophat/index.shtml; accessed Sep 2024	(Kim et al. 2013)
STAR		github.com/alexdobin/STAR; accessed Sep 2024	(Dobin et al. 2013)
Bowtie2	Splice-unaware, ungapped aligners. These tools are used for the prokaryotes, which have rare splicing events	bowtie-bio.sourceforge.net/bowtie2/index.shtml; accessed Sep 2024	(Langmead and Salzberg 2012)
BWA		github.com/lh3/bwa; accessed Sep 2024	(Li and Durbin 2009b)
Kallisto	Pseudo aligners, perform alignment and quantification of transcripts	pachterlab.github.io/kallisto/download; accessed Sep 2024	(Bray et al. 2016)
Salmon		combine-lab.github.io/salmon/getting_started/; accessed Sep 2024	(Patro, Duggal, and Kingsford 2015)

of 80%; a lower percentage indicates adapter contamination or low sequencing quality. Qualimap (García-Alcalde et al. 2012), RseQC (Wang, Wang, and Li 2012), and Picard (https://broadinstitute.github.io/picard/; accessed Sep 2024) can be utilised for ensuring the quality of mapped reads. These tools can estimate the GC content, which helps to unravel the PCR bias in sequencing and identifies nucleotide composition bias, depth of sequencing, strand specificity of data, and read distribution. Further mapping, strand specificity, and spliced reads can be visualised using IgV (Integrated Genomics Viewer) (Thorvaldsdóttir, Robinson, and Mesirov 2013).

Gene and Transcript Quantification

Different kinds of gene expression quantifiers are available and are also in development. Quantification is done at the gene and transcript level. In transcript-level quantification, all isoforms of a gene (both major and minor) are individually measured, while in gene-level quantification, counts from all transcripts of a gene are combined into a single value for that gene. For organisms with well-established, high-quality reference genomes, most experiments focus on gene-level quantification. In contrast, for organisms lacking a complete genome sequence or comprehensive annotation, transcript-level (isoform) quantification is generally preferred. Two methods, namely alignment-based and alignment-free methodologies, are followed for quantification. Alignment-based quantification typically involves mapping reads to genes or transcripts and then counting the number of reads aligned to each gene. To reduce the time-consuming mapping process, alignment-free quantification methods were developed as an alternative approach. These methods divide reads into shorter sequences called *k*-mers, map only the unique *k*-mers to the genome or transcriptome, and then use this information to quantify gene expression. Alignment-based and alignment-free quantification tools are given in Table 2.3. Accurate quantification of genes based on the number of reads mapped depends on the gene length, GC content, percentage of alignment, and other parameters like batch effect (Babarinde, Li, and Hutchins 2019).

Normalisation of Count Values

Raw count data must be normalised to achieve accurate expression levels and reduce false-positive results. This normalisation process addresses several issues: batch effects, differences in sequencing depth between samples, and variations in gene length within a sample. This step directly correlates with the identification of true differentially expressed genes (DEGs). The different normalisation methods used in RNA-seq are normalisation by library size, by read distribution, and by controls.

TABLE 2.3
Tools Used for the Quantification of Transcripts/Genes

Tool	Method	Reference
RSEM	Assign reads to isoforms in iteration utilising the expectation-maximisation (EM) algorithm	(Li and Dewey 2011)
Kallisto	Performs pseudo-alignment by using a De Bruijn graph to quantify the transcripts	(Bray et al. 2016)
Salmon	Pseudo-alignment and quantification	(Patro, Duggal, and Kingsford 2015)
htseq-count	Traditional union-exon-based methods	(Anders, Pyl, and Huber 2015)
featureCounts		(Liao, Smyth, and Shi 2014)
eXpress	The convergence speed of EM improved by utilising the online EM method, also quite a popular tool in recent days	(Forster et al. 2013)
Cufflinks	Collection of multiple packages that help in the identification of novel transcripts and perform differential expression analysis	(Trapnell et al. 2012)

Fragments per kilobase of transcript per million mapped reads (FPKM) and reads per kilobase per million mapped reads (RPKM) are methods used to normalise gene expression based on gene length. RPKM is used for single-end library samples, while FPKM is used for paired-end samples. In these methods, the total read count is first divided by one million (the scaling factor), then further divided by the length of the gene in kilobases (kb), resulting in a normalised value (per kb million). Transcripts per million (TPM) is another popular and reliable method of normalisation employed for RNA-seq analysis. This method is quite like RPKM and FPKM and paves the way for normalisation across samples. It is also the most reliable method of normalisation (Zhao et al. 2019). In addition to this, trimmed mean of m-values (TMM), UpperQuartile, DESeq, and PoissonSeq can be used for reducing the highly variable/highly expressed features in a sample. DESeq and TMM are the most reliable methods for the identification of DEGs and also for reducing false positives (Dillies et al. 2013).

Differential Gene Expression (DGE) Analysis

DGE analysis is used for the identification of DEGs under different conditions or different cell types from the count matrix (produced after normalisation). Commonly used tools, viz., edgeR (Robinson, McCarthy, and Smyth 2010), baySeq (Hardcastle and Kelly 2010), and DESeq2 (Love, Huber, and Anders 2014), have been developed, using the R language, for differential expression analysis. Normalised read counts are not necessary for differential expression analysis using these tools. The packages perform normalisation by their approach and follow the negative binomial distribution. Some more packages are available for DGE analysis, viz., NOISeq (Tarazona et al. 2011), DEGseq (Wang, Feng, et al. 2010), baySeq (Hardcastle and Kelly 2010), and EBSeq (Leng et al. 2013). When the sample size is small (without replicates or with a smaller number of replicates), DESeq2 and edgeR are poor choices for analysis. In this case, NOISeq or DEGseq can be used to reduce the noise produced by DEScq2 or edgeR. NOIseq and DEGseq use different statistical approaches for testing in smaller sample sizes: NOIseq follows an empirical distribution, while DEGseq uses a Poisson distribution (Conesa et al. 2016). NOIseq can simulate replicate values for samples that do not have replicates, which helps to avoid inaccurate *P*-value calculations. Generally, DEGs with a *P*-value less than 0.05 are considered to be significant. Additionally, the log2foldchange values filter can be used as a filter based on the specific interests and conditions of the experiment.

Functional Annotation of DEGs

Annotation of the DEG for the pathways, gene ontology, functions, and interaction with epigenetic regulators like miRNAs and lncRNAs can be retrieved from the wide variety of databases available for model organisms, viz., DAVID (Dennis et al. 2003), enrichR (Kuleshov et al. 2016), Ensembl (Cunningham et al. 2022), gene ontology (Consortium 2004), RAP-DB (Sakai et al. 2013), TAIR (Swarbreck et al. 2007), miRWalk (Sticht et al. 2018), RNAcentral (2019), KEGG (Kanehisa 2002), and Reactome (Fabregat et al. 2018). Transcriptome analysis of non-model organisms requires annotating individual transcripts to determine their functions by comparing sequence similarities across databases, viz., nr, UniProt, KEGG, GO, COG, InterPro, etc. (Hunter et al. 2009). This functional annotation helps in identifying the function of DEGs/transcripts obtained from the *de novo*–based transcriptome analysis. Diamond (Buchfink, Xie, and Huson 2015), BLAST+ (Camacho et al. 2009), Interproscan (Quevillon et al. 2005), eggNOG mapper (Huerta-Cepas et al. 2017), Omicsbox (Bioinformatics and Valencia 2019), and signalP (Teufel et al. 2022) are some of the tools that can be used for annotating the transcripts/genes in *de novo* transcriptome analysis. mirBase (Kozomara, Birgaoanu, and Griffiths-Jones 2019) is an essential database for the annotation of miRNAs. Gene and pathway enrichment analysis is yet another analysis used in transcriptome analysis to identify the overrepresented and underrepresented functions and pathways by DEGs. Gene set enrichment analysis (GSEA) and Fisher's exact test are commonly used for enrichment analysis. gProfiler

(Raudvere et al. 2019), clusterProfiler (Yu et al. 2012), KOBAS (Xie et al. 2011), and EnrichmentMap (Isserlin et al. 2014) are popular tools/web servers used for enrichment analysis.

Gene Co-Expression Network

The primary goal of gene co-expression network analysis is to identify genes with highly correlated expression patterns to better understand their functions and biological relationships (García-Ruiz et al. 2021). The systems biology method known as weighted gene co-expression network analysis (WGCNA) defines the correlation pattern of the genes in microarray or RNA-seq data. This method detects highly correlated gene clusters to identify intramodular hub genes, which are then associated with sample traits (features that are associated with the samples). WGCNA has already found applications in the field of cancer research, plant genetics, biomarker discovery, and the identification of hub genes for various traits. Normalised gene expression data are the source for the identification of hub (modular) genes. The WGCNA package, developed in R, performs weighted correlation network analysis (Langfelder and Horvath 2008). This co-expression-based network analysis requires coding skills along with expertise in bioinformatics, and web servers and packages have been developed for it, e.g., the gene whole co-expression network analysis (GWENA) (Lemoine et al. 2021) and CoExp (García-Ruiz et al. 2021). GWENA is also an R package, and it was developed as a one-stop solution for correlation analysis that includes the identification of modules, enrichment analysis, and phenotype association. Similarly, there is CoExp, a web-based tool for correlation analysis. These packages help in expanding the understanding of functions and fundamental processes in diseases or phenotypes.

An Expanded Power of RNA-seq Data

RNA-seq results serve as fundamental resources that can be integrated with various omics data, such as proteomics, genomics, ChIP-seq, ATAC-seq, whole genome bisulphite sequencing (WGBS) (methylation), and metabolomics. Additionally, interacting regulators like miRNAs, lncRNAs, and siRNAs of the genes can also be identified from RNA-seq data. Bioinformatics methods and pipelines have been developed to identify lncRNAs and miRNAs from RNA-seq and deep sequencing data. Cufflinks-Cuffdiff (Trapnell et al. 2012) and StringTie (Pertea et al. 2015) are well-known tools for the discovery of novel transcripts in the RNA-seq data. Gene fusion discovery is also an important analysis that can be performed using RNA-seq data; it helps to identify fused genes that are a result of chromosomal rearrangement.

CHIP-SEQ

ChIP-seq involves analysing DNA through high-throughput sequencing. In a ChIP assay, cells are cross-linked with proteins using chemical agents like formaldehyde. This is followed by cell disruption via sonication or enzymatic digestion to obtain chromatin fragments of 100–300 bp (Barski et al. 2007; Johnson et al. 2007; Ren et al. 2000). The DNA bound to the protein of interest, such as transcription factors (TFs), RNA polymerase, or modified histones, is enriched using specific antibodies through immunoprecipitation. Finally, the enriched DNA is purified by reversing the cross-links and then sequenced.

The key steps in ChIP-seq involve (i) quality control, (ii) mapping (iii), peak detection (iv), motif analysis, (v) Annotation, and (vi) visualisation.

QUALITY CONTROL

Quality control of the sequencing reads is the first step in ChIP-seq analysis, and it involves assessing the quality of the raw reads. FastQC is a commonly used tool to ensure that the data are of high quality and suitable for downstream analysis. Low-quality bases, adapter sequences, and reads with low mapping quality are trimmed using Cutadapt (Martin 2011) or Trimmomatic (Bolger, Lohse, and Usadel 2014).

Some of the important metrics of ChIP-seq, as suggested by the ENCODE consortium, include:

1. The mapping ratio, the read quality, and the percentage of reads sequenced from true genomic DNA. Generally, the mapping ratio of the samples sequenced using the Illumina HiSeq platform should be above 80%. However, the mapping ratio of non-DNA binding proteins, e.g., IgG, will be as low as 60%.
2. The number of non-redundant mapped reads (the read depth) depends on the genome size and the antibody signal (S) to noise (N) ratio. A minimum of 10 million uniquely mapped reads are required to analyse the sharp-mode peaks for human samples, as suggested by the ENCODE consortium. Weaker S/N ratios are often detected in broad histone marks and hence require more reads (>40 million reads) for peak calling (Figure 2.1).
3. Library complexity ranges from 0 to 1. Generally, a complexity of >0.8 is required for 10 million mapped reads, as suggested by the ENCODE consortium. Excessive PCR amplification from initial DNA leads to a lower complexity value (<0.6).
4. The normalised strand coefficient (NSC) obtained by the strand shift coefficient (SSC) profile (Anzawa, Yamagata, and Kinoshita 2020) is an S/N indicator with a recommended threshold of NSC >5.0 for sharp peaks and NSC <1.5 for broad peaks using the ChIP-seq public database for in-depth validation on multiple species. The input samples should have low S/N, and hence the NSC values should be <2. 0.
5. Background uniformity (Bu) indicates a bias in the distribution of reads in the background. But the value ranges from 0 to 1. Lower values (<0.8) indicate a biased read distribution that results in false positives in obtained peaks.
6. GC summit bias indicates the deviation occurred during immunoprecipitation and PCR amplification. The GC summit of ChIP-seq data is like that of the reference genome (~50% for humans) and an unexpected GC-rich summit (>60%) may occur due to PCR amplification bias or false-positive peaks of 'super-enriched' regions associated with cytosine-phosphate-guanine (CpG) islands.

Mapping

The sequence reads are mapped using tools such as BWA-MEM (Li and Durbin 2009b), Bowtie2 (Langmead and Salzberg 2012), SOAP (Li et al. 2008), or MAQ (Li, Ruan, and Durbin 2008). Bowtie2 performs both global and local alignment to trim low-quality bases. In addition, it supports gapped alignment and detects indels, which is essential for long or paired-end reads. Alignment sorting and deduplication is done using SAMtools (Li et al. 2009) and Picard (http://broadinstitute.github.io/picard/; accessed Sep 2024). It is necessary to observe the uniquely mapped reads obtained from the mapper. The percentage of uniquely mapped reads varies among organisms, however a recommended value of >70% is considered normal (Bailey et al. 2013). A lower percentage (<50%) is encountered as a result of excessive PCR amplification, problems with sequencing platforms, or inadequate read length.

It is important to select the appropriate number of user-settable mismatches for the NGS platform being used. Multimapped reads occur due to the frequent binding of proteins to repetitive DNA regions. Paired-end sequencing can help reduce the ambiguity of multimapping. It is essential to remove multimapped reads because most peak calling algorithms filter or ignore them, despite their potential to discover novel binding sites (Bailey et al. 2013).

Peak Detection

The key step in ChIP-seq is to detect the binding of ChIPed protein in the genome by finding a significant number of mapped reads, i.e., peaks. There are several peak calling tools such as MACS2

(Gaspar 2018), SIPeS (Wang, Xu, et al. 2010, Epic (Stovner and Sætrom 2019), BayesPeak (Spyrou et al. 2009), jMOSAiCS (Zeng et al. 2013), T-PIC (Hower, Evans, and Pachter 2011), Enriched Domain Detector (EDD) (Lund et al. 2015), and GEM (Guo, Mahony, and Gifford 2012). It is important to remove duplicate reads to ensure peak specificity. However, no tool is 100% accurate. The called regions are ranked using absolute signal (i.e., read number) or computed significance of difference (false discovery rates [FDR] threshold or *P*-values). The specific parameter setting and the algorithm used are essential criteria to determine weak or strong ChIP signals (Landt et al. 2012). An alternative to the threshold is the irreproducible discovery rate (IDR) along with motif analysis, which helps in choosing the best peak calling algorithm and parameter setting.

To ensure the reproducibility of both sequencing reads and peak calls, it is recommended to perform at least two biological replicates of the ChIP-seq experiment. The Pearson correlation coefficient (PCC) of the (mapped) read counts at each genomic position is used to measure the reproducibility of the reads (Bardet et al. 2012). The PCC should range from 0.3 to 0.4 for unrelated samples and >0.9 for replicated samples. Low-quality samples result in low values. Artifact regions showing high ChIP signals, viz., the regions near centromeres, telomeres, satellite repeats, and the blacklisted regions as given by the ENCODE and 1000 Genomes Project, should be removed before computing the PCC. The IDR analysis determines the number of peaks that satisfy a user-specified reproducibility criterion and evaluates the rank consistency of discovered peaks between replicates. The IDR, a reproducibility-based metric, provides more consistent peak identification across replicates than enrichment-based metrics such as false discovery rate (FDR) or *P*-values (Landt et al. 2012).

In addition to peak calling, it is important to detect significant differences in chromatin states between the biological conditions used in the analysis. However, the amount of noise generated during immunoprecipitation makes it difficult to detect subtle changes between conditions (Faux et al. 2021). Several tools have been developed to overcome this (Table 2.4).

Motif Analysis

Identifying DNA binding motifs is crucial for validating ChIP-seq experiments. In cases where the expected motif is unknown, successfully identifying a common motif in a large fraction of the detected peaks indicates a well-performed experiment. It also helps in identifying the DNA binding motifs of other proteins in conjunction with the ChIPed protein, thereby revealing transcriptional regulation.

Motif analysis involves several sequential steps: (a) motif discovery, which identifies recurring DNA patterns in the ChIP-seq data; (b) comparison of these discovered motifs with known DNA

TABLE 2.4
Tools Used for Differential Peak Calling Analysis

Tool	Peak Calling Tool	Normalisation	Significance Measure	Reference
MAnorm	MACS	Genome-wide MA plot	*P*-value	(Shao et al. 2012)
QChipat	BELT, MACS	Linear normalisation	*P*-value	(Liu et al. 2013)
diffReps	Sliding window based	Linear normalisation	*P*-value	(Shen et al. 2013)
DBChip	MACS	Median ratio strategy	FDR	(Liang and Keles 2012)
MultiGPS	Expectation maximisation learning scheme	edgeR	*P*-value	(Mahony et al. 2014)
DiffBind	MACS	DESeq2	*P*-value	(Stark and Brown 2011)
PePr	Window-based approach	TMM	*P*-value	(Zhang et al. 2014)
MMDiff	MACS	DESeq2	*P*-value	(Schweikert et al. 2013)
ChIPComp	MACS	Poisson distribution	Posterior probability	(Chen et al. 2015)
DiffChIPL	MACS	DESeq2	FDR	(Chen, Chen, and Lei 2022)
HOMER	Window-based approach	Library size	FDR	(homer.ucsd.edu/homer/)

TABLE 2.5
Tools Used for Motif Discovery, Comparison, Spacing, and Mapping

Tool	Motif Discovery	Motif Comparison	Motif Enrichment and Spacing	Motif Prediction and Mapping	Reference
TRANSFAC	✓				(Wingender 2008)
ChIP-Munk	✓				(Kulakovskiy et al. 2010)
CisGenome	✓	✓			(Ji et al. 2008)
CompleteMOTIFS	✓	✓			(Kuttippurathu et al. 2011)
MEME-ChIP	✓	✓	✓		(Machanick and Bailey 2011)
JASPAR			✓		(Mathelier et al. 2014)
peak-motifs	✓	✓		✓	(Thomas-Chollier et al. 2012)
Cistrome	✓	✓		✓	(Liu et al. 2011)
STAMP		✓	✓		(Mahony and Benos 2007)
TOMTOM		✓			(Gupta et al. 2007)
CentriMo			✓		(Bailey and Machanick 2012)
SpaMo			✓		(Whitington et al. 2011)
FIMO				✓	(Grant, Bailey, and Noble 2011)
PATSER				✓	(Hertz and Stormo 1999)

✓: Denotes that the tool can perform the required analysis.

motifs using specialised tools; (c) motif enrichment analysis to determine the prevalence of known DNA motifs in the dataset; (d) motif spacing analysis, which can indicate physical interactions between TFs; and (e) motif prediction, used to map and visualise motif occurrences within the ChIP-sequenced regions. The tools used for each step are mentioned in Table 2.5.

Peak Annotation

Peak annotation helps with the association of ChIP-seq peaks with functionally relevant genomic regions like transcription start sites, intergenic regions, gene promoters, etc. Peak locations (in BED or GFF file formats) and read coverage data (in WIG or bedGraph formats) are uploaded to bioinformatics tools to analyse their relationships with known genomic features (Bailey et al. 2013). There are several tools used for peak annotation such as UROPA (Kondili et al. 2017), PAVIS (Huang et al. 2013), ZINBA (Rashid et al. 2011), PeakRanger (Feng, Grossman, and Stein 2011), BEDTools (Quinlan and Hall 2010), GREAT (McLean et al. 2010), and CEAS (Shin et al. 2009). There are also Bioconductor R packages that are used for peak annotation, such as ChIPseeker (Yu, Wang, and He 2015) and ChIPpeakAnno (Zhu et al. 2010). Gene ontology enrichment analysis can be performed using Cistrome-GO (Liu et al. 2011), GREAT (McLean et al. 2010), DAVID (Dennis et al. 2003), and GSEA (Subramanian et al. 2005). When the density of sequencing reads is plotted and compared across different samples relative to specific annotated genomic features, it reveals changes in protein-binding patterns among the samples.

Peaks are detected as a point source factor, broad source factor, or a mixed source factor (Landt et al. 2012). Point source factors are found at specific genome locations, typically involving sequence-specific TFs and highly localised chromatin marks. Broad source factors identify extended genomic regions, such as chromatin marks like histone H3 lysine 9 trimethylation (H3K9me3). Mixed source factors, which detect RNA polymerase II, include both localised peaks and broader regions (Figure 2.3). Broad source and mixed source factors require more reads for accurate peak calling compared to point source factors (Landt et al. 2012).

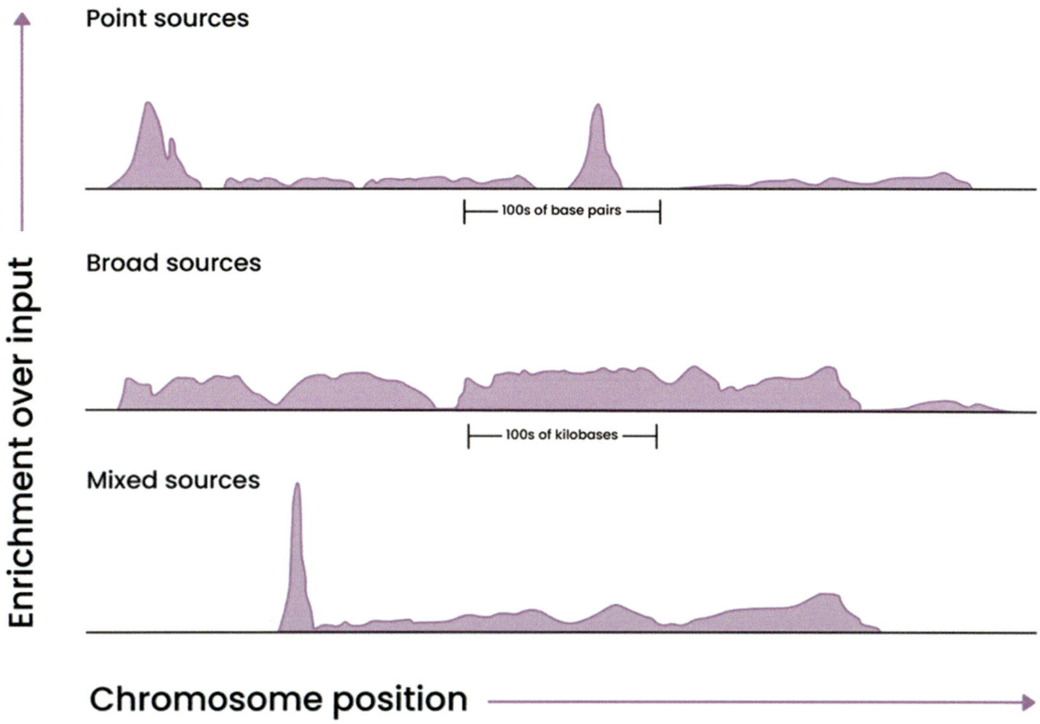

FIGURE 2.3 Different source peaks deduced in ChIP-seq analysis.

The peak calling algorithm detects peaks using a window-based method, but this approach can sometimes merge adjacent binding events incorrectly. To address this issue, several peak callers, such as PeakSplitter (Salmon-Divon et al. 2010), GPS (Guo et al. 2010), PolyaPeak (Wu and Ji 2014), and NarrowPeaks (Madrigal and Krajewski 2013), use peak shape as a clue to distinguish between those that are overlapping.

Visualisation

A visual depiction of the read distribution is required to analyse the data for which Integrated Genome Viewer (IGV) (Robinson et al. 2011) and SeqMonk (https://www.bioinformatics.babraham.ac.uk/projects/seqmonk/) are widely used. Web-based servers such as WashU Epigenome Browser (Li, Hsu, et al. 2019) and UCSC Genome Browser (Karolchik et al. 2003) help in integrating the ChIP-seq data with annotation.

ATAC-SEQ

Sequencing of ATAC-seq Libraries

ATAC-sequencing using the Illumina platform is carried out with NEXTERA-based primers and reagents. The sequencing layout for ATAC-seq is chosen based on the specific research interest; for example, paired-end sequencing is ideal for nucleosome mapping. For mammalian ATAC-seq aimed at identifying open chromatin regions and conducting differential peak analysis, it is recommended to use 25 million reads for single-end layout sequencing and 50 million reads for paired-end sequencing, with two or three biological replicates (Yan et al. 2020).

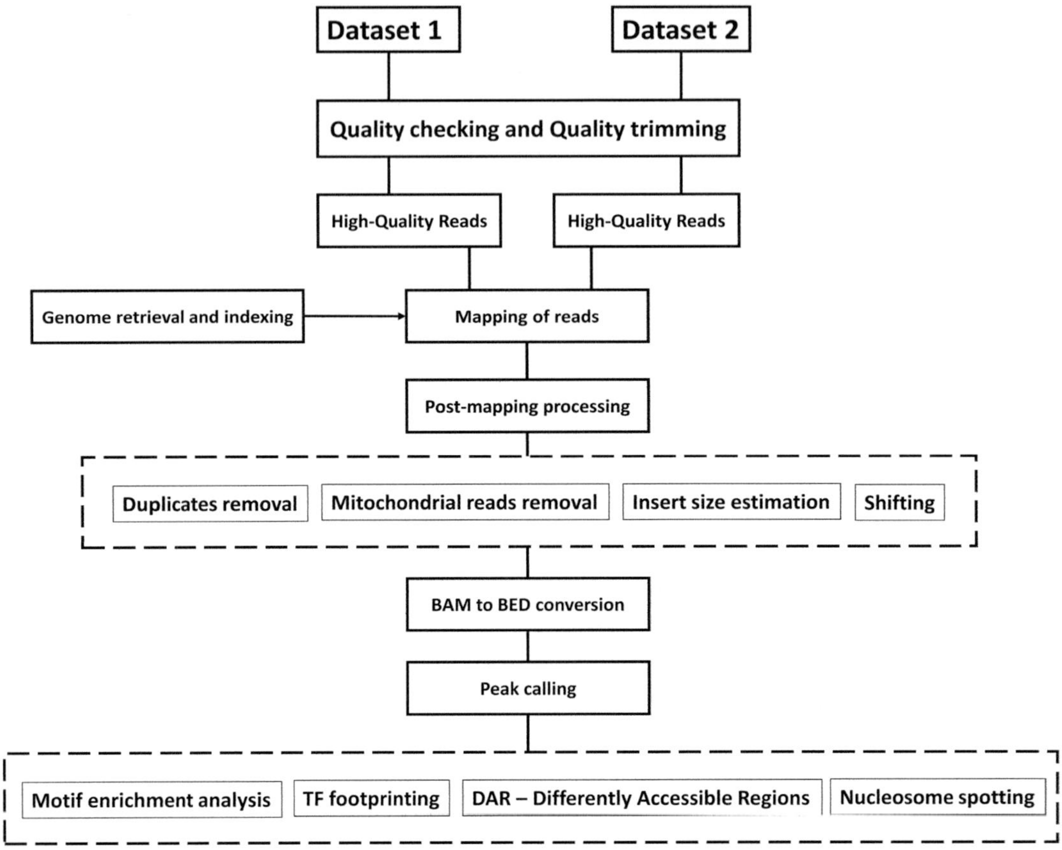

FIGURE 2.4 ATAC-seq data analysis workflow.

Processing of ATAC-seq Data in Detail

Analysis of ATAC-seq data begins with preprocessing quality assessment and ends with peak calling and annotation. The workflow is shown in Figure 2.4. The following sections describe the analysis method along with its flaws.

Preprocessing

A primary and immediate filtering step should be applied to the raw data produced by Illumina sequencing platforms. This quality control process is similar to those used in other NGS data analyses. Preprocessing of NGS data comprises quality trimming, adapter trimming, and removal of overrepresented sequences. The most common tool used for the quality assessment is FastQC (https://www.bioinformatics.babraham.ac.uk/projects/fastqc/; accessed Sep 2024). FastQC provides a graphical overview of read statistics, including low-quality bases, adapter content, GC content (a key parameter for detecting contamination), and N base content. NEXTERA adapters are commonly used in ATAC-seq and must be removed from the raw data because they are considered artifacts. Quality and adapter trimming can be performed with the available quality trimming tools, viz., Trimmomatic (Bolger et al. 2014), fastp (Chen et al. 2018), cutadapt (Martin 2011), etc.

Trimmomatic and fastp include preexisting adapter sequences for TruSeq and Nextera, simplifying the trimming process compared to other software. Removing adapters and low-quality bases improves read mapping to the genome and reduces complexity in downstream analysis.

Read Alignment to the Genome

Alignment of the DNA reads to the genome is an important step in ATAC-seq analysis, followed by peak calling. Burrows-Wheeler Aligner (BWA) (Li and Durbin 2009) and Bowtie2 (Langmead and Salzberg 2012b) are the memory- and time-efficient mapping tools for the Illumina reads, commonly used for this purpose. Indexing the reference genome reduces computational time and memory usage during mapping. Also, each mapping tool uses its own format for the genome index. In ATAC-seq analysis, a mapping percentage above 80% is considered good. Lower mapping percentages may lead to fewer identified peaks, false peaks, and can indicate sequencing errors or sample contamination. Generally, in NGS analysis, DNA fragments from the same genome should map with a high percentage, as they originate from the same reference sequence.

Post Quality Metrics

Before proceeding with downstream analysis, key parameters such as mapping percentage, non-unique mapping, and duplicate reads should be assessed. Evaluating the mapping percentage and removing reads mapped to mitochondrial DNA helps reduce the inclusion of reads from irrelevant regions. The mitochondrial genome typically shows higher coverage compared to other genomic regions due to its lack of chromatin packaging, making it more accessible to sequencing. The human and mouse genome assemblies include the mitochondrial genome, typically labelled as 'ChrM'. Removing this mitochondrial sequence from the whole genome file helps to accurately quantify peaks in the nuclear genome.

Another way to reduce the false-positive results from the analysis is by removing PCR duplicates, which also lead to false interpretation and need to be considered during preprocessing. PCR duplicates arise from over-amplification during the PCR step of short-read sequencing. This issue is common across all types of NGS analysis, including genomics and transcriptomics. PCR duplicates can be removed from the mapping file using Picard and SAMtools (Danecek et al. 2021). Reads mapped to regions blacklisted by the ENCODE project (Feingold et al. 2004) should be removed from the mapping file (BAM format).

Fragment size distribution analysis reveals nucleosome positioning in ATAC-seq data. This analysis is a key quality metric for ATAC-seq experiments. The fragsizeDist function in the ATACSeqQC package (Ou et al. 2018) can plot fragment/insert size distribution from the alignment BAM file. A high-quality ATAC-seq library typically shows a characteristic distribution pattern reflecting various chromatin states. This pattern includes nucleosome-free regions (NFRs) ranging from 36 to 100 bp, mono-nucleosomes at around 200 bp, di-nucleosomes at approximately 400 bp, and tri-nucleosomes at about 600 bp and longer fragments represent oligonucleosomes. This distribution effectively captures open chromatin regions, linker regions, and nucleosome-occupied areas, providing insights into the chromatin structure of the sample. In a successful ATAC-seq distribution plot, peak density decreases from nuclease-free regions to longer oligonucleosome (multinucleosome) regions. Peaks are enriched at the transcription start site (TSS) when DNA fragments originate from open chromatin regions (NFR). Conversely, there is a sharp decrease in peak density at the TSS when DNA fragments come from nucleosome-bound regions, with minor enrichment observed at untranslated regions (UTRs) around the TSS. The prefiltered BAM file should be assessed in order to obtain the unbiased distribution of fragment size.

Shifting of Reads for the Identification Centre of Transposon Binding

Finally, to carry out the TF footprinting and motif analysis at base-pair resolution, reads should be shifted. The Tn5 transposase, while binding as a dimer, introduces two adapters separated by 9 bp. As a result, all reads aligned to the plus strand should be shifted by +4 bp, and those aligned to the minus strand should be shifted by -5 bp. This shifting is a minor adjustment in the analysis and not affect the peak calling. Deeper analysis in TF footprinting at the base level requires shifting

TABLE 2.6
Count-, Shape-, and HMM-Based Peak Callers

Tool	Method	Reference
MACS2	Count-based	(Kumari et al. 2014)
HOMER		(Heinz et al. 2010)
epic2		(Stovner and Sætrom 2019)
F-seq		(Boyle et al. 2008)
JAMM		(Ibrahim, Lacadie, and Ohler 2015)
PICS	Shape-based	(Zhang et al. 2011)
HMMRATAC	HMM-based	(Tarbell and Liu 2019)

to accurately identify the insertion position at single-base resolution. Before shifting the reads, the BAM file must be sorted. This shifting can be achieved using bedtools (Quinlan and Hall 2010) and awk (Aho, Kernighan, and Weinberger 1979) commands. alignmentSieve (https://github.com/deeptools/deepTools; accessed Sep 2024) is another tool used for shifting reads.

Deciphering Potential Open Chromatin Region (Peak Calling and CORE Analysis)

Identification of the Tn5 insertion site (a chromatin-accessible region) is done with ATAC-seq data; this is a major and important step in the analysis (Grandi et al. 2022). Peak callers used in ATAC-seq analysis are broadly classified into three categories: count-based, shape-based, and hidden Markov model (HMM)-based (Table 2.6). Shape-based peak callers are outmoded, but count- and HMM-based methods are currently in use for peak calling. In read mapping distributions to open chromatin regions, reads are enriched more than the background read coverage, and these enriched regions are identified as peaks by peak callers. Genrich (https://github.com/jsh58/Genrich) and MACS2 (Zhang et al. 2008) are commonly used for peak calling (Grandi et al. 2022). Even though MACS2 was originally designed for ChIP-seq data analysis, most of the ATAC-seq pipelines currently use it for peak calling. Which tools to use for peak calling is highly dependent on the experimental conditions. When using MACS2 as the peak caller, the BAM file (after alignment) must be converted into a BED file. This allows the tool to efficiently use both mates of paired-end data for peak calling. If the BAM file is used as the input format, the mate pair will be discarded, and only the 5' end reads will be considered for peak calling. Some other parameters like `extsize` and `shift` are adjusted to ensure the representation of the Tn5 insertion position in the signal position. Pipelines developed for ATAC-seq data analysis often use MACS2 for peak calling and apply the parameters `--shift 100` and `--extsize 200` to accurately pinpoint Tn5 insertion sites, excluding nucleosome positions (since NFRs are less than 150-200 bp in length). The most recent update of MACS2 added the HMMRATAC module (Tarbell and Liu 2019) for peak calling in ATAC-seq data. HMMRATAC is an HMM-based module that classifies the ATAC-seq data into four signal tracks as follows: (i) short read fragments (from NFRs), (ii) mononucleosomal fragments, (iii) di-nucleosomal fragments, and (iv) tri-nucleosomal fragments. Based on the integration of these four signals, an HMM will identify the open region, nucleosomal region, and background region. And these peak callers create a BED file as output with peak regions.

Documentation of Differential Peaks and Annotation

Peaks identified in biological replicates, technical replicates, and samples should be merged to create union peaks (Figure 2.5). Peaks can be merged in two ways: using variable width or fixed width. Fixed width is the preferred method, as it simplifies the downstream analysis by reducing the need for peak normalisation. A fixed width of less than 500 bp is optimal for merging peaks (Grandi et al. 2022).

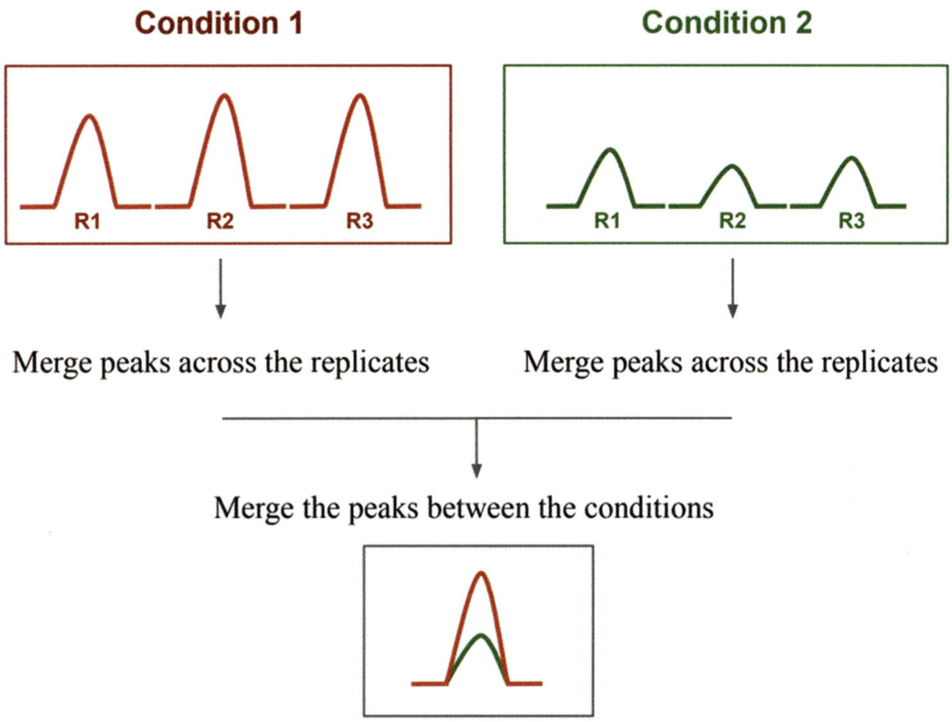

FIGURE 2.5 Image depicting the peak merging step to be conducted in ATAC-seq analysis.

Three different algorithms are used for peak merging, utilising either variable or fixed widths: (i) raw overlap with variable width, (ii) clustered overlap with fixed width, and (iii) iterative overlap peak calling (Grandi et al. 2022). The iterative overlap peak calling is performed with a Perl script using MACS2 (https://github.com/corceslab/ATAC_IterativeOverlapPeakMerging; accessed Sep 2024). This method enhances sensitivity and results in larger peaks.

Insertion events are calculated from the peaks and recorded in the counts matrix. Furthermore, these read counts will be normalised and differentially analysed using RNA-seq differential expression analysis tools, because there is no specific tool for carrying out differential accessible regions (DAR) analysis. HOMER (Heinz et al. 2010), DiffBind (Stark and Brown 2011), and DBChIP (Liang and Keles 2012) are the most used tools for DAR analysis. These tools were integrated into the RNA-seq DEG analysis packages, viz., DESeq (Anders 2010), DESeq2 (Love, Huber, and Anders 2014), edgeR (Robinson, McCarthy, and Smyth 2010a), limma, Wilcoxon, and t-test. ATAC-seq DAR analysis requires higher sensitivity than RNA-seq (Gontarz et al. 2020). The use of the tools depends entirely on the experimental design, but some key points about the packages are given in Table 2.7.

TABLE 2.7
List of Tools and Their Applications in DAR Analysis

Package	Use	Reference
DESeq	Reduces the false-positive identification. Applicable in the case of DAR with	(Anders 2010)
DESeq2	higher confidence. Specificity is more compared to other packages.	(Love et al. 2014)
edgeR	Higher sensitivity with the low incidence of replicated data and with acceptable false positives.	(Robinson, McCarthy, and Smyth 2010a)
limma	Helps to identify the signal in the lower peak intensity regions.	(Ritchie et al. 2015)

Denoising the Batch Effect

Batch effect is a common problem in omics experiments where data were generated at different times, as different biological time periods or instruments may have been used during sequencing. Larger variation within the same group can result from batch effects, directly affecting the sensitivity in differential analysis (Leek et al. 2010). BeCorrect (https://github.com/Zhang-lab/BeCorrect; accessed Sep 2024) can efficiently reduce batch effects when performing DAR analysis on ATAC-seq data signals and can be utilised for clear visualisation in genome browsers (Gontarz et al. 2020). Additionally, DESeq2 and edgeR can also be used to correct batch effects.

Track Visualisation and Peak Annotation

Visualisation of identified peaks is often considered more important than DARs, and this can be achieved with the following genome browsers: (i) UCSC (Kuhn, Haussler, and James Kent 2013), (ii) IGV (Thorvaldsdóttir, Robinson, and Mesirov 2013), and (iii) the WashU Epigenome Browser (Li et al. 2022). deepTools can be used to create a bigWig file from bamCoverage, which can then be used as a track for visualisation in genome browsers. The same tool can be employed for normalising the peaks. A bin size of 100 bp has been found to be effective (Grandi et al. 2022). In this method, peaks in different samples can be easily compared and visualised to get the peaks of biological importance, near TSS regions. Peaks are commonly annotated based on the localisation of nearby genes and gene regulators or closest genomic features (exon, *cis*-acting elements, promoter, 5′ and 3′ UTRs); this will give deeper insights into the correlation of chromatin accessibility and gene regulation. Tools used for peak annotation are given in Table 2.8. Gene ontology (GO) and pathway enrichment analysis using the Kyoto Encyclopedia of Genes and Genomes (KEGG) can be performed with enrichR (Kuleshov et al. 2016) on the identified genes and their expression.

Motif Enrichment Analysis

To discover potential TFs, motifs are analysed using motif enrichment analysis, and the collected motif information is pooled and stored in databases like JASPAR (Sandelin et al. 2004), ENCODE (de Souza 2012), and CIS-BP (Weirauch et al. 2014). These motifs are from experimental identification as well as computational prediction using *in silico* tools, viz., MEME (Bailey et al. 2015),

TABLE 2.8
Different Tools Used for the Peak Annotation

Tool	Function	Scripting Language	Reference
HOMER (annotaePeaks.pl)	Option for prioritising the feature (ex: can extract specific features like exon). Prioritises the location to be annotated	Perl	(Heinz et al. 2010)
ChIPseeker	Retrieves the function of genes nearest to the peaks and facilitates the comparison of user data with GEO databases (helps to identify the regulation of genes)	R	(Yu, Wang, and He 2015)
ChIPpeakAnno	Identifies the miRNAs, conserved elements, enriched GO terms, nearest genes, exons based on the sequence retrieved from the peak region	R	(Zhu et al. 2010)
Universal RObust Peak Annotator (UROPA)	Annotation of peaks using different target features: gene, UTR, strand, distance, up/downstream	Python/R	(Kondili et al. 2017)
Goldmine	Identifies over-/underrepresented GOs; visualisation by employing several bioinformatics resources	GUI	(Bhasin and Ting 2016)

HOMER (Heinz et al. 2010), TFBSTools (Tan and Lenhard 2016), and motifmatchr (Schep 2023). Motif enrichment analysis can be done with HOMER and MEME to identify the TFs of biological interest. The chromVAR (Schep et al. 2017) and diffTF (Berest et al. 2019) tools can be used to identify TFs of higher activity by correlating with insertion counts.

TF Footprinting

Gene expression is tightly regulated by the binding of specific TFs to the *cis*-regulatory elements in the genome. TF footprinting is a computational technique that predicts and analyses regions where TF binding protects DNA from cleavage. This method provides deeper insights into the locations and patterns of TF-DNA interactions. TF footprinting algorithms first identify the Tn5 transposase insertion sites and then analyse the surrounding DNA sequences to predict which transcription factors are likely to bind in these regions (Grandi et al. 2022). TOBIAS (TF Occupancy prediction By Investigation of ATAC-seq Signal) is a computational framework for footprinting analysis (Bentsen et al. 2020). This tool aids in identifying TFBSes and estimating changes in TF binding at specific genomic loci. HINT-ATAC (Li, Schulz, et al. 2019) is a well-known tool for footprinting TFBSes using an HMM. It also performs bias correction for ATAC-seq data. PIQ, Wellington (Piper et al. 2015), msCentipede (Raj et al. 2015), and TF footprints in active chromatin elements (TRACE) (Ouyang and Boyle 2020) are other tools that can be used for footprinting transcription factors in ATAC-seq data. Bentsen et al. (2020) discovered TF kinetics in zygotic genome activation. Their study used TOBIAS for TF footprinting and found that it performs well compared to other tools of its kind. Higher sequencing depth in ATAC-seq reduces the false positives in TF footprinting.

Nucleosome Spotting

Nucleosomes bind to DNA and alter chromatin accessibility, affecting the binding of TFs. When a nucleosome shifts from its usual position, it can suppress gene regulation by condensing the chromatin region. MNase-seq is the most commonly used method for nucleosome positioning, and several tools are available for this purpose. However, for ATAC-seq, NucleoATAC (Schep et al. 2015) and HMMRATAC are the primary tools available; during the peak calling step, HMMRATAC identifies both open chromatin regions and nucleosome positions.

Pipelines Available for ATAC-seq Data Analysis

In recent years, developing pipelines for omics data analysis has become a key focus in bioinformatics. To streamline this process, many integrated pipelines have been created, including those tailored for advancements in ATAC-seq data analysis. A list of pipelines for ATAC-seq data analysis is given in Table 2.9.

Integration of Gene Expression (RNA-seq), Chromatin Accessibility (ATAC-seq), and ChIP-seq

Instead of relying on results from a single type of omics analysis, integrating multi-omics data provides a more reliable identification of key regulators and their impact on gene expression. This approach is also valuable for uncovering the causes of various diseases and developing potential treatments, ultimately benefiting human health. Both RNA-seq and ATAC-seq data generated from samples under the same conditions enable the integration of chromatin accessibility with gene expression, facilitating the identification of key TFs and other *cis*-regulatory elements. The data thus provide insights into the regulation of gene expression and help identify novel genes involved

TABLE 2.9
List of Pipelines Developed for Analysing ATAC-seq Data

Pipeline	Features	URL	Reference
ENCODE ATAC-seq	Can run across multiple cloud platforms. Also produces a user-friendly HTML report. Supports functional annotation. Reference genome available for humans and mice.	github.com/ENCODE-DCC/atac-seq-pipeline; accessed Sep 2024	(de Souza 2012)
esATAC	End-to-end analysis from preprocessing to peak annotation. Different accessible regions analysis is not available.	www.bioconductor.org/packages/release/bioc/html/esATAC.html; accessed Sep 2024	(Wei et al. 2018)
GUAVA	GUI environment for ATAC-seq analysis and visualisation. Available for Linux and macOS. For the identification of differentially enriched ATAC-seq signals between two different conditions. The GO and Pathway analysis option is available.	github.com/MayurDivate/GUAVA; accessed Sep 2024	(Divate and Cheung 2018)
nfcore/atacseq	Installs the dependencies in containers (Docker and Singularity). Analysis starting from preprocessing to different accessibility. Motif annotation is not supported.	nf-co.re/atacseq; accessed Sep 2024	(nf-co.re/atacseq)
PEPATAC	It contains all the possible analyses that can be done with ATAC-seq data. Preprocessing to annotation of peaks and motif enrichment analysis. Tools for DAR analysis are not integrated into the pipeline. Peak callers can be changed according to the data.	github.com/databio/pepatac; accessed Sep 2024	(Smith et al. 2021)
ATAC-Pipe	Helps in creating plots from statistical data. Integrates all results from ATAC-seq analysis and produces the visualisation as charts.	github.com/QuKunLab/ATAC-pipe; accessed Sep 2024	(Zuo et al. 2019)

in certain biological or developmental conditions. Yang et al. (2019) conducted a study by integrating RNA-seq and ATAC-seq data in macrofungi for the identification of molecular mechanisms undergoing light-responsive primordial formation. Integrating ChIP (for mapping epigenetic marks) with RNA-seq provides insights into gene expression, chromatin regulators, and TFs. Integrating multiple omics data can help answer unresolved questions in biological research; this will unravel molecular mechanisms and enable researchers to test new hypotheses. At this point, there is no well-developed method for integrating these three omics data to provide efficient results. The development of methodology and bioinformatics (*in silico*) tools for integrating multi-omics data are great opportunities for bioinformaticians interested in tool development to enrich their scientific work and publish high-quality papers.

MeDIP-SEQ

Sequencing Protocol

Distinguishing methylated cytosine residues from unmethylated ones is crucial for identifying genome methylation patterns. Initially, high-performance liquid chromatography (HPLC) was used to measure overall genome methylation rates, but this method could not pinpoint the specific locations of methylated sites. To address this limitation, techniques such as restriction digestion, bisulphite conversion, and antibody-based methods were developed to identify individual methylated residues. The advent of high-throughput technologies like microarrays and genome sequencing platforms has made comprehensive methylome sequencing both feasible and reliable. The following subsections describe the most common genome-wide approaches for methylation sequencing.

Whole Genome Bisulphite Sequencing

Cytosine residues and the analogues of cytosine, such as 5-methylcytosine (5mC) and 5-hydroxymethylcytosine (5hmC), occur at significant levels in the genome, and the ability of 5mC and 5hmC to resist bisulphite treatment is the principle involved in whole genome bisulphite sequencing. Cytosine is deaminated to uracil upon treatment with bisulphite, whereas 5mC and 5hmC remain unaffected. However, a limitation of this method is its inability to distinguish between 5mC and 5hmC. 5hmC is abundant in mammalian brain cells and is presumed to be an intermediate in the demethylation pathway. Hence, quantification of 5mC and 5hmC in the genome is vital for establishing a relationship between epigenetic markers and cellular function.

Reduced Representation Bisulphite Sequencing (RRBS)

RRBS identifies genomic methylation at the single-nucleotide level. In this method, the genome is first digested with a restriction enzyme (typically MspI) and then subjected to size selection. RRBS reduces the complexity of the genome by sequencing digested fragments with greater depth. The digested fragments are added to synthesised adapters, followed by denaturation, bisulphite treatment, PCR amplification using adapter-specific primers, and sequencing. A limitation of this method is that it covers only a subset of CpG (cytosine-phosphate-guanine) sites in the genome, primarily those in CpG-rich regions, leaving many potentially methylated regions unexamined.

Methyl Binding Domain Capture Sequencing (MBDCap-seq)

While WGBS and RRBS provide an overall view of methylation sites at the genome level, their limitations are addressed by enrichment techniques like methyl-binding domain capture sequencing. This method uses methyl-binding proteins with high affinity for methylated CpG islands to enrich methylated DNA fragments. After the genome is fragmented, it is treated with these proteins, and the unmethylated sequences are washed away. The enriched fragments are then sequenced, offering 94% coverage while reducing the cost and time of sequencing the entire genome. However, this technique cannot effectively enrich and analyse genomic regions with very low CpG island density.

Methylation-Sensitive Restriction Enzyme and Sequencing

This method utilises methylation-sensitive restriction enzymes, as the name suggests. There are over 50 such enzymes available, and using rare cutters can effectively help in dissecting genome methylation sites. Methylation inhibits the activity of methylation-sensitive restriction enzymes (MREs). However, a limitation is that only methylation sites within the regions targeted by these enzymes are analysed. The use of multiple restriction enzymes can help mitigate this limitation and improve the comprehensiveness of the results.

Methylated DNA Immunoprecipitation Sequencing (MeDIP-seq)

MeDIP-seq employs antibodies specific to 5mC, which bind directly to methylated cytosine residues, regardless of neighbouring bases. To enhance antibody binding efficiency, genomic DNA is sheared and denatured. The bound DNA fragments are typically captured using magnetic beads that bind to the antibodies, followed by a washing step to remove unbound or unmethylated fragments. The methylated DNA-antibody complex is then subjected to protein denaturation, after which the DNA fragments are precipitated and purified. These fragments are then sequenced using NGS. Alternatively, the MeDIP-seq technique can be coupled with microarrays (MeDIP-chips). MeDIP-seq is particularly effective for identifying CpG islands with a low density of methylated cytosine residues. While only 2 to 3% of cytosine methylation occurs in high-density CpG islands, the majority is distributed throughout the genome and is often missed by techniques like MBDCap-seq and bisulphite sequencing. MeDIP-seq offers a comprehensive view of an organism's methylome without bias.

Data Analysis Workflow

The density of methylated CpG islands creates a bias among different genomic regions. The density of methylated residues in any CpG island determines the precipitation efficiency. For instance, a genomic region with a higher rate of methylation is enriched better compared to a CpG island with low methylation density. However, the methylation density on a CpG island does not determine its biological function. To eliminate this bias and identify key methylation sites that regulate gene expression, it is essential to use bioinformatics pipelines.

Like other sequencing data, methylome data must be subjected to three major steps to obtain biologically significant data. The following subsections describe these steps.

Data Processing and Quality Control

Experimentally derived data are obtained as enriched or depleted sequencing reads that can be mapped to specific regions of the genome. These data must be processed before visualisation or validation. All enrichment-based methylome sequencing data follow a similar computational workflow. NGS data obtained after enrichment provide information on the frequency of specific genome fragments under study. Before further analysis, the data are corrected for CpG density. Errors in performing this normalisation can affect the accuracy of absolute methylation values.

Alignment of Reads to Genome

Alignment tools such as Bowtie (Langmead and Salzberg 2012) and BWA (Li and Durbin 2009a) can be used to align methylome reads with genomic DNA.

Calculating Relative Enrichment Score

Enrichment scores are indicators of regional methylation rates. However, these data are greatly influenced by thc distribution of CpG islands in the genome. For example, a region with high CpG islands and a moderate level of methylation can outperform a region with low CpG levels and higher methylation rates. Tools such as BATMAN (Down et al. 2008), which uses Bayesian statistics, and MEDME (Pelizzola et al. 2008), which uses a logistic regression model, normalise this error. The former slows when used on larger datasets, whereas the latter requires a methylome reference to provide accurate results. MEDIPS (Lienhard et al. 2014) combines both statistical models to provide fast and accurate results.

Tools such as MeQA (Huang et al. 2012) and MeDUSA (the toolkit comes with packages necessary for methylome data analysis) help in semi-automating the process by serving as a step between BWA and MEDIPS. Quality control can be performed using Repitools, a R/Bioconductor package designed exclusively for the enrichment of datasets.

Statistical Analysis

The normalised data obtained from processing are provided in the form of a table called the CpG methylation table, which represents the absolute methylation level for all the samples. This forms a dataset for further analysis. Like data in whole genome analyses, where a variant calling format (VCF), methylome analysis uses the methylation call format (MCF) for further downstream processing and visualisation.

The MCF tables are then converted into formats that aid the visualisation process. BigBed and bigWig are some of the formats used by visualisation tools that make spotting the CpG island on the genome possible. The data in the BED or WIG format are later uploaded to be used with the corresponding genome on web-available or standalone genome browsers. This helps with understanding the local methylation rate of a gene or a selected portion of the genome. For a more global-level understanding of DNA methylation, statistical packages can be used to create box plots, scatter plots, or the Hilbert curve for visualisation.

Methylation mapping can also be visualised as differences in methylation patterns between two distinct samples or populations. R packages can be used for this but none with an easy-to-use interface. This makes it relatively difficult to do without using command-line tools.

The visualised data provide an outline of how the methylation pattern is distributed throughout the genome. For a better understanding, local or region-specific differential methylation rates must be analysed. The biological effect of methylation is independent of the number of C residues methylated on a particular CpG island. While advanced tools help in detecting methylation rates, the underlying biological function could also be due to the methylation of a single C residue between the control and the sample under study. Bisulphite sequencing uses a methylation table to understand differential rates, while MeDIP-seq and other enrichment-based methods use the read count data to analyse the variation in methylation. Differential methylation rates can be analysed using a simple student's t-test, which compares the methylation rates of C residues between test and control samples. Other advanced methods include the mixture model, Shannon entropy, and logistic M values, which can be used to obtain more accurate results. However, these analytical tools should be chosen based on the sample under study. To use these tools, understanding the nature and pattern of methylated samples is essential. To study the methylation pattern at multiple sites, it is essential to include a correction factor for the multiple hypotheses formulated. Alternatively, among the multiple methylated C residue sites on a selected CpG island, only a few strongly methylated sites tend to show significant differences. The remaining sites are often found to be false positives.

Data Validation

The statistical significance of a differentially methylated region does not necessarily confirm the biological significance of the sample. Experimental validation, accompanied by statistical significance, enables a better understanding of the differential DNA methylation patterns in the sample. The first step is to rank methylation sites based on statistical data. Following this, the methylation sites are manually compared using a genome browser to ensure that they do not occur in repeat regions or microsatellites, as they may not then contribute to biological significance. In addition, we ensure that there are no unusually high methylation sites in random genomic fragments. This is because such very high methylation rates might result in false positives. Perform a quality check using plots such as volcano plots, Manhattan plots, or Q-Q plots to identify biased or false-positive data.

Once the data are verified *in silico*, an experimental confirmation is performed. Two approaches are used: one is to carry out methylation analyses using a new set of samples (i.e., using biological replication); the other is to use the same sample and analyse its differential methylation patterns using other techniques. This is performed to determine the technical accuracy of the method used to study the patterns. Therefore, it is not essential to validate global patterns; validation is generally performed for specific site. MethMarker (Schüffler et al. 2009) is a tool used extensively for validating such data. This tool helps in designing specific analyses to validate methylation sites based on methods such as bisulphite sequencing and methylation-specific PCR.

Tools such as EpiExplorer (Halachev et al. 2012), Galaxy (Blankenberg et al. 2010), and Genomic Hyperbrowser (Sandve et al. 2010) help identify the biological significance of a methylation site by comparing the site with the genomic data available in public databases. Furthermore, pathway analysis tools can aid in a better understanding of biological trends based on differential methylation rates.

CONCLUSIONS

In this chapter, we extensively discussed bioinformatics analyses that can be performed on epigenetics data. In the NGS era, most biological experiments are accompanied by high-throughput sequencing data. It is therefore very important to understand the quality of the data and the protocols to be used for their analysis and interpretation. This chapter provides an overview of all the methodologies involved in sequencing experiments, such as RNA-seq, ChIP-seq, ATAC-seq, and MeDIP-seq. We presented a step-by-step workflow and the tools associated with each step for data analysis, from preprocessing to obtaining results. A challenging segment in NGS data analysis, viz., integration of multi-omics data, was also reviewed and discussed. With the availability of a plethora of bioinformatics servers and software, keep in mind that the decision on which tool to use for a specific analysis will depend on the objective of the experiment. We hope that this chapter will be useful for beginners in epigenetic data analysis and for the experts who develop tools, pipelines, and databases.

REFERENCES

Aho, Alfred V, Brian W Kernighan, and Peter J Weinberger. 1979. "Awk – A pattern scanning and processing language." *Software: Practice and Experience* 9 (4):267–279.

Anaïs F, Bardet, Qiye He, Julia Zeitlinger, and Alexander Stark. 2012. "A computational pipeline for comparative ChIP-seq analyses." *Nature Protocols* 7 (1): 45–61.

Anders, Simon. 2010. "Analysing RNA-seq data with the DESeq package." Molecular Biology 43 (4):1–17.

Anders, Simon, Paul Theodor Pyl, and Wolfgang Huber. 2015. "HTSeq – A python framework to work with high-throughput sequencing data." *Bioinformatics* 31 (2):166–169.

Anzawa, Hayato, Hitoshi Yamagata, and Kengo Kinoshita. 2020. "Theoretical characterisation of strand cross-correlation in ChIP-seq." *BMC Bioinformatics*, 21: 1–19.

Artem, Barski, Suresh, Cuddapah, Cui, Kairong, Roh, Tae-Young, Schones, Dustin E, Wang, Zhibin, Wei, Gang, Chepelev Iouri, and Keji, Zhao. 2007. "High-resolution profiling of histone methylations in the human genome." *Cell* 129 (4): 823–837.

Babarinde, Isaac A, Yuhao Li, and Andrew P Hutchins. 2019. "Computational methods for mapping, assembly and quantification for coding and non-coding transcripts." *Computational and Structural Biotechnology Journal* 17:628–637.

Bailey, Timothy L, James Johnson, Charles E Grant, and William S Noble. 2015. "The MEME suite." *Nucleic Acids Research* 43 (W1):W39–W49.

Bailey, Timothy, Pawel Krajewski, Istvan Ladunga, Celine Lefebvre, Qunhua Li, Tao Liu, Pedro Madrigal, Cenny Taslim, and Jie Zhang. 2013. "Practical guidelines for the comprehensive analysis of ChIP-seq data." *PLOS Computational Biology* 9 (11):e1003326.

Bailey, Timothy L, and Philip Machanick. 2012. "Inferring direct DNA binding from ChIP-seq." *Nucleic Acids Research* 40 (17):e128.

Bentsen, Mette, Philipp Goymann, Hendrik Schultheis, Kathrin Klee, Anastasiia Petrova, René Wiegandt, Annika Fust, Jens Preussner, Carsten Kuenne, and Thomas Braun. 2020. "ATAC-seq footprinting unravels kinetics of transcription factor binding during zygotic genome activation." *Nature Communications* 11 (1):4267.

Berest, Ivan, Christian Arnold, Armando Reyes-Palomares, Giovanni Palla, Kasper Dindler Rasmussen, Holly Giles, Peter-Martin Bruch, Wolfgang Huber, Sascha Dietrich, and Kristian Helin. 2019. "Quantification of differential transcription factor activity and multiomics-based classification into activators and repressors: diffTF." *Cell Reports* 29 (10):3147–3159. e12.

Bersanelli, Matteo, Ettore Mosca, Daniel Remondini, Enrico Giampieri, Claudia Sala, Gastone Castellani, and Luciano Milanesi. 2016. "Methods for the integration of multi-omics data: Mathematical aspects." *BMC Bioinformatics* 17:167–177.

Bhasin, Jeffrey M, and Angela H Ting. 2016. "Goldmine integrates information placing genomic ranges into meaningful biological contexts." *Nucleic Acids Research* 44 (12):5550–5556.

Bioinformatics, BioBam, and Spain Valencia. 2019. "OmicsBox-Bioinformatics made easy." *March* 3:2019.

Birol, Inanç, Shaun D Jackman, Cydney B Nielsen, Jenny Q Qian, Richard Varhol, Greg Stazyk, Ryan D Morin, Yongjun Zhao, Martin Hirst, and Jacqueline E Schein. 2009. "*De novo* transcriptome assembly with ABySS." *Bioinformatics* 25 (21):2872–2877.

Blankenberg, D, G Von Kuster, N Coraor, G Ananda, R Lazarus, M Mangan, A Nekrutenko, and J Taylor. 2010. "Galaxy: A web-based genome analysis tool for experimentalists." *Current Protocols in Molecular Biology* 89(Suppl):19.10.1–19.10.21. doi: 10.1002/0471142727.mb1910s89

Bolger, Anthony M, Marc Lohse, and Bjoern Usadel. 2014. "Trimmomatic: A flexible trimmer for Illumina sequence data." *Bioinformatics* 30 (15):2114–2120.

Boyle, Alan P, Justin Guinney, Gregory E Crawford, and Terrence S Furey. 2008. "F-seq: A feature density estimator for high-throughput sequence tags." *Bioinformatics* 24 (21):2537–2538.

Bray, Nicolas L, Harold Pimentel, Páll Melsted, and Lior Pachter. 2016. "Near-optimal probabilistic RNA-seq quantification." *Nature Biotechnology* 34 (5):525–527.

Bruning, Oskar, Wendy Rodenburg, Conny T van Oostrom, Martijs J Jonker, Mark de Jong, Rob J Dekker, Han Rauwerda, Wim A Ensink, Annemieke de Vries, and Timo M Breit. 2014. "A range finding protocol to support design for transcriptomics experimentation: Examples of in-vitro and in-vivo murine UV exposure." *PLOS One* 9 (5):e97089.

Buchfink, Benjamin, Chao Xie, and Daniel H Huson. 2015. "Fast and sensitive protein alignment using DIAMOND." *Nature Methods* 12 (1):59–60.

Bushmanova, Elena, Dmitry Antipov, Alla Lapidus, and Andrey D Prjibelski. 2019. "rnaSPAdes: A de novo transcriptome assembler and its application to RNA-seq data." *GigaScience* 8 (9):giz100.

Camacho, Christiam, George Coulouris, Vahram Avagyan, Ning Ma, Jason Papadopoulos, Kevin Bealer, and Thomas L Madden. 2009. "BLAST+: Architecture and applications." *BMC Bioinformatics* 10:1–9.

Chen, Yang, Shue Chen, and Elissa P Lei. 2022. "DiffChIPL: A differential peak analysis method for high-throughput sequencing data with biological replicates based on limma." *Bioinformatics* 38 (17):4062–4069.

Chen, Li, Chi Wang, Zhaohui Qin, and Hao Wu. 2015. "ChIPComp: A novel statistical method for quantitative comparison of multiple ChIP-seq datasets." *Bioinformatics* 31(12): 1889–1896.

Chen, Shifu, Yanqing Zhou, Yaru Chen, and Jia Gu. 2018. "Fastp: An ultra-fast all-in-one FASTQ preprocessor." *Bioinformatics* 34 (17):i884–i890.

Conesa, Ana, Pedro Madrigal, Sonia Tarazona, David Gomez-Cabrero, Alejandra Cervera, Andrew McPherson, Michał Wojciech Szcześniak, Daniel J Gaffney, Laura L Elo, and Xuegong Zhang. 2016. "A survey of best practices for RNA-seq data analysis." *Genome Biology* 17 (1):1–19.

Consortium, Gene Ontology. 2004. "The gene ontology (GO) database and informatics resource." *Nucleic Acids Research* 32 (suppl_1):D258–D261.

Cunningham, Fiona, James E Allen, Jamie Allen, Jorge Alvarez-Jarreta, M Ridwan Amode, Irina M Armean, Olanrewaju Austine-Orimoloye, G Andrey If, Azov Ruth, and Barnes Bennett. 2022. "Ensembl 2022." *Nucleic Acids Research* 50 (D1):D988–D995.

Danecek, Petr, James K Bonfield, Jennifer Liddle, John Marshall, Valeriu Ohan, Martin O Pollard, Andrew Whitwham, Thomas Keane, Shane A McCarthy, and Robert M Davies. 2021. "Twelve years of SAMtools and BCFtools." *GigaScience* 10 (2):giab008.

de Souza, Natalie. 2012. "The ENCODE project." *Nature Methods* 9 (11):1046–1046.

Dennis, Glynn, Brad T Sherman, Douglas A Hosack, Jun Yang, Wei Gao, H Clifford Lane, and Richard A Lempicki. 2003. "DAVID: Database for annotation, visualisation, and integrated discovery." *Genome Biology* 4 (9):1–11.

Dillies, Marie-Agnès, Andrea Rau, Julie Aubert, Christelle Hennequet-Antier, Marine Jeanmougin, Nicolas Servant, Céline Keime, Guillemette Marot, David Castel, and Jordi Estelle. 2013. "A comprehensive evaluation of normalisation methods for Illumina high-throughput RNA sequencing data analysis." *Briefings in Bioinformatics* 14 (6):671–683.

Divate, Mayur, and Edwin Cheung. 2018. "GUAVA: A graphical user interface for the analysis and visualisation of ATAC-seq data." *Frontiers in Genetics* 9:250.

Dobin, Alexander, Carrie A Davis, Felix Schlesinger, Jorg Drenkow, Chris Zaleski, Sonali Jha, Philippe Batut, Mark Chaisson, and Thomas R Gingeras. 2013. "STAR: Ultrafast universal RNA-seq aligner." *Bioinformatics* 29 (1):15–21.

Down, Thomas A, Vardhman K Rakyan, Daniel J Turner, Paul Flicek, Heng Li, Eugene Kulesha, Stefan Graef, Nathan Johnson, Javier Herrero, and Eleni M Tomazou. 2008. "A Bayesian deconvolution strategy for immunoprecipitation-based DNA methylome analysis." *Nature Biotechnology* 26 (7):779–785.

Fabregat, Antonio, Steven Jupe, Lisa Matthews, Konstantinos Sidiropoulos, Marc Gillespie, Phani Garapati, Robin Haw, Bijay Jassal, Florian Korninger, and Bruce May. 2018. "The reactome pathway knowledgebase." *Nucleic Acids Research* 46 (D1):D649–D655.

Faux, Thomas, Kalle T Rytkönen, Mehrad Mahmoudian, Niklas Paulin, Sini Junttila, Asta Laiho, and Laura L Elo. 2021. "Differential ATAC-seq and ChIP-seq peak detection using ROTS." *NAR Genomics and Bioinformatics* 3 (3):lqab059.

Feingold, EA, PJ Good, MS Guyer, S Kamholz, L Liefer, K Wetterstrand, FS Collins, TR Gingeras, D Kampa, and EA Sekinger. 2004. "The ENCODE (ENCyclopedia of DNA elements) project." *Science* 306 (5696):636–640.

Feng, Xin, Robert Grossman, and Lincoln Stein. 2011. "PeakRanger: A cloud-enabled peak caller for ChIP-seq data." *BMC Bioinformatics* 12:1–11.

Forster, Samuel C, Alexander M Finkel, Jodee A Gould, and Paul J Hertzog. 2013. "RNA-eXpress annotates novel transcript features in RNA-seq data." *Bioinformatics* 29 (6):810–812.

García-Alcalde, Fernando, Konstantin Okonechnikov, José Carbonell, Luis M Cruz, Stefan Götz, Sonia Tarazona, Joaquín Dopazo, Thomas F Meyer, and Ana Conesa. 2012. "Qualimap: Evaluating next-generation sequencing alignment data." *Bioinformatics* 28 (20):2678–2679.

García-Ruiz, Sonia, Ana L Gil-Martínez, Alejandro Cisterna, Federico Jurado-Ruiz, Regina H Reynolds, Mark R Cookson, John Hardy, Mina Ryten, and Juan A Botía, and NABEC. 2021. "CoExp: A web tool for the exploitation of co-expression networks." *Frontiers in Genetics* 12:630187.

Gaspar, John M. 2018. "Improved peak-calling with MACS2." *BioRxiv*:496521.

Gontarz, Paul, Shuhua Fu, Xiaoyun Xing, Shaopeng Liu, Benpeng Miao, Viktoriia Bazylianska, Akhil Sharma, Pamela Madden, Kitra Cates, and Andrew Yoo. 2020. "Comparison of differential accessibility analysis strategies for ATAC-seq data." *Scientific Reports* 10 (1):10150.

Grabherr, Manfred G, Brian J Haas, Moran Yassour, Joshua Z Levin, Dawn A Thompson, Ido Amit, Xian Adiconis, Lin Fan, Raktima Raychowdhury, and Qiandong Zeng. 2011. Trinity: Reconstructing a full-length transcriptome without a genome from RNA-seq data." *Nature Biotechnology* 29 (7):644.

Grandi, Fiorella C, Hailey Modi, Lucas Kampman, and M Ryan Corces. 2022. "Chromatin accessibility profiling by ATAC-seq." *Nature Protocols* 17 (6):1518–1552.

Grant, Charles E, Timothy L Bailey, and William Stafford Noble. 2011. "FIMO: Scanning for occurrences of a given motif." *Bioinformatics* 27 (7):1017–1018.

Guo, Yuchun, Georgios Papachristoudis, Robert C Altshuler, Georg K Gerber, Tommi S Jaakkola, David K Gifford, and Shaun Mahony. 2010. "Discovering homotypic binding events at high spatial resolution." *Bioinformatics* 26 (24):3028–3034.

Guo, Yuchun, Shaun Mahony, and David K Gifford. 2012. "High resolution genome wide binding event finding and motif discovery reveals transcription factor spatial binding constraints." *PLOS Computational Biology* 8(8): e1002638

Gupta, Shobhit, John A Stamatoyannopoulos, Timothy L Bailey, and William Stafford Noble. 2007. "Quantifying similarity between motifs." *Genome Biology* 8 (2):1–9.

Halachev, Konstantin, Hannah Bast, Felipe Albrecht, Thomas Lengauer, and Christoph Bock. 2012. "EpiExplorer: Live exploration and global analysis of large epigenomic datasets." *Genome Biology* 13 (10):1–14.

Handy, Diane E, Rita Castro, and Joseph Loscalzo. 2011. "Epigenetic modifications: Basic mechanisms and role in cardiovascular disease." *Circulation* 123 (19):2145–2156.

Hardcastle, Thomas J, and Krystyna A Kelly. 2010. "baySeq: Empirical Bayesian methods for identifying differential expression in sequence count data." *BMC Bioinformatics* 11:1–14.

Heinz, Sven, Christopher Benner, Nathanael Spann, Eric Bertolino, Yin C Lin, Peter Laslo, Jason X Cheng, Cornelis Murre, Harinder Singh, and Christopher K Glass. 2010. "Simple combinations of lineage-determining transcription factors prime cis-regulatory elements required for macrophage and B cell identities." *Molecular Cell* 38 (4):576–589.

Hertz, Gerald Z, and Gary D. Stormo. 1999. "Identifying DNA and protein patterns with statistically significant alignments of multiple sequences." *Bioinformatics (Oxford, England)* 15 (7):563–577.

Hower, Valerie, Steven N Evans, and Lior Pachter. 2011. "Shape-based peak identification for ChIP-Seq." *BMC Bioinformatics* 12 (1):1–9.

Huang, Jinyan, Valérie Renault, J Sengenes, Nizar Touleimat, S Michel, M Lathrop, and Jorg Tost. 2012. "MeQA: A pipeline for MeDIP-seq data quality assessment and analysis." *Bioinformatics* 28 (4):587–588.

Huang, Weichun, Rasiah Loganantharaj, Bryce Schroeder, David Fargo, and Leping Li. 2013. "Pavis: A tool for peak annotation and visualisation." *Bioinformatics* 29 (23):3097–3099.

Huerta-Cepas, Jaime, Kristoffer Forslund, Luis Pedro Coelho, Damian Szklarczyk, Lars Juhl Jensen, Christian Von Mering, and Peer Bork. 2017. "Fast genome-wide functional annotation through orthology assignment by eggNOG-mapper." *Molecular Biology and Evolution* 34 (8):2115–2122.

Hunter, Sarah, Rolf Apweiler, Teresa K Attwood, Amos Bairoch, Alex Bateman, David Binns, Peer Bork, Ujjwal Das, Louise Daugherty, and Lauranne Duquenne. 2009. "InterPro: The integrative protein signature database." *Nucleic Acids Research* 37 (suppl_1):D211–D215.

Ibrahim, Mahmoud M, Scott A Lacadie, and Uwe Ohler. 2015. "JAMM: A peak finder for joint analysis of NGS replicates." *Bioinformatics* 31 (1):48–55.

Isserlin, Ruth, Daniele Merico, Veronique Voisin, and Gary D Bader. 2014. "Enrichment Map – A Cytoscape app to visualise and explore OMICs pathway enrichment results." *F1000Research* 3: 141.

Jaenisch, Rudolf, and Adrian Bird. 2003. "Epigenetic regulation of gene expression: How the genome integrates intrinsic and environmental signals." *Nature Genetics* 33 (3):245–254.

Ji, Hongkai, Hui Jiang, Wenxiu Ma, David S Johnson, Richard M Myers, and Wing H Wong. 2008. "An integrated software system for analysing ChIP-chip and ChIP-seq data." *Nature Biotechnology* 26 (11): 1293–1300.

Johnson, DS, A Mortazavi, RM Myers, and B Wold. 2007. "Genome-wide mapping of in vivo protein-DNA interactions." *Science* 316 (5830): 1497–1502.

Kanehisa, Minoru. 2002. "The KEGG database." 'In silico' simulation of biological processes: Novartis Foundation Symposium 247.

Karolchik, Donna, Robert Baertsch, Mark Diekhans, Terrence S Furey, Angie Hinrichs, YT Lu, Krishna M Roskin, Matt Schwartz, Charles W Sugnet, and Daryl J Thomas. 2003. "The UCSC genome browser database." *Nucleic Acids Research* 31 (1):51–54.

Kim, Daehwan, Geo Pertea, Cole Trapnell, Harold Pimentel, Ryan Kelley, and Steven L Salzberg. 2013. "TopHat2: Accurate alignment of transcriptomes in the presence of insertions, deletions and Gene fusions." *Genome Biology* 14 (4):1–13.

Kim, Daehwan, Joseph M Paggi, Chanhee Park, Christopher Bennett, and Steven L Salzberg. 2019. "Graph-based genome alignment and genotyping with HISAT2 and HISAT-genotype." *Nature Biotechnology* 37 (8):907–915.

Kim, Daehwan, Li Song, Florian P Breitwieser, and Steven L Salzberg. 2016. "Centrifuge: Rapid and sensitive classification of metagenomic sequences." *Genome Research* 26 (12):1721–1729.

Kondili, Maria, Annika Fust, Jens Preussner, Carsten Kuenne, Thomas Braun, and Mario Looso. 2017. "UROPA: A tool for universal RObust peak annotation." *Scientific Reports* 7 (1):2593.

Kozomara, Ana, Maria Birgaoanu, and Sam Griffiths-Jones. 2019. "miRBase: From microRNA sequences to function." *Nucleic Acids Research* 47 (D1):D155–D162.

Kuhn, Robert M, David Haussler, and W James Kent. 2013. "The UCSC genome browser and associated tools." *Briefings in Bioinformatics* 14 (2):144–161.

Kulakovskiy, Ivan V, VA Boeva, Alexander V Favorov, and Vsevolod J Makeev. 2010. "Deep and wide digging for binding motifs in ChIP-seq data." *Bioinformatics* 26 (20):2622–2623.

Kuleshov, Maxim V, Matthew R Jones, Andrew D Rouillard, Nicolas F Fernandez, Qiaonan Duan, Zichen Wang, Simon Koplev, Sherry L Jenkins, Kathleen M Jagodnik, and Alexander Lachmann. 2016. "Enrichr: A comprehensive Gene set enrichment analysis web server 2016 update." *Nucleic Acids Research* 44 (W1):W90–W97.

Kumari, R., R. Kumar, Consortium Open Source Drug Discovery, and A. Lynn. 2014. "g_mmpbsa – A GROMACS tool for high-throughput MM-PBSA calculations." *Journal of Chemical Information and Modeling* 54 (7):1951–1962. doi: 10.1021/ci500020m

Kuttippurathu, Lakshmi, Michael Hsing, Yongchao Liu, Bertil Schmidt, Douglas L Maskell, Kyungjoon Lee, Aibin He, William T Pu, and Sek Won Kong. 2011. "CompleteMOTIFs: DNA motif discovery platform for transcription factor binding experiments." *Bioinformatics* 27 (5):715–717.

Landt, Stephen G, Georgi K Marinov, Anshul Kundaje, Pouya Kheradpour, Florencia Pauli, Serafim Batzoglou, Bradley E Bernstein, Peter Bickel, James B Brown, and Philip Cayting. 2012. "ChIP-seq guidelines and practices of the ENCODE and modENCODE consortia." *Genome Research* 22 (9):1813–1831.

Langfelder, Peter, and Steve Horvath. 2008. "WGCNA: An r package for weighted correlation network analysis." *BMC Bioinformatics* 9 (1):1–13.

Langmead, B., and S. L. Salzberg. 2012. "Fast gapped-read alignment with Bowtie 2." *Nature Methods* 9 (4): 357–9. doi: 10.1038/nmeth.1923

Leek, Jeffrey T, Robert B Scharpf, Héctor Corrada Bravo, David Simcha, Benjamin Langmead, W Evan Johnson, Donald Geman, Keith Baggerly, and Rafael A Irizarry. 2010. "Tackling the widespread and critical impact of batch effects in high-throughput data." *Nature Reviews Genetics* 11 (10):733–739.

Lemoine, Gwenaëlle G, Marie-Pier Scott-Boyer, Bathilde Ambroise, Olivier Périn, and Arnaud Droit. 2021. "GWENA: Gene Co-expression networks analysis and extended modules characterization in a single Bioconductor package." *BMC Bioinformatics* 22 (1):1–20.

Leng, Ning, John A Dawson, James A Thomson, Victor Ruotti, Anna I Rissman, Bart MG Smits, Jill D Haag, Michael N Gould, Ron M Stewart, and Christina Kendziorski. 2013. "EBSeq: An empirical bayes hierarchical model for inference in RNA-seq experiments." *Bioinformatics* 29 (8):1035–1043.

Liang, Kun, and Sündüz Keles. 2012. "DBChIP: Detecting differential binding of transcription factors with ChIP-seq." *Bioinformatics* 28 (1):121–122.

Liao, Yang, Gordon K Smyth, and Wei Shi. 2014. "featureCounts: An efficient general purpose program for assigning sequence reads to genomic features." *Bioinformatics* 30 (7):923–930.

Li, Bo, and Colin N Dewey. 2011. "RSEM: Accurate transcript quantification from RNA-seq data with or without a reference genome." BMC Bioinformatics 12 (1):16.

Li, Daofeng, Deepak Purushotham, Jessica K Harrison, Silas Hsu, Xiaoyu Zhuo, Changxu Fan, Shane Liu, Vincent Xu, Samuel Chen, and Jason Xu. 2022. "WashU epigenome browser update 2022." *Nucleic Acids Research* 50 (W1):W774–W781.

Li, Daofeng, Silas Hsu, Deepak Purushotham, Renee L Sears, and Ting Wang. 2019. "WashU epigenome browser update 2019." *Nucleic Acids Research* 47 (W1):W158–W165.

Li, Heng, and Richard Durbin. 2009. "Fast and accurate short read alignment with Burrows–Wheeler transform." *Bioinformatics* 25 (14):1754–1760.

Li, Heng, Bob Handsaker, Alec Wysoker, Tim Fennell, Jue Ruan, Nils Homer, Gabor Marth, Goncalo Abecasis, Richard Durbin, and Genome Project Data Processing Subgroup. 2009. "The sequence alignment/map format and SAMtools." *Bioinformatics* 25 (16):2078–2079.

Li, Heng, Jue Ruan, and Richard Durbin. 2008. "Mapping short DNA sequencing reads and calling variants using mapping quality scores." *Genome Research* 18 (11):1851–1858.

Li, Ruiqiang, Yingrui Li, Karsten Kristiansen, and Jun Wang. 2008. "SOAP: Short oligonucleotide alignment program." *Bioinformatics* 24 (5):713–714.

Li, Xinmin, and Cun-Yu Wang. 2021. "From bulk, single-cell to spatial RNA sequencing." *International Journal of Oral Science* 13 (1):36.

Li, Zhijian, Marcel H Schulz, Thomas Look, Matthias Begemann, Martin Zenke, and Ivan G Costa. 2019. "Identification of transcription factor binding sites using ATAC-seq." *Genome Biology* 20:1–21.

Lienhard, Matthias, Christina Grimm, Markus Morkel, Ralf Herwig, and Lukas Chavez. 2014. "MEDIPS: Genome-wide differential coverage analysis of sequencing data derived from DNA enrichment experiments." *Bioinformatics* 30 (2):284–286.

Liu, Bin, Jimmy Yi, Aishwarya Sv, Xun Lan, Yilin Ma, Tim HM Huang, Gustavo Leone, and Victor X Jin. 2013. "QChIPat: A quantitative method to identify distinct binding patterns for two biological ChIP-seq samples in different experimental conditions." *BMC Genomics* 14:1–11.

Liu, Tao, Jorge A Ortiz, Len Taing, Clifford A Meyer, Bernett Lee, Yong Zhang, Hyunjin Shin, Swee S Wong, Jian Ma, and Ying Lei. 2011. "Cistrome: An integrative platform for transcriptional regulation studies." *Genome Biology* 12 (8):1–10.

Love, Michael I, Wolfgang Huber, and Simon Anders. 2014. "Moderated estimation of fold change and dispersion for RNA-seq data with DESeq2." *Genome Biology* 15 (12):1–21.

Lowe, Rohan, Neil Shirley, Mark Bleackley, Stephen Dolan, and Thomas Shafee. 2017. "Transcriptomics technologies." *PLOS Computational Biology* 13 (5):e1005457.

Lund, Eivind G, Isabelle Duband-Goulet, Anja Oldenburg, Brigitte Buendia, and Philippe Collas. 2015. "Distinct features of lamin A-interacting chromatin domains mapped by ChIP-sequencing from sonicated or micrococcal nuclease-digested chromatin." *Nucleus* 6 (1):30–39.

Machanick, Philip, and Timothy L Bailey. 2011. "MEME-ChIP: Motif analysis of large DNA datasets." *Bioinformatics* 27 (12):1696–1697.

Madrigal, Pedro, and Pawel Krajewski. 2013. "Package 'NarrowPeaks'." https://www.bioconductor.org/packages//2.10/bioc/html/NarrowPeaks.html

Mahony, Shaun, Matthew D Edwards, Esteban O Mazzoni, Richard I Sherwood, Akshay Kakumanu, Carolyn A Morrison, Hynek Wichterle, and David K Gifford. 2014. "An integrated model of multiple-condition ChIP-seq data reveals predeterminants of Cdx2 binding." *PLOS Computational Biology* 10 (3):e1003501.

Mahony, Shaun, and Panayiotis V Benos. 2007. "STAMP: A web tool for exploring DNA-binding motif similarities." *Nucleic Acids Research* 35 (suppl_2):W253–W258.

Martin, Marcel. 2011. "Cutadapt removes adapter sequences from high-throughput sequencing reads." *EMBnet Journal* 17 (1):10–12.

Mathelier, Anthony, Xiaobei Zhao, Allen W Zhang, François Parcy, Rebecca Worsley-Hunt, David J Arenillas, Sorana Buchman, Chih-yu Chen, Alice Chou, and Hans Ienasescu. 2014. "JASPAR 2014: An extensively expanded and updated open-access database of transcription factor binding profiles." *Nucleic Acids Research* 42 (D1):D142–D147.

McLean, Cory Y, Dave Bristor, Michael Hiller, Shoa L Clarke, Bruce T Schaar, Craig B Lowe, Aaron M Wenger, and Gill Bejerano. 2010. "GREAT improves functional interpretation of *cis*-regulatory regions." *Nature Biotechnology* 28 (5):495–501.

Nakato, Ryuichiro, and Toyonori Sakata. 2021. "Methods for ChIP-seq analysis: A practical workflow and advanced applications." *Methods* 187:44–53.

O'Connor, Lance M, Blake A O'Connor, Su Bin Lim, Jialiu Zeng, and Chih Hung Lo. 2023. "Integrative multi-omics and systems bioinformatics in translational neuroscience: A data mining perspective." *Journal of Pharmaceutical Analysis* 13:836–850.

Ou, Jianhong, Haibo Liu, Jun Yu, Michelle A Kelliher, Lucio H Castilla, Nathan D Lawson, and Lihua Julie Zhu. 2018. "ATACseqQC: A bioconductor package for post-alignment quality assessment of ATAC-seq data." *BMC Genomics* 19:1–13.

Ouyang, Ningxin, and Alan P Boyle. 2020. "TRACE: Transcription factor footprinting using chromatin accessibility data and DNA sequence." *Genome Research* 30 (7):1040–1046.

Patro, Rob, Geet Duggal, and Carl Kingsford. 2015. "Salmon: Accurate, versatile and ultrafast quantification from RNA-seq data using lightweight-alignment." *bioRxiv* 10:021592.

Pelizzola, Mattia, Yasuo Koga, Alexander Eckehart Urban, Michael Krauthammer, Sherman Weissman, Ruth Halaban, and Annette M Molinaro. 2008. "MEDME: An experimental And Analytical methodology for the estimation of DNA methylation levels based on microarray derived MeDIP-enrichment." *Genome Research* 18 (10):1652–1659.

Pertea, Mihaela, Geo M Pertea, Corina M Antonescu, Tsung-Cheng Chang, Joshua T Mendell, and Steven L Salzberg. 2015. "StringTie enables improved reconstruction of a transcriptome from RNA-seq reads." *Nature Biotechnology* 33 (3):290–295.

Piper, Jason, Salam A Assi, Pierre Cauchy, Christophe Ladroue, Peter N Cockerill, Constanze Bonifer, and Sascha Ott. 2015. "Wellington-bootstrap: Differential DNase-seq footprinting identifies cell-type determining transcription factors." *BMC Genomics* 16 (1):1–8.

Quevillon, Emmanuel, Ville Silventoinen, Sharmila Pillai, Nicola Harte, N Mulder, Rolf Apweiler, and Rodrigo Lopez. 2005. "InterProScan: protein domains identifier." *Nucleic Acids Research* 33 (suppl_2):W116–W120.

Quinlan, Aaron R, and Ira M Hall. 2010. "BEDTools: A flexible suite of utilities for comparing genomic features." *Bioinformatics* 26 (6):841–842.

Raghavan, Venket, Louis Kraft, Fantin Mesny, and Linda Rigerte. 2022. "A simple guide to *de novo* transcriptome assembly and annotation." *Briefings in Bioinformatics* 23 (2):bbab563.

Raj, Anil, Heejung Shim, Yoav Gilad, Jonathan K Pritchard, and Matthew Stephens. 2015. "msCentipede: Modeling heterogeneity across genomic sites and replicates improves accuracy in the inference of transcription factor binding." *PLOS One* 10 (9):e0138030.

Rashid, Naim U, Paul G Giresi, Joseph G Ibrahim, Wei Sun, and Jason D Lieb. 2011. "ZINBA integrates local covariates with DNA-seq data to identify broad and narrow regions of enrichment, even within amplified genomic regions." *Genome Biology* 12:1–20.

Raudvere, Uku, Liis Kolberg, Ivan Kuzmin, Tambet Arak, Priit Adler, Hedi Peterson, and Jaak Vilo. 2019. "g: Profiler: a web server for functional enrichment analysis and conversions of gene lists (2019 update)." *Nucleic Acids Research* 47 (W1):W191–W198.

Ren, Bing, François Robert, John J Wyrick, Oscar Aparicio, Ezra G Jennings, Itamar Simon, Julia Zeitlinger, Jorg Schreiber, Nancy Hannett, and Elenita Kanin. 2000. "Genome-wide location and function of DNA binding proteins." *Science* 290 (5500):2306–2309.

Ritchie, Matthew E, Belinda Phipson, DI Wu, Yifang Hu, Charity W Law, Wei Shi, and Gordon K Smyth. 2015. "Limma powers differential expression analyses for RNA-sequencing and microarray studies." *Nucleic Acids Research* 43 (7):e47–e47.

Robinson, James T, Helga Thorvaldsdóttir, Wendy Winckler, Mitchell Guttman, Eric S Lander, Gad Getz, and Jill P Mesirov. 2011. "Integrative genomics viewer." *Nature Biotechnology* 29 (1):24–26.

Robinson, M. D., D. J. McCarthy, and G. K. Smyth. 2010. "edgeR: A Bioconductor package for differential expression analysis of digital gene expression data." *Bioinformatics* 26 (1):139–140. doi: 10.1093/bioinformatics/btp616

Sahinyan, Korin, Darren M Blackburn, Marie-Michelle Simon, Felicia Lazure, Tony Kwan, Guillaume Bourque, and Vahab D Soleimani. 2022. "Application of ATAC-Seq for genome-wide analysis of the chromatin state at single myofiber resolution." *eLife* 11:e72792.

Sakai, Hiroaki, Sung Shin Lee, Tsuyoshi Tanaka, Hisataka Numa, Jungsok Kim, Yoshihiro Kawahara, Hironobu Wakimoto, Ching-chia Yang, Masao Iwamoto, and Takashi Abe. 2013. "Rice Annotation Project Database (RAP-DB): an integrative and interactive database for rice genomics." *Plant and Cell Physiology* 54 (2):e6.

Salmon-Divon, Mali, Heidi Dvinge, Kairi Tammoja, and Paul Bertone. 2010. "PeakAnalyzer: Genome-wide annotation of chromatin binding and modification loci." *BMC Bioinformatics* 11:1–12.

Sandelin, Albin, Wynand Alkema, Pär Engström, Wyeth W Wasserman, and Boris Lenhard. 2004. "JASPAR: An open-access database for eukaryotic transcription factor binding profiles." *Nucleic Acids Research* 32 (suppl_1):D91–D94.

Sandve, Geir K, Sveinung Gundersen, Halfdan Rydbeck, Ingrid K Glad, Lars Holden, Marit Holden, Knut Liestøl, Trevor Clancy, Egil Ferkingstad, and Morten Johansen. 2010. "The genomic HyperBrowser: Inferential genomics at the sequence level." *Genome Biology* 11:1–12.

Schep, Alicia. 2023. *motifmatchr: Fast Motif Matching in R*. R package version 1.24.0, https://bioconductor.org/packages/motifmatchr.

Schep, Alicia N, Beijing Wu, Jason D Buenrostro, and William J Greenleaf. 2017. "chromVAR: Inferring transcription-factor-associated accessibility from single-cell epigenomic data." *Nature Methods* 14 (10): 975–978.

Schep, Alicia N, Jason D Buenrostro, Sarah K Denny, Katja Schwartz, Gavin Sherlock, and William J Greenleaf. 2015. "Structured nucleosome fingerprints enable high-resolution mapping of chromatin architecture within regulatory regions." *Genome Research* 25 (11):1757–1770.

Schüffler, Peter, Thomas Mikeska, Andreas Waha, Thomas Lengauer, and Christoph Bock. 2009. "MethMarker: User-Friendly design and optimization of gene-specific DNA methylation assays." *Genome Biology* 10 (10):1–10.

Schulz, Marcel H, Daniel R Zerbino, Martin Vingron, and Ewan Birney. 2012. "Oases: Robust de novo RNA-seq assembly across the dynamic range of expression levels." *Bioinformatics* 28 (8):1086–1092.

Schweikert, Gabriele, Botond Cseke, Thomas Clouaire, Adrian Bird, and Guido Sanguinetti. 2013. "MMDiff: Quantitative testing for shape changes in ChIP-seq data sets." *BMC Genomics* 14 (1):1–17.

Shao, Zhen, Yijing Zhang, Guo-Cheng Yuan, Stuart H Orkin, and David J Waxman. 2012. "MAnorm: A robust model for quantitative comparison of ChIP-seq data sets." *Genome Biology* 13:1–17.

Shen, Li, Ning-Yi Shao, Xiaochuan Liu, Ian Maze, Jian Feng, and Eric J Nestler. 2013. "diffReps: Detecting differential chromatin modification sites from ChIP-seq data with biological replicates." *PLOS One* 8 (6):e65598.

Shi, Dong-Qiao, Iftikhar Ali, Jun Tang, and Wei-Cai Yang. 2017. "New insights into 5hmC DNA modification: Generation, distribution and function." *Frontiers in Genetics* 8:100.

Shin, Hyunjin, Tao Liu, Arjun K Manrai, and X Shirley Liu. 2009. "CEAS: *cis*-regulatory element annotation system." *Bioinformatics* 25 (19):2605–2606.

Simão, Felipe A, Robert M Waterhouse, Panagiotis Ioannidis, Evgenia V Kriventseva, and Evgeny M Zdobnov. 2015. "BUSCO: Assessing genome assembly and annotation completeness with single-copy orthologs." *Bioinformatics* 31 (19):3210–3212.

Smith, Jason P, M Ryan Corces, Jin Xu, Vincent P Reuter, Howard Y Chang, and Nathan C Sheffield. 2021. "PEPATAC: An optimized pipeline for ATAC-seq data Analysis with serial alignments." *NAR Genomics and Bioinformatics* 3 (4):lqab101.

Spyrou, Christiana, Rory Stark, Andy G Lynch, and Simon Tavaré. 2009. "BayesPeak: Bayesian analysis of ChIP-seq data." *BMC Bioinformatics* 10:1–17.

Stark, Rory, Gordon Brown (2011). DiffBind: differential binding analysis of ChIP-Seq peak data. http://bioconductor.org/packages/release/bioc/vignettes/DiffBind/inst/doc/DiffBind.pdf.

Sticht, Carsten, Carolina De La Torre, Alisha Parveen, and Norbert Gretz. 2018. "miRWalk: An online resource for prediction of microRNA binding sites." *PLOS One* 13 (10):e0206239.

Stovner, Endre Bakken, and Pål Sætrom. 2019. "epic2 efficiently finds diffuse domains in ChIP-seq data." *Bioinformatics* 35 (21):4392–4393.

Subramanian, Aravind, Pablo Tamayo, Vamsi K Mootha, Sayan Mukherjee, Benjamin L Ebert, Michael A Gillette, Amanda Paulovich, Scott L Pomeroy, Todd R Golub, and Eric S Lander. 2005. "Gene Set enrichment analysis: A knowledge-based approach for interpreting genome-wide expression profiles." *Proceedings of the National Academy of Sciences* 102 (43):15545–15550.

Swarbreck, David, Christopher Wilks, Philippe Lamesch, Tanya Z Berardini, Margarita Garcia-Hernandez, Hartmut Foerster, Donghui Li, Tom Meyer, Robert Muller, and Larry Ploetz. 2007. "The Arabidopsis information resource (TAIR): Gene Structure and function annotation." *Nucleic Acids Research* 36 (suppl_1):D1009–D1014.

Tan, Ge, and Boris Lenhard. 2016. "TFBSTools: An R/bioconductor package for transcription factor binding site analysis." *Bioinformatics* 32 (10):1555–1556.

Tarazona, Sonia, Fernando García, Alberto Ferrer, Joaquín Dopazo, and Ana Conesa. 2011. "NOIseq: A RNA-seq differential expression method robust for sequencing depth biases." *EMBnet Journal* 17 (B):18–19.

Tarbell, Evan D, and Tao Liu. 2019. "HMMRATAC: A hidden Markov ModeleR for ATAC-seq." *Nucleic Acids Research* 47 (16):e91–e91.

Teufel, Felix, José Juan Almagro Armenteros, Alexander Rosenberg Johansen, Magnús Halldór Gíslason, Silas Irby Pihl, Konstantinos D Tsirigos, Ole Winther, Søren Brunak, Gunnar von Heijne, and Henrik Nielsen. 2022. "SignalP 6.0 predicts all five types of signal peptides using protein language models." *Nature Biotechnology* 40 (7):1023–1025.

The RNAcentral Consortium. 2019. "RNAcentral: A hub of information for non-coding RNA sequences." *Nucleic Acids Research* 47 (D1):D221–D229.

Thomas-Chollier, Morgane, Carl Herrmann, Matthieu Defrance, Olivier Sand, Denis Thieffry, and Jacques Van Helden. 2012. "RSAT peak-Motifs: Motif analysis in full-size ChIP-seq datasets." *Nucleic Acids Research* 40 (4):e31–e31.

Thorvaldsdóttir, Helga, James T Robinson, and Jill P Mesirov. 2013. "Integrative genomics viewer (IGV): High-performance genomics data visualisation and exploration." *Briefings in Bioinformatics* 14 (2):178–192.

Trapnell, Cole, Adam Roberts, Loyal Goff, Geo Pertea, Daehwan Kim, David R Kelley, Harold Pimentel, Steven L Salzberg, John L Rinn, and Lior Pachter. 2012. "Differential gene and transcript expression analysis of RNA-seq experiments with TopHat and cufflinks." *Nature Protocols* 7 (3):562–578.

Wang, Congmao, Jie Xu, Dasheng Zhang, Zoe A Wilson, and Dabing Zhang. 2010. "An effective approach for identification of in vivo protein–DNA binding sites from paired-end ChIP-seq data." *BMC Bioinformatics* 11:1–8.

Wang, Liguo, Shengqin Wang, and Wei Li. 2012. "RSeQC: Quality control of RNA-seq experiments." *Bioinformatics* 28 (16):2184–2185.

Wang, Likun, Zhixing Feng, Xi Wang, Xiaowo Wang, and Xuegong Zhang. 2010. "DEGseq: An R package for identifying DEGs from RNA-seq data." *Bioinformatics* 26 (1):136–138.

Wei, Zheng, Wei Zhang, Huan Fang, Yanda Li, and Xiaowo Wang. 2018. "esATAC: An easy-to-use systematic pipeline for ATAC-seq data analysis." *Bioinformatics* 34 (15):2664–2665.

Weirauch, Matthew T, Ally Yang, Mihai Albu, Atina G Cote, Alejandro Montenegro-Montero, Philipp Drewe, Hamed S Najafabadi, Samuel A Lambert, Ishminder Mann, and Kate Cook. 2014. "Determination and inference of eukaryotic transcription factor sequence specificity." *Cell* 158 (6):1431–1443.

Whitington, Tom, Martin C Frith, James Johnson, and Timothy L Bailey. 2011. "Inferring transcription factor complexes from ChIP-seq data." *Nucleic Acids Research* 39 (15):e98–e98.

Wingender, Edgar. 2008. "The TRANSFAC project as an example of framework technology that supports the analysis of genomic regulation." *Briefings in Bioinformatics* 9 (4):326–332.

Wood, Derrick E, Jennifer Lu, and Ben Langmead. 2019. "Improved metagenomic analysis with kraken 2." *Genome Biology* 20:1–13.

Wu, Hao, and Hongkai Ji. 2014. "PolyaPeak: Detecting transcription factor binding sites from ChIP-seq using peak shape information." *PLOS One* 9 (3):e89694.

Xie, Chen, Xizeng Mao, Jiaju Huang, Yang Ding, Jianmin Wu, Shan Dong, Lei Kong, Ge Gao, Chuan-Yun Li, and Liping Wei. 2011. "KOBAS 2.0: A web server for annotation and identification of enriched pathways and diseases." *Nucleic Acids Research* 39 (suppl_2):W316–W322.

Xie, Yinlong, Gengxiong Wu, Jingbo Tang, Ruibang Luo, Jordan Patterson, Shanlin Liu, Weihua Huang, Guangzhu He, Shengchang Gu, and Shengkang Li. 2014. "SOAPdenovo-trans: *de novo* transcriptome assembly with short RNA-seq reads." *Bioinformatics* 30 (12):1660–1666.

Yan, Feng, David R Powell, David J Curtis, and Nicholas C Wong. 2020. "From reads to insight: A hitchhiker's guide to ATAC-seq data analysis." *Genome Biology* 21:1–16.

Yang, Chi, Lu Ma, Donglai Xiao, Zhenghe Ying, Xiaoling Jiang, and Yanquan Lin. 2019. "Integration of ATAC-seq and RNA-seq identifies key genes in light-induced primordia formation of *Sparassis* latifolia." *International Journal of Molecular Sciences* 21 (1):185.

Yang, Xi, Di Liu, Fei Liu, Jun Wu, Jing Zou, Xue Xiao, Fangqing Zhao, and Baoli Zhu. 2013. "HTQC: A fast quality control toolkit for Illumina sequencing data." *BMC Bioinformatics* 14 (1):1–4.

Yu, Guangchuang, Li-Gen Wang, and Qing-Yu He. 2015. "ChIPseeker: An R/Bioconductor package for ChIP peak annotation, comparison and visualisation." *Bioinformatics* 31 (14):2382–2383.

Yu, Guangchuang, Li-Gen Wang, Yanyan Han, and Qing-Yu He. 2012. "clusterProfiler: An r package for comparing biological themes among Gene clusters." *Omics: A Journal of Integrative Biology* 16 (5):284–287.

Zeng, Xin, Rajendran Sanalkumar, Emery H Bresnick, Hongda Li, Qiang Chang, and Sündüz Keleş. 2013. "jMOSAiCS: Joint analysis of multiple ChIP-seq datasets." *Genome Biology* 14 (4):1–18.

Zhang, Yanxiao, Yu-Hsuan Lin, Timothy D Johnson, Laura S Rozek, and Maureen A Sartor. 2014. "PePr: A peak-calling prioritization pipeline to identify consistent or differential peaks from replicated ChIP-seq data." *Bioinformatics* 30 (18):2568–2575.

Zhang, Yong, Tao Liu, Clifford A Meyer, Jérôme Eeckhoute, David S Johnson, Bradley E Bernstein, Chad Nusbaum, Richard M Myers, Myles Brown, and Wei Li. 2008. "Model-based analysis of ChIP-seq (MACS)." *Genome Biology* 9 (9):1–9.

Zhang, Xuekui, Gordon Robertson, Martin Krzywinski, Kaida Ning, Arnaud Droit, Steven Jones, and Raphael Gottardo. 2011. "PICS: Probabilistic inference for ChIP-seq." *Biometrics* 67 (1):151–163.

Zhao, Zhongming, Yulin Dai, Chi Zhang, Ewy Mathé, Lai Wei, and Kai Wang. 2019. The International Conference on Intelligent Biology and Medicine (ICIBM) 2019: Bioinformatics Methods and Applications for Human Diseases. Springer.

Zhu, Lihua J, Claude Gazin, Nathan D Lawson, Hervé Pagès, Simon M Lin, David S Lapointe, and Michael R Green. 2010. "ChIPpeakAnno: A bioconductor package to annotate ChIP-seq and ChIP-chip data." *BMC Bioinformatics* 11:1–10.

Zuo, Zuqi, Yonghao Jin, Wen Zhang, Yichen Lu, Bin Li, and Kun Qu. 2019. "ATAC-pipe: General analysis of genome-wide chromatin accessibility." *Briefings in Bioinformatics* 20 (5):1934–1943.

3 Streamlining Next-Generation Sequencing Data Analysis with Nextflow and nf-core Pipelines

Sandeep Mallya, Nikhil Gunala, Sai Kiran KS, Sateesh Peri, and Bobby Paul

NEXT-GENERATION SEQUENCING (NGS): TECHNOLOGY EVOLUTION

NGS is a high-throughput DNA sequencing technique that generates millions of short reads in a single experiment. The availability of NGS technologies has significantly increased the amount of sequence data available and has transformed the approach to genome sequencing in biomedical sciences [1]. The emergence of NGS technologies and computational workflows for data analysis has greatly reduced the cost of sequencing to about $1,000 per genome [2].

Sequencing platforms such as Illumina and Ion Torrent are currently the leading choices for NGS due to their high throughput and accuracy and the wide range of applications they can be used for in research and clinical settings. Illumina NGS technology relies on reversible terminator chemistry and the sequential addition of nucleotides to generate short DNA fragments that can then be sequenced by synthesis [3]. It is commonly used for applications such as whole genome sequencing (WGS), whole transcriptome shotgun sequencing (WTSS/RNA-seq), and ChIP-seq. Ion Torrent NGS technology, on the other hand, uses the detection of hydrogen ions released during DNA synthesis and is suitable for applications including WGS, targeted sequencing, and metagenomics [4].

Currently, the third-generation sequencing technology is expected to revolutionise genome sequencing due to its ability to generate very long reads of up to thousands of base pairs [5]. It is also known to have substantially higher error rates compared to earlier technologies, complicating the analysis of the resulting data. Despite the high error rates, these technologies have demonstrated the ability to outperform existing methods in applications that can tolerate higher error rates, such as structural variant calling, at low depths of sequencing [6]. Pacific Biosciences (PacBio) and Oxford Nanopore Technologies (ONT) are the leading third-generation sequencing technologies that are known for their ability to generate very long reads (up to 50,000 base pairs or more) with high accuracy [7]. PacBio relies on single-molecule sequencing and is used for applications including *de novo* genome sequencing, structural variation detection, and long-range epigenetic analysis [8], whereas ONT involves the passage of DNA through a nanopore and the detection of changes in ionic current. This technology can be applied to a range of applications such as whole genome sequencing, transcriptome sequencing, and metagenomics [9].

APPLICATIONS OF NGS TECHNOLOGY

NGS technologies have their own unique applications in research and clinical settings. There are different types of sequencing, such as WGS, whole exome sequencing (WES), targeted gene(s), different types of RNAs, proteins interacting with DNA, methylated genomic regions, and DNA from different environments, and each has its own specific goals and needs. Factors such as goals of the project, size and complexity of genome, and type and quality of the sample determine the choice of the technology employed in sequencing [10]. DNA sequencing is used primarily for detecting genetic variations, especially for clinically relevant mutations. WGS gives us the output of all the

DOI: 10.1201/9781003354062-3

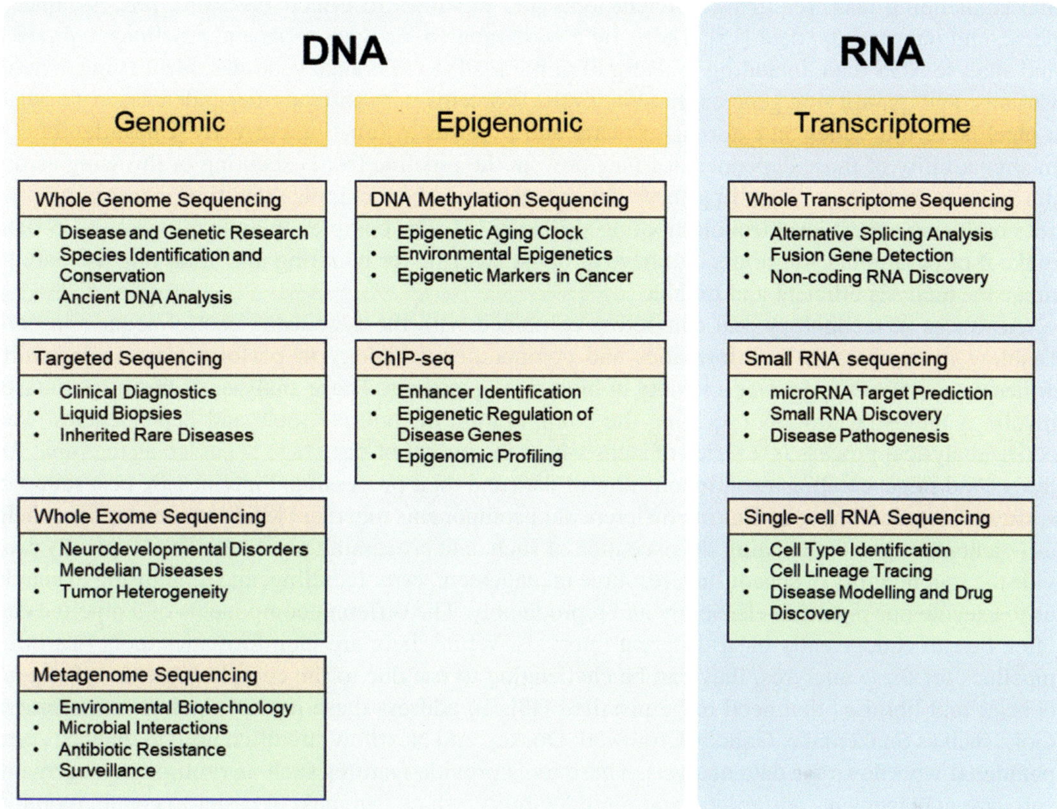

FIGURE 3.1 Chart representing the various types of NGS techniques and the areas of their application.

coding and non-coding regions and can identify genetic variations at high resolution but is limited by the cost and time required for data generation [11]. On the other hand, exome and targeted sequencing are cost-effective and efficient alternatives that provide us with a more efficient way to generate and analyse data that can be used in lieu of whole genome sequencing for a wide range of applications, including genetic diagnosis, ancestry testing, evolutionary studies, and personalised medicine [12]. RNA-seq, a technique useful for studying gene expression and regulation, identifying transcript isoforms, and understanding the functional role of coding and non-coding RNA, provides us with a comprehensive view of an organism's transcriptome [13]. Technologies such as ChIP-seq are used to study the binding sites and regulatory roles of specific proteins and can be used to identify regulatory elements and understand the mechanisms of gene regulation [10,14]. As a culture-independent method, the metagenome sequencing technique is widely used to amplify DNA from different environments that can be bioinformatically classified into bacteria/archaea or fungi respectively. Types and applications of various NGS technologies are depicted in Figure 3.1.

CHALLENGES OF NGS DATA ANALYSIS

The popularity of the technology and the reduced cost of data generation has provided NGS with a wide range of applications in biology and medicine. The explosion in data generation has led to NGS being routinely used in research as well as diagnostic settings. Advances in information technology revolutionised Big Data analysis and reduced the daunting task in data analysis for biologists. However, managing and analysing the vast amount of data generated by NGS is a complex and challenging task, requiring efficient tools and pipelines to effectively store, process, query, parse, and interpret the data [15]. There are a wide range of analysis tools and pipelines to process and analyse NGS data, including tools for aligning reads to reference genomes, identifying genetic variants, and quantifying gene expression levels. Problems with software dependencies arise when a pipeline or tool relies on external software and libraries to function correctly. Unavailability or incompatibility of these dependencies may prevent the pipeline from executing or throwing errors during execution. Also, installing these dependencies can be a complex and time-consuming process requiring expert-level troubleshooting skills [16,17]. The complexity and volume of NGS data make it of paramount importance to have a robust pipeline for handling and analysing the data to make the analysis efficient and reliable.

Advances in technology and computing combined with the decreasing cost of sequencing are enabling researchers in bioinformatics and computational biology to perform large-scale multi-omics data analyses to answer a variety of biological questions. These analyses of this large dataset involve a multistep process requiring the configuration of multiple tools and dependencies. The entire analytical process is a series of steps where the output of one stage is passed as the input for the second stage, creating a continuous flow of data and their processing. Pipelines are best suited to address these challenges by linking different data components together [18]. Pipeline managers help us in automating and managing the execution of such data processing pipelines. They typically provide the functionality of scheduling, resource management, error handling, and monitoring to enable us to execute our pipelines efficiently and reproducibly. The different components of a pipeline can often be run concurrently or at different intervals. While there are bioinformatics tools that offer pipelines for these analyses, they can be challenging to use due to the complex systems of dependencies and libraries that need to be installed [18]. To address these issues, pipeline management tools, such as Snakemake, Galaxy, Cromwell, Docker, and Nextflow, are often used to manage computational workflows for data analysis. These tools provide features such as runtime management, software management, portability, and reproducibility, which can make it easier to create, monitor, execute, and share pipelines. Pipeline management tools can streamline the data analysis and overcome the challenges of managing complex processes with numerous software dependencies [19]. A comparison of Nextflow with other pipeline management tools is shown in Table 3.1.

TABLE 3.1
Comparison of Pipeline Management Tools

Feature	Snakemake	Cromwell	Docker	Galaxy Tools	Nextflow
Workflow definition language	Python-based DSL	WDL	Dockerfile	XML	Groovy-based DSL
Parallelisation	Rule-based	Multiple backend execution systems	Containerised	Data-driven	Data parallelism
Reproducibility and portability	Good compatibility with various computational platforms	Integration with data and analysis tools	Encapsulates software dependencies	Vast collection of prebuilt tools and workflows	Active community and extensive collection of reusable workflows
Familiarity with tool required	Python	WDL	Containerisation concepts	Web-based interface	Nextflow DSL
Setup complexity	Relatively simple	Complex	Relatively simple	Simple	Simple
Customisation and extensibility	Limited	Limited	Limited	Limited	Extensive
Integration with external tools	Limited	Limited	Limited	Limited	Extensive

OVERVIEW OF NEXTFLOW AND ITS FEATURES

Pipeline management tools help us streamline the execution of data processing workflows efficiently and reproducibly with minimal error. Bioinformatics analyses often involve multiple steps such as data preprocessing, analysis, and visualisation. Nextflow is an open-source bioinformatics workflow manager that can be used for the development and execution of reproducible data analysis pipelines. It is designed to make it easy to write and run pipelines that can be scaled and adapted to different computing environments. The software is written in the Groovy programming language and uses a simple and expressive syntax for defining pipelines and processes [20].

Nextflow enables the deployment of workflows on various platforms, including Kubernetes, AWS Batch, Google Cloud Life Sciences, and high-performance computing (HPC) schedulers, as well as local execution. Further, its domain specific language (DSL) allows for the easy creation of sophisticated parallel and dynamic operations, particularly on clouds and clusters. It also has built-in support for container technologies like Conda, Docker, Podman, Singularity, and Modules, which help manage dependencies within the workflow. These containers allow for the isolation of processes from the primary platform they were built and run on and enable the easy integration of pre-existing pipelines written in languages like Bash, R, and Python [21].

The key features of Nextflow include the ability to quickly prototype, reuse existing scripts and tools, and ensure reproducibility. It also integrates with GitHub, allowing for the creation of self-contained pipelines with version management and reproducibility capabilities. The abstraction layer provides Nextflow with the ability to run pipelines on multiple platforms, including cloud services. It is based on the dataflow programming model, which simplifies the writing of complex distributed pipelines and allows for inherent parallelisation that can scale up or scale out. It has resume functionality that tracks intermediate results during pipeline execution and allows for the resumption of pipelines from the last successfully completed step [20].

BASIC CONCEPTS OF NEXTFLOW

Nextflow is designed around the popularity of the Linux platform in the field of data science. Linux offers various command-line and scripting tools that can be easily linked together to effectively handle complex data processing tasks. Nextflow goes further by enabling the creation of complex program interactions and a high-level parallel computing environment based on the dataflow programming model. The three main parts of Nextflow workflows are processes, channels, and workflows, as discussed in the following section [22].

PROCESSES, CHANNELS, AND WORKFLOWS

In Nextflow, a process is a defined unit of work that performs a specific task. A process can be as simple as a single command or script, or it can be a complex series of steps that involve multiple inputs, outputs, and dependencies. A process is defined in a Nextflow script using the process keyword followed by a name and a block of code. For example:

```
process fastqc {
    input:
        file fastq_file
    output:
        file html_report
    script:
        """
        fastqc $fastq_file -o .
        """
}
```

This process takes a single input, a FASTQ file, and produces a single output, an HTML report generated by the `fastqc` tool. The code to be executed by the process is defined in the script block, which can contain one or more commands.

Data may only be transferred between processes via the asynchronous queues known as channels. A channel is a data flow mechanism in Nextflow that allows processes to communicate with each other. Channels are defined in a Nextflow script using the `channel` keyword followed by a name and a data type. For example:

```
channel fastq_files
```

This defines a channel called `fastq_files` that can be used to pass FASTQ files between processes. A process can read from a channel using the `from` keyword and write to a channel using the `into` keyword, as shown in the following example:

```
process fastqc {
    input:
        file fastq_file
    output:
        file html_report
    script:
        """
        fastqc $fastq_file -o .
        """
}

fastq_files = Channel.fromFile("*.fastq")
fastq_files.read().map { fastq_file ->
    fastqc(fastq_file)
}.into(html_reports)
```

This code defines a channel called `fastq_files` that reads all FASTQ files in the current directory, and a process called `fastqc` that takes a FASTQ file as input and produces an HTML output [23].

Essentially in Nextflow, processes are defined by specifying the inputs and outputs for a specific task. Channels are used to control the flow of data between processes, and the workflow section explicitly defines the interaction between processes and the overall execution flow of the pipeline. Nextflow allows end-users to define processes, which are the specific commands or scripts that need to be executed, and the executors determine how the processes are actually run on the target system. By default, processes are run on the local computer, but Nextflow also supports running processes on HPCs or cloud platforms. This separation between the functional logic of the pipeline and the execution platform makes it possible to write a pipeline once and then run it on different systems without modifying the workflow. Nextflow includes built-in support for popular batch schedulers and cloud platforms like Sun Grid Engine, SLURM, AWS Batch, and Kubernetes [20].

NF-CORE AND ITS RELEVANCE FOR NGS DATA ANALYSIS

nf-core is a collection of community-built pipelines created using Nextflow and can be found on the nf-core home page (https://nf-co.re). nf-core is meant to be a resource for best practices in pipeline development, with a focus on reproducibility and ease of use, and is aimed at three main audiences: facilities, single users, and developers. Facilities can use nf-core pipelines to provide automated, optimised workflows that ensure the reproducibility of results for their users. Single users can benefit from the easy-to-use, portable, and well-documented pipelines. Developers can also use nf-core to write their own pipelines using ready-made templates and helper tools. nf-core pipelines are designed to be easy to install and run, with clear documentation and a consistent directory structure, and are extensively tested and validated to produce reliable results. In addition to the pipelines, nf-core provides tools and resources for developers to help them create high-quality pipelines that meet the standards of the nf-core community [24].

As on May 1, 2024, there are 73 pipelines that are currently available as part of nf-core for analysis of RNA sequencing, variant calling, ChIP-seq, single-cell RNA sequencing, metagenomics, and other sequencing data.

INSTALLING AND RUNNING NEXTFLOW AND NF-CORE PIPELINES

To install Nextflow on system, one needs to have Bash 3.2 or later and Java 11 or later (up to 18) installed. Nextflow can be used on any system that is compatible with POSIX, such as Linux or macOS, and does not require any additional installation steps. On Windows, it can be run through Windows Subsystem for Linux (WSL).

The following command can be used to install Nextflow:

```
wget -qO- https://get.nextflow.io | bash

## Or using curl
curl -s https://get.nextflow.io | bash
```

This will download and install the latest version of Nextflow in the current directory and then add the Nextflow `bin` directory to `$PATH` environment variable to run the `nextflow` command from any location.

Nextflow can also be installed through Bioconda. Note that one needs to have Bioconda installed (https://bioconda.github.io/).

```
conda create -name myenv
conda activate myenv
conda install -c bioconda nextflow
```

This will create a new environment for Conda, activate the Bioconda environment, and then install Nextflow in the environment from the Bioconda channel, which will allow the use of the `nextflow` command from any directory.

Now that Nextflow is set up, nf-core pipelines can be set up.

Note that nf-core is typically distributed as Conda packages and installed using Bioconda.

```
conda install -c bioconda nf-core
```

This will install the latest version of nf-core in the environment.

To install a specific nf-core pipeline, such as the RNA-seq pipeline, execute the following command:

```
conda install -c bioconda nf-core-rnaseq
```

This will install the RNA-seq pipeline in the environment, and we can then use the pipeline by running the `nextflow` command with the appropriate pipeline name, as in the following example:

```
nextflow run nf-core/rnaseq
```

At this point, the nf-core RNA-seq pipeline is installed and is ready for analysis. nf-core pipelines use the built-in support for software packaging that Nextflow offers and can work with Docker and Singularity, and most pipelines also support Conda.

EXAMPLES OF USING NEXTFLOW AND NF-CORE PIPELINES FOR NGS ANALYSIS

Before getting started, a checklist will be run through to ensure that everything needed is available to proceed with the installation of the nf-core pipeline. This may include verifying the installation of required dependencies, such as a compatible version of Python or Conda, and confirming that the necessary access privileges to install packages on the system are in place. An active internet connection is also required to fetch all pipeline requirements.

Certain software dependencies must be installed, such as Nextflow and one of the following container platforms such as Docker, Singularity, or Conda. The user can test that everything is working correctly by executing the `nextflow run hello` command. If there are no errors thrown, then proceed with choosing a pipeline to execute using the command `nf-core list`. The next step is to configure Nextflow and specify the pipeline-specific parameters. It is recommended to use Docker or Singularity for reproducibility, although Conda is also supported.

Last step is to test the pipe by executing the following command:

```
nextflow run nf-core/<pipeline_name> -profile test,docker --outdir <OUTDIR>
```

More information about the available pipelines is available in the nf-core pipeline documentation.

Getting Ready to Run an nf-core Pipeline

nf-core provides a set of tools to help users run and develop pipelines. These tools allow end user to list all the available pipelines and their versions, along with information about the versions installed locally. There are also commands to assist with downloading pipelines for offline use. The `--help` option is used to list the subcommands.

The most important subcommand is the `list` command that gives the output of all available nf-core pipelines along with information about the version with local installation. The output includes the latest version number and release date, as well as details last pulled the pipeline locally using Nextflow with the latest version.

The command runs as follows:

```
nf-core list
```

This generates the following output:

```
nf-core/tools version 2.4.1 - https://nf-co.re
There is a new version of nf-core/tools available! (2.7.2)
```

Pipeline Name	Stars	Latest Release	Released	Last Pulled	Have latest release?
fetchngs	73	1.9	2 weeks ago	-	-
rnaseq	549	3.10	2 weeks ago	2 years ago	Yes (v3.1)
methylseq	96	2.3.0	2 weeks ago	-	-
ampliseq	101	2.4.1	4 weeks ago	-	-
airrflow	23	2.4.0	4 weeks ago	-	-
mhcquant	21	2.4.0	1 months ago	-	-
atacseq	127	2.0	1 months ago	-	-
sarek	214	3.1.1	1 months ago	-	-
eager	82	2.4.6	2 months ago	-	-
coproid	7	1.1.1	2 months ago	-	-
cutandrun	36	3.0	2 months ago	-	-
smrnaseq	45	2.1.0	2 months ago	1 years ago	No (v1.1.0)
nascent	5	2.0.0	2 months ago	-	-
hgtseq	14	1.0.0	2 months ago	-	-
hlatyping	36	2.0.0	3 months ago	-	-
demultiplex	18	1.0.0	3 months ago	-	-
scrnaseq	67	2.1.0	3 months ago	5 months ago	No (v2.0.0)
chipseq	135	2.0.0	3 months ago	-	-
isoseq	5	1.1.1	3 months ago	-	-
mag	109	2.2.1	4 months ago	-	-
epitopeprediction	22	2.1.0	5 months ago	-	-
viralrecon	86	2.5	6 months ago	-	-

This output shows that the latest version of the RNA-seq pipeline is installed [25].

Executing an nf-core RNA-seq Pipeline

RNA-seq pipeline on nf-core is a pipeline for analysing the RNA-seq data having a reference genome and annotation. The data in this pipeline are processed in five stages. In stage one, the QC is estimated by FastQC [26] and then UMI barcode is extracted using UMI-tools [27], followed by the trimming of the adapter using TrimGalore [28]; then quality is checked by removing any contamination of the genome using BBSplit [29], and SortMeRNA [30] is used to remove ribosomal RNA. The second stage, which includes the alignment and quantification stages, is where the sequence alignment as well as quantification of expression levels is done using STAR [31], Salmon [32], RSEM [33], and HISAT2 [34]. The next stage is optional and includes pseudo-alignment and quantification stages where the data are quantified by Salmon. The fourth stage of postprocessing includes sorting and indexing of alignment using SAMtools [35], then deduplication is done using UMI-tools dedup, and Picard MarkDuplicates [36] is used to mark the duplicate reads. The final QC stage includes visualisation, such as heat maps and dendrograms followed by MultiQC [37] to generate the single HTML report, which includes all the output from the supported tools, i.e., SortMeRNA, RSEM, SAMtools, FastQC, and featureCounts [38]. With the system configured, the analysis is ready to run. The execution of the pipeline can be initiated by running the following piece of code.

```
nextflow run nf-core/rnaseq --input samplesheet.csv --outdir <OUTDIR>
--genome GRCh37 -profile docker
```

This command runs the pipeline using Docker on all the input `fastq` files mentioned in the `samplesheet.csv` file using the default set of tools and parameters. Here the reads are aligned to the GRCh37 version of the genome that is downloaded, and the output files are saved to the `<OUTDIR>` directory. This is the most basic command, and additional parameters can also be added for customisation.

Creating a Sample Sheet

This is a four-column comma-separated file with information on sample IDs, the name of `fastq` files, and the strandedness information. We can add information about the single- and paired-end `fastq` files. The file consists of a header line followed by other information, and the full construct of a sample sheet is as follows:

```
sample,fastq_1,fastq_2,strandedness
CONTROL_REP1,AEG588A1_R1_001.fastq.gz,AEG588A1_R2_001.fastq.gz,forward
CONTROL_REP2,AEG588A2_R1_001.fastq.gz,AEG588A2_R2_001.fastq.gz,forward
CONTROL_REP3,AEG588A3_R1_001.fastq.gz,AEG588A3_R2_001.fastq.gz,forward
```

The sample sheet has a header row followed by sample information. The first column is the custom sample name given to each of the sequencing runs. The second and third columns should be the full path to the FastQ files for read pair 1 and 2, respectively. It is to be noted that the FastQ files must be in compressed formats with the extension `.fastq.gz`. If the sample file is a single-end file, then the path to the FastQ file is to be entered in the second column, and the third column can be left blank. The fourth column is for the sample strand specificity.

Reference Genome Options

A FASTA and GTF file are required for the alignment and annotation of the sample files. Either a path to the locally available reference genome files can be provided as a parameter, or the pipeline will download the files from AWS iGenomes directly. The parameter `--save_reference`, when used, saves the downloaded genome locally, making it available for future runs. This also saves time and space, as building the genome index is a time-consuming process and need not be repeated every run.

Choice of Aligner and Other Alignment Options

The default choice of aligner by the pipeline is STAR, and the quantification is done by SALMON. The end-user can also use HISAT2 to align by adding an additional parameter `--aligner hisat2` to the original command. Similarly, RSEM can also be used for quantification instead of SALMON. It is recommended to check the pipeline documentation for specific parameters that can be used with the command while executing the pipeline. Note also that the QC and other downstream analysis steps are independent of the choice of aligners.

The following is a fully functional command for executing the pipeline using the locally saved genome:

```
nextflow run nf-core/rnaseq --input samplesheet.csv --outdir /path/to/
output --genome GRCh37 -profile docker --max_cpus 32 --skip_stringtie 0
--star_index /path/to/STARIndex/ --fasta /path/to/genome.fa --gtf /path/
to/genes.gtf -resume
```

In this command, options passed to the pipeline are as follows:

1. `--input samplesheet.csv`: This specifies the input sample sheet, a CSV file that lists the samples to be processed and their associated file paths.
2. `--outdir /path/to/output`: This specifies the output directory where the pipeline will write its results.
3. `--genome GRCh37`: This specifies the genome build to be used by the pipeline, in this case GRCh37.
4. `-profile docker`: This specifies that the pipeline should be run using Docker to provide the necessary software dependencies.
5. `--max_cpus 32`: This limits the number of CPUs to be used by the pipeline to 32.
6. `--skip_stringtie 0`: This is a flag to skip the StringTie step of the pipeline; by default, it is set to 0, meaning it will not be skipped. StringTie is used for assembling transcripts and to estimate their abundance.
7. `--star_index /path/to/STARIndex/`: This specifies the path to the STAR index for genome alignment.
8. `--fasta /path/to/genome.fa`: This specifies the path to the FASTA file for genome reference.
9. `--gtf /path/to/genes.gtf`: This specifies the path to the GTF file for gene annotation
10. `-resume`: This tells Nextflow to resume a previous execution of the pipeline. It will check the checkpoint directory and use the information to continue from where it last ran. This is not a mandatory parameter, but it is suggested for use, as it will prevent the pipeline from starting from the beginning, saving time and resources.

In summary, this command successfully completes the running of the nf-core/RNA-seq pipeline with defined input sample sheet, output directory, genome, profile, `max_cpus`, `skip_stringtie`, `star_index`, `fasta`, and `gtf` files. Also, the pipeline has the option of resuming its execution.

Results Directory

Typically, all the output files are written to the directory specified in the command with the `--outdir` option. Additionally, a `work/` directory gets created that stores all the intermediate files generated during the pipeline execution. Log files are generated that have the details on each of the processes executed during the analysis. These logs can be used for debugging and troubleshooting issues that may have occurred during the run. Typically, all the necessary files are saved in the results directory specified previously. The file types and their location are discussed in Table 3.2. The output files

TABLE 3.2
List of the Directories Generated by the RNA-seq Pipeline as Output and Its Contents, Including the Type of Files and the Contents of the Output Files

Directory	File Formats	Description
`fastqc`	HTML and ZIP	Quality statistics of individual samples
`multiqc`	HTML	Contains read alignment, read mapping statistics and plots in an HTML file format.
`pipeline_info`	Text and HTML	Contains the execution report and metadata related to pipeline execution.
`salmon`	TSV and RData	Contains the merged counts and quantified expression values in transcripts per million (TPM). RData files contain the DESeqDataSet object that can be used for differential expression analysis.
`star_salmon`	BAM and TSV	Contains aligned `.bam files`, indices, and optionally gene- and isoform-level count/TPM matrices.
`trimgalore`	Text	Contains the preprocessing statistics for all the sample files.

include all those necessary for further downstream analysis, and they are labelled and arranged in a directory structure that makes it easy to use them.

CONCLUSIONS AND FUTURE DIRECTIONS

The development of the Nextflow framework to write and execute reproducible data analysis pipelines and the subsequent development of nf-core, a collection of pipelines created using Nextflow, have provided a standardised and open-source framework for pipeline development and execution. This has enabled the standardisation of pipelines due to strict implementation guidelines, making it easy for researchers to compare results from different studies and facilitating collaboration. The open-source nature of Nextflow makes it available to all to use, modify, and distribute, allowing researchers access to high-quality and well-maintained pipelines without having to develop an in-house framework. Community efforts for developing and maintaining pipelines allow continuous improvements with fewer bugs. The platform-independent nature of the nf-core pipelines allows them to be deployed on local workstations or on a cloud, as well as on high-performance computing clusters, allowing flexibility in terms of computational resource availability. These pipelines are designed to be reproducible, enabling the results to be independently verified and replicated by others and eliminating inconsistencies creeping in by using different versions of software or different parameters.

While the nf-core pipelines have many advantages, a few potential drawbacks should be considered. First up, the nf-core pipelines are designed to be highly modular and customisable, and their use may require a certain level of programming expertise. A few of the nf-core pipelines may require a significant amount of computational resources and time to run. Due to the open-source nature of the development, there may be limited support and insufficient documentation compared to some of the very popular tools or their commercial equivalents. Frequent updates to the pipeline may lead to compatibility issues with the older versions of the tools installed that may be difficult to troubleshoot. Overall, nf-core has made it easier for researchers to access, use, and share high-quality NGS analysis pipelines enabling efficiency and accuracy in research.

REFERENCES

1. Goodwin, S., McPherson, J. D. & McCombie, W. R. Coming of age: Ten years of next-generation sequencing technologies. *Nat. Rev. Genet.* **17**, 333–351 (2016).
2. Slatko, B. E., Gardner, A. F. & Ausubel, F. M. Overview of next-generation sequencing technologies. *Curr. Protoc. Mol. Biol.* **122**, e59 (2018).

3. Bentley, D. R. *et al.* Accurate whole human genome sequencing using reversible terminator chemistry. *Nature* **456**, 53–59 (2008).
4. Rusk, N. Torrents of sequence. *Nat. Methods* **8**, 44–44 (2011).
5. Heather, J. M. & Chain, B. The sequence of sequencers: The history of sequencing DNA. *Genomics* **107**, 1–8 (2016).
6. Alvarez, J. R., Skachkov, D., Massey, S. E., Kalitsov, A. & Velev, J. P. DNA/RNA transverse current sequencing: Intrinsic structural noise from neighboring bases. *Front. Genet.* **6**, 213–213 (2015).
7. Dijk, E. L., van, Jaszczyszyn, Y., Naquin, D. & Thermes, C. The third revolution in sequencing technology. *Trends Genet.* **34**, 666–681 (2018).
8. Rhoads, A. & Au, K. F. PacBio sequencing and its applications. *Genom. Proteom. Bioinform.* **13**, 278–289 (2015).
9. Lu, H., Giordano, F. & Ning, Z. Oxford Nanopore MinION sequencing and genome assembly. *Genom. Proteom. Bioinform.* **14**, 265–279 (2016).
10. Rabbani, B., Nakaoka, H., Akhondzadeh, S., Tekin, M. & Mahdieh, N. Next generation sequencing: Implications in personalized medicine and pharmacogenomics. *Mol. Biosyst.* **12**, 1818–1830 (2016).
11. Alfares, A. *et al.* Whole-genome sequencing offers additional but limited clinical utility compared with reanalysis of whole-exome sequencing. *Genet. Med.* **20**, 1328–1333 (2018).
12. Platt, C. D. *et al.* Efficacy and economics of targeted panel versus whole-exome sequencing in 878 patients with suspected primary immunodeficiency. *J. Allergy Clin. Immunol.* **147**, 723–726 (2021).
13. Wang, Z., Gerstein, M. & Snyder, M. RNA-seq: A revolutionary tool for transcriptomics. *Nat. Rev. Genet.* **10**, 57–63 (2009).
14. Visel, A. *et al.* ChIP-seq accurately predicts tissue-specific activity of enhancers. *Nature* **457**, 854–858 (2009).
15. Gullapalli, R. R., Desai, K. V., Santana-Santos, L., Kant, J. A. & Becich, M. J. Next generation sequencing in clinical medicine: Challenges and lessons for pathology and biomedical informatics. *J. Pathol. Inform.* **3**, 40 (2012).
16. Roy, S. *et al.* Standards and guidelines for validating next-generation sequencing bioinformatics pipelines: A joint recommendation of the association for molecular pathology and the college of American pathologists. *J. Mol. Diagn.* **20**, 4–27 (2018).
17. SoRelle, J. A., Wachsmann, M. & Cantarel, B. L. Assembling and validating bioinformatic pipelines for next-generation sequencing clinical assays. *Arch. Pathol. Lab. Med.* **144**, 1118–1130 (2020).
18. Leipzig, J. A review of bioinformatic pipeline frameworks. *Brief. Bioinform.* **18**, 530–536 (2017).
19. Wratten, L., Wilm, A. & Göke, J. Reproducible, scalable, and shareable analysis pipelines with bioinformatics workflow managers. *Nat. Methods* **18**, 1161–1168 (2021).
20. Di Tommaso, P. *et al.* Nextflow enables reproducible computational workflows. *Nat. Biotechnol.* **35**, 316–319 (2017).
21. Spišaková, V., Hejtmánek, L. & Hynšt, J. Nextflow in bioinformatics: Executors performance comparison using genomics data. *Future Gener. Comput. Syst.* **142**, 328–339 (2023).
22. Introduction to bioinformatics workflows with Nextflow and nf-core: Summary and setup. https://carpentries-incubator.github.io/workflows-nextflow/. Accessed 30 Sep 2024.
23. Nextflow 22.10.4 documentation. https://www.nextflow.io/docs/latest/. Accessed 30 Jan 2023.
24. Ewels, P. A. *et al.* The nf-core framework for community-curated bioinformatics pipelines. *Nat. Biotechnol.* **38**, 276–278 (2020).
25. Patel, H. *et al.* nf-core/RNA-Seq: nf-core/RNA-Seq v3.10.1 - Plastered Rhodium Rudolph. (2023) doi:10.5281/ZENODO.1400710
26. Andrews, S. Babraham bioinformatics - FastQC A quality control tool for high throughput sequence data. https://www.bioinformatics.babraham.ac.uk/projects/fastqc/. Accessed 30 Jan 2023.
27. Smith, T., Heger, A. & Sudbery, I. UMI-tools: Modeling sequencing errors in unique molecular identifiers to improve quantification accuracy. *Genome Res.* **27**, 491–499 (2017).
28. Krueger, F. Babraham bioinformatics - Trim Galore! https://www.bioinformatics.babraham.ac.uk/projects/trim_galore/. Accessed 30 Jan 2023.
29. Bushnell, B. BBMap: A fast, accurate, splice-aware aligner. https://www.osti.gov/biblio/1241166 (2014).
30. Kopylova, E., Noé, L. & Touzet, H. SortMeRNA: Fast and accurate filtering of ribosomal RNAs in metatranscriptomic data. *Bioinformatics* **28**, 3211–3217 (2012).
31. Dobin, A. *et al.* STAR: Ultrafast universal RNA-seq aligner. *Bioinforma. Oxf. Engl.* **29**, 15–21 (2013).
32. Patro, R., Duggal, G., Love, M. I., Irizarry, R. A. & Kingsford, C. Salmon provides fast and bias-aware quantification of transcript expression. *Nat. Methods* **14**, 417–419 (2017).

33. Li, B. & Dewey, C. N. RSEM: Accurate transcript quantification from RNA-seq data with or without a reference genome. *BMC Bioinformatics* **12**, 323 (2011).
34. Kim, D., Paggi, J. M., Park, C., Bennett, C. & Salzberg, S. L. Graph-based genome alignment and genotyping with HISAT2 and HISAT-genotype. *Nat. Biotechnol.* **37**, 907–915 (2019).
35. Li, H. *et al.* The sequence Alignment/Map format and SAMtools. *Bioinformatics* **25**, 2078–2079 (2009).
36. Broad Institute. Picard Tools. http://broadinstitute.github.io/picard/. Accessed 30 Jan 2023.
37. Ewels, P., Magnusson, M., Lundin, S. & Käller, M. MultiQC: Summarize analysis results for multiple tools and samples in a single report. *Bioinformatics* **32**, 3047–3048 (2016).
38. Liao, Y., Smyth, G. K. & Shi, W. featureCounts: An efficient general purpose program for assigning sequence reads to genomic features. *Bioinformatics* **30**, 923–930 (2014).

4 Best Practices for Variant Calling Using the Genome Analysis Toolkit

V. P. Snijesh, K. Shivakumar, K. M. Punya, and Ashitha S. Niranjana Murthy

INTRODUCTION

Next-generation sequencing (NGS) has revolutionised the field of genomics since its introduction over a decade ago. It represents a high-throughput and cost-effective method for sequencing DNA and RNA, allowing researchers to decode and analyse genetic information on a scale not previously achievable [1]. The growth of NGS technology has been remarkable, driving significant advancements in various fields and contributing to numerous scientific breakthroughs and medical discoveries. NGS has found numerous applications across different areas of research and medicine [2–6]. In the field of genomics, it has facilitated large-scale projects such as the Human Genome Project (HGP), which aimed to sequence the entire human genome [7–9]. NGS has enabled the exploration of genomic variations, including single nucleotide polymorphisms (SNPs), insertions, deletions, and structural variations, which play critical roles in understanding genetic diseases and population genetics.

In the field of clinical diagnostics, NGS has transformed genetic testing by enabling comprehensive and rapid analysis of patient samples. It has become a valuable tool for diagnosing genetic disorders, identifying disease-causing mutations, and guiding treatment decisions. NGS-based tests are now routinely used for carrier screening, prenatal testing, newborn screening, and oncology diagnostics. Beyond genomics, NGS has extended its applications to other areas of research, such as transcriptomics and epigenetics [10,11]. RNA sequencing (RNA-seq) allows the quantification and profiling of gene expression, leading to a deeper understanding of cellular functions, developmental processes, and disease mechanisms [12–14]. Epigenetic modifications, such as DNA methylation and histone modifications, can also be analysed using NGS techniques, providing insights into gene regulation and epigenetic alterations associated with diseases [15].

The growth of NGS has been fuelled by continuous advancements in sequencing technologies, bioinformatics tools, and data analysis methods. The introduction of new platforms and improvements in sequencing chemistry and read lengths have enhanced the accuracy, speed, and cost-effectiveness of NGS. Moreover, the development of robust bioinformatics pipelines and computational algorithms has enabled the efficient processing and interpretation of vast amounts of sequencing data.

In this chapter, we discuss the current best practices for variant calling in clinical sequencing studies, with a particular emphasis on variant calling for inherited disorders. Variant calling is a fundamental step in the analysis of NGS data that involves identifying genetic variations in a sample relative to a reference genome. Accurate and reliable variant calling is crucial for understanding the genetic basis of diseases, identifying potential drug targets, and developing personalised therapies. The Genome Analysis Toolkit (GATK) is a widely used software package for variant calling in NGS data. Developed by the Broad Institute, GATK provides a comprehensive set of tools for variant discovery, genotyping, and annotation [16].

DOI: 10.1201/9781003354062-4

The software is highly customisable and can be tailored to specific experimental designs and sequencing platforms.

RESOURCE BUNDLE AND BIOLOGICAL SAMPLES

The resource bundle refers to a collection of reference files and known variant datasets that are commonly used in genomic analysis. The GATK resource bundle is a curated set of high-quality reference and annotation files provided by the Broad Institute to aid researchers in their genomic analysis pipelines. The resource bundle includes reference genome in FASTA format; variant-calling resources such as dbSNP, HapMap, and 1000 Genomes Project (1KGP) variants; annotation resources that provide functional information about genomic elements, which include information about gene locations, transcripts, protein-coding regions, and other genomic features; and index files that are necessary for efficient data retrieval and querying. These index files enable quick access to specific genomic regions, improving the performance of read alignment and variant-calling algorithms. The GATK resource bundle is regularly updated and maintained by the Broad Institute to ensure the availability of the most up-to-date and relevant genomic references and annotations. Researchers can download the resource bundle and incorporate it into their analysis pipelines, simplifying the process of obtaining and using essential genomic data. By providing a comprehensive set of reference and annotation files, the GATK resource bundle streamlines the setup and execution of genomic analysis workflows. It ensures consistent and standardised analysis across different projects and facilitates reproducibility and comparability of results within the genomics community.

The European Nucleotide Archive (ENA) [17] and Sequence Read Archive (SRA) [18] are key databases for storing and sharing biological data, particularly high-throughput sequencing and omics data. ENA serves as a comprehensive repository for nucleotide sequences, accepting data from various sequencing technologies and providing a platform for archiving, retrieval, and analysis. It encompasses not only sequence data but also metadata, annotations, and associated information. On the other hand, SRA, maintained by the National Center for Biotechnology Information (NCBI) [19], specialises in raw sequencing data storage and access, including short and long reads from NGS platforms. Researchers can submit their data to SRA for public availability, promoting data sharing and facilitating its reuse for diverse research purposes. In the current study, we used a trio dataset consisting of three samples (father, mother, and child) from the Simons Simplex Collection Project (PRJNA167318). The samples were downloaded using wget and subsequently considered for downstream analyses.

```
wget ftp.sra.ebi.ac.uk/vol1/fastq/SRR130/006/SRR1301246/SRR1301246_1.
fastq.gz
wget ftp.sra.ebi.ac.uk/vol1/fastq/SRR130/006/SRR1301246/SRR1301246_2.
fastq.gz
wget ftp.sra.ebi.ac.uk/vol1/fastq/SRR130/007/SRR1301247/SRR1301247_1.
fastq.gz
wget ftp.sra.ebi.ac.uk/vol1/fastq/SRR130/007/SRR1301247/SRR1301247_2.
fastq.gz
wget ftp.sra.ebi.ac.uk/vol1/fastq/SRR130/008/SRR1301248/SRR1301248_1.
fastq.gz
wget ftp.sra.ebi.ac.uk/vol1/fastq/SRR130/008/SRR1301248/SRR1301248_2.
fastq.gz
```

In this protocol series, we use terms as mentioned in Table 4.1, and the demo of running the script will be shown only in one sample. The usage of certain attributes with their path are given in Table 4.1. The overall workflow described in this protocol is shown in Figure 4.1. The requirements for running variant-calling analysis for optimal performance are given in supplementary File 1.

TABLE 4.1
The Usage of Attributes with Their Path

Attribute	Path
`$REFERENCE`	`bundle/Homo_sapiens_assembly38.fasta`
`$MillsIndels`	`bundle/Mills_and_1000G_gold_standard.indels.hg38.vcf.gz`
`$DBSNP`	`bundle/dbsnp_138.hg38.vcf.gz`
`$G1000`	`bundle/1000G_phase1.snps.high_confidence.hg38.vcf.gz`
`$hapmap`	`bundle/hapmap_3.3.hg38.vcf.gz`
`$omni`	`bundle/1000G_omni2.5.hg38.vcf.gz`
`$gnomad`	`bundle/af-only-gnomad.hg38.vcf.gz`

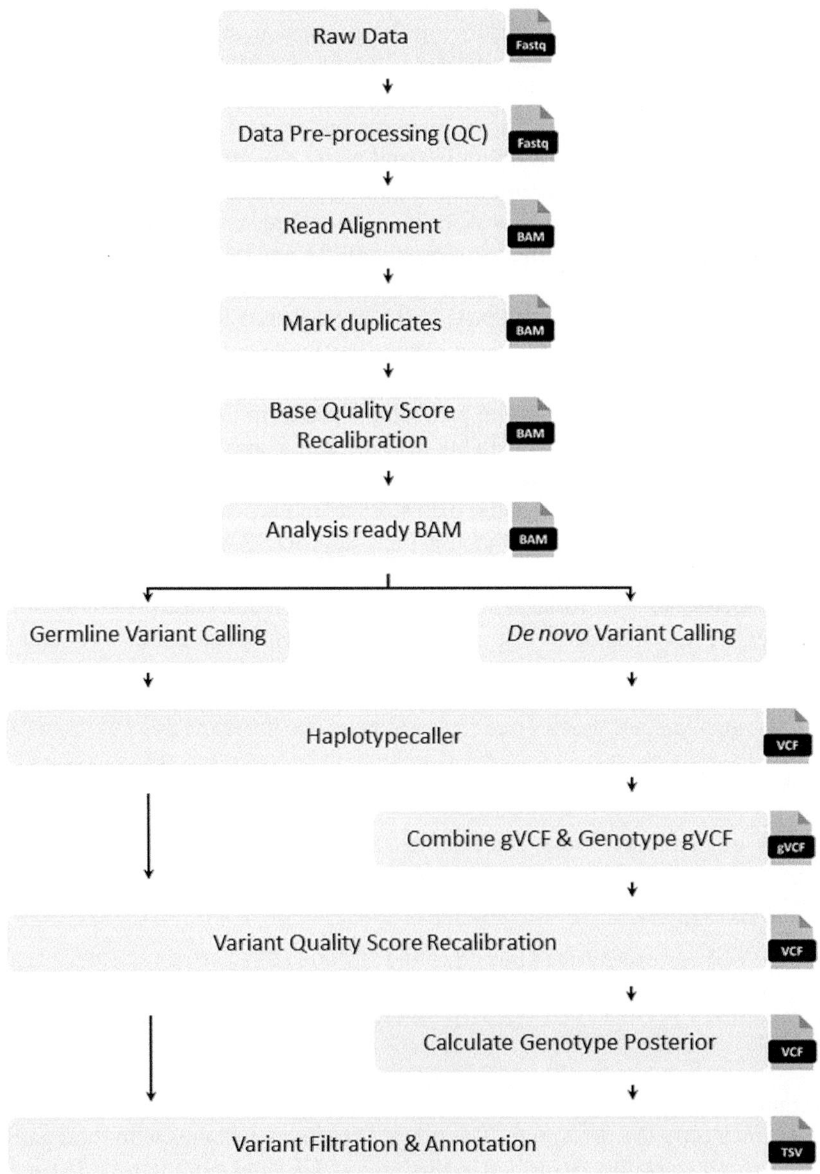

FIGURE 4.1 The overall workflow used in this protocol.

QUALITY CONTROL

Quality control (QC) involves assessing the quality of sequencing data to identify and remove low-quality reads. The goal of QC is to minimise noise and errors in the downstream analysis and improve the accuracy and reliability of variant calls. The first step in QC is to assess the quality of the raw sequencing data. This involves examining the sequencing metrics, such as read length, sequencing depth, and base quality scores, to identify any issues that may affect the quality of the data. Common quality issues include low read depth, adapters, high error rates, and base quality score compression. These issues can result in false-positive variant calls, which can lead to erroneous conclusions. The next step is to remove low-quality reads and trim adapters and low-quality bases from the sequencing data. This step is critical for improving the accuracy and reliability of variant calls. Several tools are available for quality control, including FastQC [20], Trimmomatic [21], TrimGalore [22,23]. and Cutadapt [24], among others. These tools help to remove low-quality reads and trim adapters and low-quality bases from the sequencing data. In this chapter, we used FastQC [20] for quality check and TrimGalore for removal of adapters and low-quality reads in the sample. The following commands were used to run QC and remove noise in the sample:

```
# Performing the quality control of the samples
fastqc \
  -t 2 \
  data/SRR1301246_1.fastq data/SRR1301246_2.fastq \
  -o /path/to/output/

# Trim Galore to trim adapters and low-quality bases
trim_galore \
  --paired \
  --quality 30 \
  --phred33 \
  -o data/trimmed \
  --illumina \
  data/SRR1301246_1.fastq data/SRR1301246_2.fastq
```

Parameters of fastqc: `-t` = number of files to run in parallel; `-o` = the directory to store the results.

Parameters of trim_galore: `--paired` = to mention input files are paired end; `--quality` = the Phred score cutoff to trim the samples; `-phred33` = to use ASCII+33 quality scores as Phred scores; `-o` = the directory to store the results; `--illumina` = to mention adapters are from Illumina.

ALIGNING THE READS TO THE REFERENCE GENOME

Alignment is a critical step in variant calling that involves mapping the sequencing reads to a reference genome to identify variations in the sequence. The goal of alignment is to map the reads accurately and efficiently to the reference genome while minimising false-positive and false-negative variant calls. The first step in alignment is to select an appropriate aligner tool. There are many aligner tools available, each with its own advantages and disadvantages. Some commonly used aligner tools include BWA [25], Bowtie [26], and HISAT2 [27], among others. BWA is a versatile and widely used bioinformatics tool specifically designed for the alignment of genomic DNA. Its adaptability makes it a popular choice for various DNA alignment tasks. HISAT, on the other hand, is uniquely tailored for spliced alignment, making it particularly well suited for RNA-seq data analysis, where the consideration of exons and introns is essential. Unlike BWA and HISAT, Bowtie is not designed for spliced alignment but excels in short read mapping for genomic DNA analysis. Each tool serves distinct purposes in the realm of bioinformatics, allowing researchers to choose the most suitable option based on the specific requirements of their genomic or transcriptomic analyses.

Each tool has its own parameters and options that can be optimised to improve the accuracy and efficiency of the alignment.

The next step is to prepare the reference genome and the sequenced reads for alignment. This involves indexing the reference genome and creating a sequence dictionary to enable faster and more efficient alignment of reads to the genome. GATK requires the main FASTA file to be accompanied by a dictionary file ending `in.dict` and an index file ending `in.fai`, as it allows efficient random access to the reference bases. Earlier, generating `a.dict` file was performed using the Picard tool, but it has been integrated into the GATK program. The alignment process generates an alignment file, which contains information about the alignment quality, read position, and read depth. The primary output of alignment is a Sequence Alignment Map (SAM) file which is sorted and converted to a Binary Alignment Map (BAM) file. Sorting SAM files organises the data by coordinates, improving efficiency. Converting SAM to BAM reduces file size and facilitates downstream analysis. Both processes are crucial for efficient storage and analysis of high-throughput data. In this chapter, we use the BWA aligner to map the reads to the reference genome build version hg38. The following commands were used to prepare the reference genome and read alignment:

```
# Generate BWA reference
bwa index \
  -a \
  bwtsw $REFERENCE

# Generate fasta index
samtools faidx $REFERENCE

# Create sequence dictionary
gatk CreateSequenceDictionary \
  -R $REFERENCE \
  -O Homo_sapiens_assembly38.dict

# Aligning the paired end reads to the reference genome
bwa mem \
  -M \
  -t 8 \
  -R "@RG\tID:SRR1301246\tSM:SRR1301246\tLB:SRR1301246\tPL:Illumina" \
  $REFERENCE \
  data/SRR1301246_1.fastq data/SRR1301246_2.fastq > SRR1301246_aligned.sam

# Sort and Convert SAM to BAM file
samtools view \
  -S -b SRR1301246_aligned.sam > SRR1301246_aligned_sorted.bam
```

Parameters of BWA index: `-a` = algorithm for constructing BWT index; `bwtsw` = algorithm implemented in BWT-SW for the whole human genome.

Parameters of BWA mem: `-M` = mark shorter split hits as secondary for GATK compatibility; `-t` = number of threads; `-R` = read group (RG), a metadata attribute associated with sequencing read.

Parameters GATK: `-R` = reference in FASTA format; `-O` = outputs sequence dictionary.

Parameters of samtools: `-S` = sort the SAM file; `-b` = convert to BAM format.

IDENTIFYING DUPLICATE READS IN SEQUENCING DATA

MarkDuplicates is a program in the GATK suite of tools that identifies and marks duplicate reads in sequencing data. Duplicate reads can arise during library preparation or sequencing due to amplification bias, polymerase chain reaction (PCR) errors, or sequencing errors, and they can introduce biases into downstream analyses. Marking duplicates is, therefore, an essential step in preparing high-quality sequencing data for variant calling and downstream analyses. MarkDuplicates works by comparing the start position, end position, and orientation of each read pair to identify duplicate reads. It uses an algorithm that takes into account the mapping quality, the length of the reads, and other factors to minimise the number of false positives and false negatives. MarkDuplicates also takes into account optical duplicates, which are duplicate reads that arise from imaging artefacts on the sequencing instrument. Once duplicate reads are identified, MarkDuplicates marks them with a flag in the read header, indicating that they should be treated as duplicates in downstream analyses. Marking duplicates does not remove them from the data, but it allows downstream programs to filter or ignore them if necessary. MarkDuplicates is particularly important in applications such as exome sequencing and whole-genome sequencing, where the depth of coverage is often high, and duplicate reads can represent a significant fraction of the data. Failure to mark duplicates can result in false-positive variant calls, decreased sensitivity, and decreased specificity in downstream analyses. MarkDuplicates can also help to identify and diagnose technical issues during sequencing and library preparation.

The output file in BAM format (`SRR1301246_aligned_sorted.bam`) generated from the alignment step will be given as input to MarkDuplicates function in GATK. The command used to perform MarkDuplicates is as follows:

```
# Mark duplicate reads in sequencing data
gatk MarkDuplicates \
  -I SRR1301246_aligned_sorted.bam \
  -M SRR1301246_markdups_gatk_metrics.txt \
  -O SRR1301246_MarkDup.bam \
  -VS LENIENT
```

Parameters of MarkDuplicates: `-I` = input file in SAM or BAM format; `-M` = outputs duplication metrics; `-O` = output duplicate marked file; `-VS` = setting stringency to improve performance (`STRICT/LENIENT/SILENT`).

RECALIBRATION OF BASE QUALITY SCORES

Variant-calling algorithms rely greatly on the quality score assigned to the individual base calls in each sequence read. The quality score indicates the reliability of the observation in reflecting the true biological state of the aligned base. A low-quality score suggests uncertainty in accurately identifying the base, reducing our confidence in its accuracy. Therefore, base calls with higher-quality scores are considered more trustworthy and receive greater reliance in the analysis process. BaseRecalibrator is a tool in the GATK that is used to recalibrate base quality scores in high-throughput sequencing data. Accurate base quality scores are essential for the identification of genetic variants and the interpretation of sequencing data, but these scores can be influenced by several factors, such as machine-specific errors, sequence context, and reference genome mismatches. BaseRecalibrator addresses these issues by creating a model that estimates the probability of a base being miscalled, based on several covariates, such as read group, cycle, and dinucleotide context. The process of recalibration involves two steps. The first step is to generate a recalibration table using the BaseRecalibrator tool. This table contains information about the relationship between the observed quality scores and the true error rates and is generated by analysing a set of known variant sites in the reference genome. The second step is to apply the recalibration table to the sequencing data using the ApplyBQSR tool. This tool uses the recalibration table to adjust the base quality scores in the sequencing data, resulting in more accurate base quality scores that can improve downstream analyses. BaseRecalibrator can be used to improve the accuracy of variant calling, particularly for rare or low-frequency variants, as well as for the identification of somatic mutations in cancer samples. It is also useful for reducing false-positive calls caused by sequencing artefacts and improving the performance of downstream analyses such as variant annotation and functional analysis. The output (`SRR1301246_MarkDup.bam`) generated from the MarkDuplicates step were given as the input to the BaseRecalibrator function in GATK. The following commands were used for BaseRecalibrator and ApplyBQSR with the parameters described below:

```
# Generate a recalibration table using the BaseRecalibrator
gatk BaseRecalibrator \
  -I SRR1301246_MarkDup.bam \
  -R $REFERENCE
  -O recal_data.table
  --known-sites $MillsIndels \
  --known-sites $G1000 \
  --known-sites $DBSNP

# Apply the recalibration table to the data
gatk ApplyBQSR \
  -I SRR1301246_MarkDup.bam \
  -R $REFERENCE \
  -bqsr recal_data.table \
  -O SRR1301246.bqsr.bam
```

Parameters of BaseRecalibrator: `-I` = input file in BAM format; `-R` = the reference genome; `-O` = name of file to write recalibration table; `-known-sites` = use databases of known polymorphic sites which will be used to exclude regions around known polymorphisms.

Parameters of ApplyBQSR: `-I` = input file in BAM format; `-R` = reference genome; `-bqsr` = name of the recalibrated table; `-O` = name of the output file to store recalibrated data.

The final output of the analysis will be an analysis-ready BAM file. The scripts can be run on other samples in a similar way. So, the final output for the three files will be `SRR1301246.bqsr.bam`, `SRR1301247.bqsr.bam`, and `SRR1301248.bqsr.bam`.

VARIANT CALLING FROM ANALYSIS-READY BAM FILES

UnifiedGenotyper and HaplotypeCaller are variant-calling tools in GATK that are used for variants in NGS datasets. While UnifiedGenotyper is not recommended anymore, currently, HaplotypeCaller is the widely used variant caller due to its ability to identify variants in complex regions and accurately call variants with high sensitivity and specificity. HaplotypeCaller uses a probabilistic model to identify haplotypes, which are sets of variants that are closely linked on a chromosome. This approach allows the tool to identify variants in regions that are difficult to call using traditional variant-calling methods, such as those with high levels of genetic variation or regions with low read depth [28].

The HaplotypeCaller algorithm works by first identifying potential variant sites in the sequence data using a local *de novo* assembly method. It then uses these candidate variants to construct haplotypes, which are used to generate a likelihood model for each read at each position in the genome. This model incorporates information about the quality of the base calls, the read depth at each position, and the probability of observing each possible genotype at the position. Once the likelihood models have been generated, HaplotypeCaller uses a Bayesian approach to calculate the posterior probability of each genotype at each position in the genome. This approach takes into account the uncertainty associated with each genotype call and the probability of sequencing errors, resulting in highly accurate genotype calls.

HaplotypeCaller also incorporates a number of additional features to improve the accuracy and reliability of variant calls. For example, it includes a read-backed phasing algorithm that uses information from read pairs to determine the haplotype phase of variants. It also includes a local realignment algorithm that corrects for alignment errors around indels, improving the accuracy of variant calls in regions with high levels of variation.

UnifiedGenotyper is another variant-calling tool in GATK that was widely used in the past but has since been replaced by HaplotypeCaller due to its limitations. Some of the main limitations of UnifiedGenotyper are:

Difficulty in detecting indels: UnifiedGenotyper was not optimised for detecting indels, which can result in lower sensitivity and specificity for indel calls.

Inability to handle complex variants: UnifiedGenotyper may not be able to accurately call complex variants, such as those involving multiple nucleotide substitutions and indels in close proximity.

Performance issues: UnifiedGenotyper can be computationally intensive and may require a lot of computational resources, especially for large datasets.

Limited scalability: UnifiedGenotyper may not scale well for large-scale sequencing projects, which can result in longer turnaround times for variant calling.

Dependency on population-based data: UnifiedGenotyper requires a large population-based dataset for accurate variant calling, which may not be available for all projects.

Overall, while UnifiedGenotyper was a valuable tool for variant calling in the past, its limitations have led to its replacement by HaplotypeCaller, which offers improved accuracy, sensitivity, and scalability. Also, HaplotypeCaller's ability to accurately identify variants in complex regions and its high sensitivity and specificity make it an essential tool for many sequencing applications, such as genome-wide association studies, disease gene mapping, and cancer genomics.

The final output generated from recalibration of base quality scores is provided as the input (`SRR1301246.bqsr.bam`, `SRR1301247.bqsr.bam`, and `SRR1301248.bqsr.bam`) to HaplotypeCaller. Here, we demonstrate germline variant calling from a sample as well as joint analysis of multiple samples like cohort data to derive *de novo* variants.

Calling *De Novo* Variants from the Trio Cohort

The HaplotypeCaller tool is employed to identify potential variant sites within individual samples and store the outcomes in the Genomic Variant Call Format (gVCF). The gVCF format retains variant sites while grouping non-variant sites into blocks based on genotype quality during the calling process. This approach compresses the variant call format (VCF) file without sacrificing any sites, enabling efficient joint analysis in subsequent stages. To call *de novo* variants, we require the family information as a pedigree file and gVCF of all samples in the trio set (father, mother, and affected child). The format of the pedigree (PED) file is given in Table 4.2. The PED files must contain six mandatory columns with Family ID, Individual ID, Paternal ID, Maternal ID, Sex (1 = male; 2 = female; other = unknown), and Phenotype (–9 missing, 0 missing, 1 unaffected, 2 affected). The PED file for the family is stored in a file named `trio.ped`.

TABLE 4.2
The Format of the PED File

#fam.id	ind.id	pat.id	mat.id	sex	phen
fam001	SRS596246	SRS596246	SRS596248	1	1
fam001	SRS596247	0	0	1	2
fam001	SRS596248	0	0	2	1

To call *de novo* variants from the trio cohort, the gVCF files for the trio samples were first generated by using HaplotypeCaller separately and combined into a multisample gVCF file through CombineGVCFs. Then the raw variants were called by GenotypeGVCFs, and the utility MakeSitesOnlyVcf was used to generate a simplified VCF file that includes only the variant sites without any sample-specific genotype information. Next these variants were recalibrated using VariantRecalibrator and ApplyVQSR functions. Finally, after deriving the posterior probabilities of the genotypes by CalculateGenotypePosteriors, the low-quality genotypes were filtered out by VariantFiltration, and the *de novo* single nucleotide variants (SNVs) were annotated by VariantAnnotator. The following commands are used to call *de novo* variants from analysis-ready BAM files:

```
# gVCF Calling from SRR1301246 HaplotypeCaller
gatk HaplotypeCaller \
-R $REFERENCE \
-D $DBSNP \
-I SRR1301246.bqsr.bam \
-O SRR1301246.g.vcf.gz \
-ERC GVCF

# gVCF Calling from SRR1301247 HaplotypeCaller
gatk HaplotypeCaller \
-R $REFERENCE \
-D $DBSNP \
-I SRR1301247.bqsr.bam \
-O SRR1301247.g.vcf.gz \
-ERC GVCF

# gVCF Calling from SRR1301248 HaplotypeCaller
gatk HaplotypeCaller \
-R $REFERENCE \
-D $DBSNP \
-I SRR1301248.bqsr.bam \
-O SRR1301248.g.vcf.gz \
-ERC GVCF
```

Parameters of HaplotypeCaller: `-R` = reference genome; `-D` = dbSNP file; `-I` = recalibrated BAM as the input file; `-O` = name of the output file with variants (gVCF); `-ERC` = mode for emitting reference confidence scores (other options: `NONE/BP_RESOLUTION/GVCF`)

Merging gVCF to a Single Consolidated File

CombineGVCFs is a utility within GATK used to merge multiple gVCF files into a single consolidated file. During the merging process, CombineGVCFs combines the variant information from each gVCF file, retaining the genotype likelihoods, allele counts, and other relevant variant annotations. CombineGVCFs also supports the handling of regions with overlapping variant calls, where it performs intelligent merging to consolidate the variant information without duplication. This helps to ensure accuracy and eliminate redundancy in the resulting combined gVCF file. The output of the CombineGVCFs command is a single consolidated gVCF file containing all the variant sites from the input files. The following command is used for CombineGVCFs:

```
# Merging gVCF to single consolidated file using CombineGVCFs
gatk CombineGVCFs \
  -R $REFERENCE \
  -V SRR1301246.g.vcf.gz \
  -V SRR1301247.g.vcf.gz \
```

```
  -V SRR1301248.g.vcf.gz \
  -O trio.g.vcf.gz
```

Parameters of CombineGVCFs: `-R` = reference genome; `-V` = input in gVCF format; `-O` = combined gVCF output file.

Joint Genotyping on Merged gVCF Files

GenotypeGVCFs is a command used by GATK to perform joint genotyping on merged gVCF files. By utilising the information from the GVCF files, GenotypeGVCFs determines the genotypes and allele frequencies for each variant site across the entire dataset. The joint genotyping process performed by GenotypeGVCFs involves considering the shared genomic regions and combining the individual genotype likelihoods from the GVCF files to generate the most accurate and comprehensive genotype calls. During the joint genotyping process, GenotypeGVCFs apply sophisticated statistical models to estimate the likelihoods of various genotype combinations at each variant site. It incorporates information such as allele counts, base qualities, and read alignments to make informed genotype calls. The output of GenotypeGVCFs is a consolidated VCF file containing genotype information for each variant site across all the samples. This file includes the genotypes, genotype qualities, and other relevant annotations. GenotypeGVCFs offer flexibility in handling different ploidy levels and can handle datasets with diploid, haploid, or polyploid genomes.

MakeSitesOnlyVcf is another command in GATK to create a simplified VCF file that includes only the variant sites without any sample-specific genotype information. This utility is particularly useful when researchers are interested in focusing solely on the presence or absence of variants across different samples, rather than individual genotypes or allele frequencies. By excluding the sample-specific genotype information, MakeSitesOnlyVcf streamlines the VCF file, reducing its size and complexity. MakeSitesOnlyVcf takes as input a VCF file that contains both variant and non-variant sites and generates a new VCF file that retains only the variant sites. The following commands are used for GenotypeGVCFs and MakeSitesOnlyVcf:

```
# Performing joint genotyping on merged gVCF files using GenotypeGVCFs
gatk GenotypeGVCFs \
  -R $REFERENCE \
  -V trio.g.vcf.gz \
  -D $DBSNP \
  -O trio.vcf.gz

# Make VCF with only the variant sites using MakeSitesOnlyVcf
gatk MakeSitesOnlyVcf \
  -I trio.vcf.gz \
  -O trio.sites.vcf.gz
```

Parameters of GenotypeGVCFs: `-R` = reference genome; `-V` = combined gVCF as the input file; `-D` = dbSNP file; `-O` = name of the output variant file (VCF).

Parameters of MakeSitesOnlyVcf: `-I` = input as the combined VCF file; `-O` = name of the output file with site-level information.

Variant Quality Score Recalibration

VariantRecalibrator and ApplyVQSR are two key components of the GATK used for variant quality score recalibration (VQSR) in genomic data analysis. These tools work together to enhance the accuracy and reliability of variant calls by recalibrating variant quality scores based on various features and known truth sets.

VariantRecalibrator is the initial step in the process. It takes as input a variant call set and a set of known variant truth data. The tool performs a statistical analysis by considering multiple annotations associated with the variants, such as read depth, mapping quality, allele frequency, and various others. VariantRecalibrator builds a machine learning model, often using Gaussian mixture models or support vector machines, to determine the likelihood that a variant is true or false based on these annotations. The model is trained on the known truth data, enabling it to assign more accurate quality scores to the variants. ApplyVQSR is the subsequent step after VariantRecalibrator. It applies the recalibration model generated by VariantRecalibrator to the original variant call set. ApplyVQSR takes into account the recalibrated quality scores and applies filtering thresholds to classify variants as either 'PASS' or 'FAIL' based on their quality. Variants that pass the filtering thresholds are considered to be of higher quality and more likely to be true positives.

By utilising VariantRecalibrator and ApplyVQSR in tandem, researchers can refine the variant quality scores and improve the accuracy of variant calling. This process is particularly beneficial when working with large-scale genomic datasets or when dealing with samples from diverse populations. It is important to note that both tools require appropriate training sets and high-quality known truth data to achieve optimal recalibration performance. The choice and configuration of annotations used by VariantRecalibrator, as well as the specific filtering thresholds set by ApplyVQSR, should be carefully tailored to the analysis objectives and dataset characteristics. The process of recalibration takes place in two steps, that is, we need to run the VariantRecalibrator and ApplyVQSR commands on SNPs and INDELs separately. The commands used for performing VariantRecalibrator and ApplyVQSR are given in the following subsections.

Variant Recalibration of SNPs

```
# Performing variant recalibration of SNP
gatk VariantRecalibrator \
  -R $REFERENCE \
  -V trio.sites.vcf.gz \
  --resource hapmap,known=false,training=true,truth=true,prior=15.0:$hap
map \
  --resource omni,known=false,training=true,truth=false,prior=12.0:$omni \
  --resource 1000G,known=false,training=true,truth=false,prior=10.0
:$G1000 \
  --resource dbsnp,known=true,training=false,truth=false,prior=2.0:$DBSNP \
  -an QD -an MQ -an MQRankSum -an ReadPosRankSum -an FS -an SOR \
  -mode SNP \
  -O SNP.recal \
  --tranches-file SNP.tranches \
  --rscript-file SNP.plots.R

# ApplyVQSR on recalibrated SNP
gatk ApplyVQSR \
  -R $REFERENCE \
  -V trio.sites.vcf.gz \
  --mode SNP \
  --truth-sensitivity-filter-level 99.0 \
  --recal-file SNP.recal \
  --tranches-file SNP.tranches \
  -O recalibrated_snps.vcf
```

Variant Recalibration of INDELs

```
# Performing variant recalibration of INDEL
gatk VariantRecalibrator \
  -R $REFERENCE \
  -V recalibrated_snps.vcf \
  --resource hapmap,known=false,training=true,truth=true,prior=15.0:$hap
map \
```

```
  --resource omni,known=false,training=true,truth=false,prior=12.0:$omni \
  --resource 1000G,known=false,training=true,truth=false,prior=10.0
:$G1000 \
  --resource dbsnp,known=true,training=false,truth=false,prior=2.0:$DBSNP \
  -an QD -an MQ -an MQRankSum -an ReadPosRankSum -an FS -an SOR \
  -mode INDEL \
  -O INDEL.recal \
  --tranches-file INDEL.tranches \
 --rscript-file INDEL.plots.R

# ApplyVQSR on recalibrated INDEL
gatk ApplyVQSR \
  -R $REFERENCE \
  -V recalibrated_snps.vcf \
  --mode INDEL \
  --truth-sensitivity-filter-level 99.0 \
  --recal-file INDEL.recal \
  --tranches-file INDEL.tranches \
  -O recalibrated_snps_indels.vcf
```

Parameters of VariantRecalibrator: `-R` = reference genome; `-V` = input variants in VCF format; `-resource` = list of sites for which to apply a prior probability of being correct but which are not used by the algorithm; `-an` = annotation files which will be used for calculation; `-mode` = type of recalibration mode to use; `-O` = write the output recal file; `-tranches-file` = output tranches file used by ApplyRecalibration; `-rscript-file` = output rscript file generated by the VQSR to aid in visualisation.

Parameters of ApplyVQSR: `-R` = reference genome; `-V` = variants in VCF format; `-mode` = type of recalibration mode to use; `-truth-sensitivity-filter-level` = truth sensitivity level at which to start filtering; `-tranches-file` = input tranches file describing where to cut the data; `-O` = annotated variants with VQSLOD value will be written into the VCF file.

Estimate the Posterior Distribution of Genotypes

CalculateGenotypePosteriors is a statistical method used to estimate the probability or posterior distribution of genotypes given observed genetic data. It is commonly employed in genetic research, particularly in fields such as population genetics, medical genetics, and evolutionary biology. This method allows researchers to make inferences about the underlying genotypes of individuals based on observed genetic markers or traits.

The genotype of an individual refers to the genetic composition at a specific locus or set of loci. In diploid organisms like humans, each individual carries two copies of each chromosome, and thus two alleles at each locus. The combination of alleles determines the genotype. For example, at a particular locus, an individual can be homozygous (having two identical alleles) or heterozygous (having two different alleles).

To calculate genotype posteriors, several factors need to be taken into account, such as the observed genotype data, the allele frequencies in the population, and the assumptions about inheritance patterns using Bayesian statistics.

Here is a general outline of the steps involved in calculating genotype posteriors:

1. *Define the genetic markers:* Determine the specific markers or loci that will be used to infer genotypes. These markers can be single nucleotide polymorphisms (SNPs), microsatellites, or other genetic variants.
2. *Collect genotype data:* Obtain genotype data by genotyping individuals at the selected markers. Such data can be in the form of individual genotypes (e.g., AA, AB, BB) or allele frequencies.

3. *Estimate allele frequencies:* Estimate the frequencies of alleles at each marker in the population. This can be done using a variety of methods, such as direct counting or maximum likelihood estimation based on a representative sample.
4. *Model assumptions:* Specify the assumptions about the mode of inheritance, population structure, and other relevant factors. These assumptions will influence the calculation of genotype posteriors.
5. *Calculate likelihoods:* Calculate the likelihood of the observed genotype data given each possible genotype combination for an individual. The likelihood is determined by the allele frequencies and the assumed mode of inheritance.
6. *Calculate prior probabilities:* Assign prior probabilities to each genotype combination based on prior knowledge, assumptions, or population-level data. These priors represent the initial beliefs about the likelihood of each genotype.
7. *Apply Bayes' theorem:* Combine the likelihoods and priors using Bayes' theorem to calculate the posterior probabilities of each genotype combination. The posterior probability represents the updated belief about the genotype given the observed data.
8. *Normalise posteriors:* Normalise the posterior probabilities so that they sum to 1, ensuring they represent a valid probability distribution.
9. *Interpretation and analysis:* Analyse the posterior probabilities to make inferences about the genotypes of individuals. This can include identifying the most likely genotype, calculating genotype frequencies in a population, or comparing genotypes between different groups.

It is important to note that the accuracy of genotype posterior estimation depends on the quality of the data, the appropriateness of the assumptions, and the representativeness of the population sample. Additionally, incorporating additional information, such as linkage disequilibrium or phenotypic data, can enhance the accuracy and precision of genotype posterior calculations.

CalculateGenotypePosteriors is a valuable tool for estimating the probability or posterior distribution of genotypes based on observed genetic data. By combining genotype likelihoods, prior probabilities, and Bayesian inference, researchers can gain insights into genetic variations, population structure, and heritability of traits. This method has broad applications in various genetic research fields and contributes to our understanding of the genetic basis of complex traits and diseases. The following command is used to perform CalculateGenotypePosteriors:

```
# Performing joint genotyping on merged gVCF files using GenotypeGVCFs
gatk CalculateGenotypePosteriors \
 -V recalibrated_snps_indels.vcf \
 -ped trio.ped \
 --supporting-callsets $gnomad \
 -O trio.vcf
```

Parameters of CalculateGenotypePosteriors: `-V` = input recalibrated VCF; `-ped` = pedigree file; `-supporting-callsets` = other call sets to use in generating genotype posteriors; `-O` = filename of output variants.

Filter Low Confidence Variants

VariantFiltration, a component of GATK, is utilised to filter through variants in NGS data, eliminating potential false positives and producing a reliable variant subset for downstream analyses. It operates by implementing user-defined filters based on quality metrics like read depth, genotype quality, mapping quality, strand bias, and allele frequency. The tool functions in either hard-filtering mode, which excludes variants failing the filters from the output VCF file, or soft-filtering mode, where filter criteria are included as annotations. Users commonly set filtering thresholds, such as a minimum read depth of 10 or a maximum strand bias of 0.01, through the `-filter-expression` option, assigning a filter tag like `my_filter` to flagged variants for subsequent identification.

Overall, VariantFiltration proves instrumental in refining NGS variant data, finding utility across applications like variant discovery, population genetics, and disease gene mapping due to its adaptability and customisable filtering capacities.

```
# Performing VariantFiltration
gatk VariantFiltration \
 -R $REFERENCE \
 -V trio.vcf \
 --filterExpression "GQ<20" \
 --filterName "lowGQ" \
 -O trio_filtered.vcf
```

Parameters of VariantFiltration: `-R` = reference genome; `-V` = input variants in VCF format; `-filterExpression` = expressions used from INFO fields to filter; `-filter-name` = names of the filter; `-O` = filename of output variants.

De Novo Variant Calling Using PossibleDeNovo

PossibleDeNovo is an analysis approach that can be performed using the VariantAnnotator tool in the GATK. VariantAnnotator is one of the modules available within GATK, specifically designed for annotating VCF files with various types of annotations. These annotations provide additional information about the detected variants, aiding in the interpretation of their potential impact on genes and their association with genetic diseases. When utilising VariantAnnotator in GATK, the PossibleDeNovo analysis aims to identify *de novo* mutations or variants in an individual that are absent in their parent's genomes. This analysis is particularly useful in identifying potential causative variants in genetic diseases that may have originated as *de novo* mutations.

The following steps outline how VariantAnnotator in GATK can be used for PossibleDeNovo analysis:

1. *Prepare input files:* Ensure that the VCF file containing variant calls for the individual of interest is available and properly formatted for VariantAnnotator.
2. *Set up annotations:* Determine the specific annotations to be used for the PossibleDeNovo analysis. GATK offers a variety of built-in annotations that can be employed, such as DepthOfCoverage, GenotypeSummaries, InbreedingCoeff, and others. Select annotations that are relevant to *de novo* mutation analysis.
3. *Run VariantAnnotator:* Execute the VariantAnnotator module in GATK, specifying the input VCF file, reference genome, desired annotations, and any additional parameters required for the analysis. The output is a VCF file containing the original variant calls annotated with the specified annotations.
4. *Filter and interpret results:* Analyse the annotated VCF file to identify potential *de novo* variants. Focus on variants that are present in the individual of interest but absent in the parental genomes. These variants have the potential to be *de novo* mutations and may warrant further investigation to assess their impact on genes or potential association with genetic diseases.

By applying the PossibleDeNovo analysis using VariantAnnotator in GATK, we can effectively identify and prioritise *de novo* variants in individuals, facilitating the identification of disease-causing mutations and advancing our understanding of the genetic basis of various disorders. The following command is used for running PossibleDeNovo using VariantAnnotator:

```
# Performing VariantAnnotator with PossibleDeNovo
gatk VariantAnnotator \
 -R $REFERENCE \
 -V trio_filtered.vcf \
```

```
-A PossibleDeNovo \
-ped trio.ped \
-O trio_filtered.denovos.vcf
```

Parameters of VariantAnnotator: `-R` = reference genome; `-V` = variants in VCF format; `-A` = one or more specific annotations to add to variant; `-ped` = pedigree file; `-O` = filename of output variants.

The existence of a *de novo* mutation in the output file `trio_filtered.denovos.vcf` will have the mutations marked as hiConfDeNovo or loConfDeNovo if they are called *de novo* mutations by the GATK pipeline.

Calling Germline Variants from a Single Sample

HaplotypeCaller possesses the ability to concurrently identify SNPs and insertions/deletions (indels) by performing local *de novo* assembly of haplotypes within an active region. In simpler terms, when the program detects a region displaying indications of variation, it disregards the current mapping data and performs a complete reassembly of the reads within that specific region. This particular step is specifically designed to enhance sensitivity, thereby minimising the occurrence of false negatives, which refers to the failure to identify genuine genetic variants. There are not many differences between variants calling for *de novo* and single sample. The core command used to call variants from a single sample is as follows:

```
# Calling variants from single sample using HaplotypeCaller
gatk HaplotypeCaller \
 -R $REFERENCE \
 -I SRR1301246.bqsr.bam \
 -D $DBSNP \
 -O SRR1301246_raw.vcf.gz
```

Parameters of HaplotypeCaller: `-R` = reference genome; `-I` = recalibrated BAM file; `-D` = dbSNP file; `-O` = filename of output variants.

Further, steps VariantRecalibrator, ApplyVQSR, and VariantFiltration are performed as explained in *de novo* variant calling. The following is an example of a VCF file:

```
#CHROM POS     ID         REF    ALT     QUAL FILTER INFO
20     14370   rs6054257  G      A       29   PASS   NS=3;DP=20;AF=0.5;DB;H2
20     17330   .          T      A       3    q10    NS=3;DP=88;AF=0.017
```

FUNCTIONAL ANNOTATION OF VARIANTS

There are several variant annotation tools available that are widely used in genomic research and clinical applications. These tools provide valuable information about the functional impact of genetic variants and aid in the interpretation of genomic data. Here are some popular variant annotation tools:

SnpEff (SNP Effect) [29]: SnpEff is a widely used variant annotation tool that predicts the functional consequences of genetic variants. It integrates variant information with genomic annotations, such as gene models and regulatory elements, to determine the potential effects on genes, proteins, and non-coding regions.

ANNOVAR (Annotate Variation) [30]: ANNOVAR is a versatile variant annotation tool that annotates genetic variants based on various databases and algorithms. It supports a wide range of annotations, including functional impact, conservation scores, population frequencies, and disease associations. ANNOVAR is highly customisable, allowing researchers to incorporate custom annotation sources.

VEP (Variant Effect Predictor) [31]: VEP is a popular variant annotation tool developed by the Ensembl project. It provides comprehensive annotations for genetic variants, including information on gene and transcript features, protein domains, regulatory elements, and conservation scores. VEP supports a wide range of organisms and integrates with multiple databases and prediction algorithms.

VAAST (Variant Annotation, Analysis, and Search Tool) [32]: VAAST is a powerful tool specifically designed for disease gene identification in rare genetic disorders. It integrates variant annotation with statistical analysis to prioritise potentially disease-causing variants based on family-based or population-based studies. VAAST considers variant functional impact, conservation, inheritance patterns, and phenotype information.

VEST (Variant Effect Scoring Tool) [33]: VEST is a tool that focuses on predicting the functional impact of missense variants. It employs machine learning algorithms trained on large-scale variant data to assign a score indicating the likelihood of a variant being deleterious. VEST can aid in prioritising potentially pathogenic variants for further investigation.

CADD (Combined Annotation-Dependent Depletion) [34]: CADD is a widely used tool that combines multiple annotations and machine learning methods to estimate the deleteriousness of genetic variants. It integrates diverse annotations, including conservation scores, protein structure information, and functional genomics data, to provide a comprehensive ranking of variant pathogenicity.

These are just a few examples of the variant annotation tools available in the field of genomics. Each tool has its own strengths, features, and compatibility with different annotation sources and databases. The choice of annotation tool depends on the specific research needs, data types, and analysis goals of the researcher or clinician.

Variant annotation is a crucial step in genomic analysis that involves interpreting and understanding the functional impact of genetic variants. With the advent of high-throughput sequencing technologies, a vast amount of genetic data is generated, necessitating efficient tools to annotate and prioritise variants for further investigation. One popular tool used for variant annotation is SnpEff.

The core functionality of SnpEff relies on its ability to integrate variant information with genomic annotations. It achieves this by leveraging a database of known gene models and regulatory elements, such as transcription factor binding sites and microRNA target sites. These annotations are typically obtained from well-curated reference genome databases like RefSeq or Ensembl.

The annotation process with SnpEff involves several steps:

1. *Database preparation*: Before annotating variants, SnpEff requires a prebuilt database specific to the reference genome being used. This database contains genomic annotations like gene models, transcripts, and functional elements.
2. *Variant annotation*: Once the database is prepared, SnpEff takes a VCF file containing genomic variants as input. It compares the genomic positions of the variants with the reference genome and determines the affected genes and transcripts.
3. *Functional prediction*: After identifying the affected genes and transcripts, SnpEff predicts the functional consequences of the variants. It categorises the variants based on their impact, such as missense mutations, frameshifts, splice site alterations, or non-coding changes. SnpEff also provides information about the potential effect on protein structure, protein domains, and regulatory regions.
4. *Annotation output*: SnpEff generates an annotated VCF file as output, which includes additional information about the variants and their functional effects. This annotated file helps researchers prioritise variants for downstream analysis and interpret their potential biological relevance.

SnpEff offers several additional features and capabilities to enhance variant annotation and analysis. These include the following:

Support for various organisms: SnpEff is not limited to human genome analysis but can annotate variants in a wide range of organisms, including model organisms and agriculturally important species.
Integration with other tools: SnpEff can be seamlessly integrated with other bioinformatics tools, such as variant callers, downstream analysis pipelines, and visualisation tools, to facilitate comprehensive genomic analysis.
Custom annotations: Researchers can extend SnpEff's functionality by incorporating custom annotations specific to their research interests. This allows the inclusion of additional information, such as population frequencies, conservation scores, or disease associations, in the annotation output.

In short, SnpEff is a powerful variant annotation tool that aids in the interpretation and prioritisation of genetic variants. By leveraging genomic annotations and functional predictions, SnpEff provides valuable insights into the potential impact of variants on genes, proteins, and regulatory elements. Its versatility, accuracy, and ability to integrate with other tools make it an indispensable component of genomic analysis workflows. The following commands were used to perform variant annotation using SnpEff:

```
# Downloading the database for the analysis
snpeff download GRCh38.86

# Performing variant annotation using snpeff
snpeff GRCh38.86 \
  -o gatk \
  -canon -t -lof \
  SRR1301246.vcf > SRR1301246.ann.vcf
```

Parameters of snpeff: `-o` = for GATK compatibility; `-canon` = Only use canonical transcripts; `-lof` = add loss of function (LOF) and Nonsense mediated decay (NMD) tags; `-t` = to use multiple threads.

SnpEff produces three output files:

1. An HTML file that contains summary statistics about the variants and their annotations.
2. An annotated VCF file.
3. A text file summarising the number of variant types per gene.

LIMITATIONS OF GATK

GATK is a widely used software package for variant discovery and analysis in genomic data. While GATK is widely used for variant calling and analysis, it has certain limitations when it comes to copy number variation (CNV) calling [35–37]. Here are some of the limitations of GATK in CNV calling:

CNV detection: GATK's primary focus is on single nucleotide variants (SNVs) and small insertions and deletions (indels). It does not have specific algorithms or tools dedicated to CNV detection. While GATK provides some functionality for CNV analysis, it may not offer the same level of sensitivity or accuracy as specialised CNV detection tools.
Sensitivity to variant size: GATK's algorithms are optimised for detecting small variants, typically in the range of a few base pairs to a few hundred base pairs. CNVs, on the other

hand, involve larger genomic segments, ranging from kilobases to megabases. GATK may not be as sensitive in identifying larger-scale CNVs or complex rearrangements.

Lack of specialised models: GATK does not employ specific models or algorithms designed specifically for CNV calling. CNV analysis often requires specialised statistical models and algorithms that account for factors like read depth, regional biases, and mappability. GATK's general-purpose variant calling models may not be as effective in capturing the complexities of CNV analysis.

Read depth-based methods: GATK's approach to CNV calling primarily relies on read depth analysis, where deviations in coverage across the genome are used to identify potential CNVs. While read depth analysis can provide some information about CNVs, it may be limited in its ability to accurately detect and characterise complex CNV events, such as nested or overlapping CNVs.

Lack of comprehensive annotation: GATK's annotation capabilities for CNVs may be limited compared to specialised CNV analysis tools. CNV calling often requires detailed annotation information, such as gene content, functional impact, and overlap with known CNV databases. GATK's annotation functionality may not provide the same level of comprehensive information as dedicated CNV annotation tools.

Performance with low-coverage data: GATK's CNV calling performance may be compromised when working with low-coverage sequencing data. CNV detection typically requires higher coverage depths to accurately capture variations in read depth across the genome. In low-coverage scenarios, the sensitivity and specificity of GATK's CNV calling may be reduced.

Some of the other drawbacks when it comes to performance and computational resources are as follows:

Computational resource requirements: GATK can be computationally demanding, requiring substantial computational resources, including memory and processing power. Large-scale analyses or datasets may require high-performance computing infrastructure to run efficiently.

Complexity of installation and usage: GATK has a steep learning curve, especially for users who are new to bioinformatics. It requires a significant investment of time and effort to understand the various command-line tools, workflows, and best practices associated with GATK. This can be a barrier for researchers without a strong bioinformatics background.

Cost: GATK is an open-source software, but some of its functionalities and best practices are available only to users who purchase a GATK licence from the Broad Institute. This licensing model may limit accessibility for researchers or institutions with limited financial resources.

Limited support for non-human genomes: While GATK is primarily designed for human genome analysis, its functionality and performance on non-human genomes can be limited. The best practices and optimised workflows provided by GATK may not be as comprehensive or well established for non-human species.

Sensitivity to sequencing and mapping errors: GATK's variant calling algorithms are sensitive to errors introduced during sequencing and mapping steps. While GATK provides various tools and filters to mitigate false-positive variant calls, the inherent errors in sequencing and mapping can still affect the accuracy of variant calls.

Limited scalability for certain analyses: GATK's performance may decrease when dealing with large-scale datasets or specific types of analyses. For example, joint variant calling across a large cohort or whole-genome sequencing data can be challenging due to the increased computational requirements and longer processing times.

Dependency on high-quality reference genomes and annotations: GATK heavily relies on accurate and up-to-date reference genome assemblies and annotations. If the reference genome used is incomplete or contains errors, it can affect the accuracy of variant calling and downstream analyses.

Limited integration with other analysis tools: GATK has its own suite of tools and workflows, but integrating it with other bioinformatics tools or custom pipelines can be challenging. This can be a limitation when researchers want to combine GATK with specialised tools or incorporate additional analyses into their workflows.

It is important to note that many of these limitations are not unique to GATK but are inherent challenges in genomic data analysis. Researchers should carefully consider these limitations and assess whether GATK is the most suitable tool for their specific research needs, taking into account factors such as dataset size, computational resources, cost, and available support.

CONCLUSIONS

The GATK stands as a crucial player in the era of NGS. Its probabilistic model, meticulous preprocessing, and innovative HaplotypeCaller algorithm enhance the accuracy of variant calling. However, proficiency in bioinformatics is essential for optimal results. GATK's filtering and annotation tools further refine variant calls, providing valuable functional insights. Its widespread adoption underscores its significance in the scientific community. GATK's journey exemplifies the synergy between advanced technologies and expert analysis, shaping our comprehension of genetics and its applications in medicine and research. As genomics continues to shape diverse domains, GATK remains an indispensable guide, illuminating the path towards accurate variant detection and comprehensive genomic insights.

REFERENCES

1. Jamuar SS, Tan E-C. Clinical application of next-generation sequencing for mendelian diseases. Hum Genomics. 2015;9(1):1–6.
2. Singh S, Snijesh VP, Jannet Vennila J. Rheumatoid arthritis candidate genes identification by investigating core and periphery interaction structures. Comput Intell Med Informatics. 2015;6:87–96.
3. Snijesh VP, Matchado MS, Singh S. Classifying rheumatoid arthritis gene network signatures for identifying key regulatory molecules and their altered pathways by adopting network biology approach. Gene Reports. 2018;13:199–211.
4. Singh S, Vennila JJ, Snijesh VP, George G, Sunny C. Implying analytic measures for unravelling rheumatoid arthritis significant proteins through drug–target interaction. Interdiscip Sci Comput Life Sci. 2016;8:122–31.
5. Poudel G, Tolland MG, Hughes TP, Pagani IS. Mechanisms of resistance and implications for treatment strategies in chronic myeloid leukaemia. Cancers (Basel). 2022;14(14):3300.
6. Rajarajan S, Snijesh VP, Anupama CE, Nair MG, Mavatkar AD, Naidu CM, et al. An androgen receptor regulated gene score is associated with epithelial to mesenchymal transition features in triple negative breast cancers. Transl Oncol. 2023;37:101761.
7. Collins FS, Morgan M, Patrinos A. The human genome project: Lessons from large-scale biology. Science. 2003;295(5561):1817.
8. Olson MV. The human genome project. Proc Natl Acad Sci. 1993;90(10):4338–44.
9. Watson JD. The human genome project: Past, present, and future. Science (80-). 1990;248(4951):44–9.
10. Nair MG, Patil S, Mukherjee G, Kumar RV, et al. miR-18a mediates immune evasion in ER-positive breast cancer through wnt signaling. Cells. 2022;11(10):1672.
11. Kamps R, Brandão RD, van den Bosch BJ, Paulussen ADC, Xanthoulea S, Blok MJ, et al. Next-generation sequencing in oncology: Genetic diagnosis, risk prediction and cancer classification. Int J Mol Sci. 2017;18(2):308.
12. Bima AIH, Elsamanoudy AZ, Albaqami WF, Khan Z, Parambath SV, Al-Rayes N, et al. Integrative system biology and mathematical modeling of genetic networks identifies shared biomarkers for obesity and diabetes. Math Biosci Eng. 2022;19(3):2310–29.
13. George G, Parambath SV, Lokappa SB, Varkey J. Construction of Parkinson's disease marker-based weighted protein-protein interaction network for prioritization of co-expressed genes. Gene. 2019;697:67–77.

14. Alsulaimany FA, Zabermawi NMO, Almukadi H, Parambath SV, Shetty PJ, Vaidyanathan V, et al. Transcriptome-based molecular networks uncovered interplay between druggable genes of CD8+ T cells and changes in immune cell landscape in patients with pulmonary tuberculosis. Front Med. 2022;8:812857.
15. Arora I, Tollefsbol TO. Computational methods and next-generation sequencing approaches to analyze epigenetics data: Profiling of methods and applications. Methods. 2021;187:92–103.
16. der Auwera GA, Carneiro MO, Hartl C, Poplin R, Del Angel G, Levy-Moonshine A, et al. From FastQ data to high-confidence variant calls: The genome analysis toolkit best practices pipeline. Curr Protoc Bioinforma. 2013;43(1):10–1.
17. Leinonen R, Akhtar R, Birney E, Bower L, Cerdeno-Tárraga A, Cheng Y, et al. The European nucleotide archive. Nucleic Acids Res. 2010;39(suppl_1):D28–D31.
18. Leinonen R, Sugawara H, Shumway M, Collaboration INSD. The sequence read archive. Nucleic Acids Res. 2010;39(suppl_1):D19–D21.
19. Wheeler DL, Barrett T, Benson DA, Bryant SH, Canese K, Chetvernin V, et al. Database resources of the national center for biotechnology information. Nucleic Acids Res. 2006;34(suppl_1):D173–D180.
20. Andrews S. FastQC: A quality control tool for high throughput sequence data. Babraham Bioinformatics, Babraham Institute, Cambridge, United Kingdom; 2010.
21. Bolger AM, Lohse M, Usadel B. Trimmomatic: A flexible trimmer for Illumina sequence data. Bioinformatics. 2014;30(15):2114–20.
22. Nimbalkar VP, Snijesh VP, Rajarajan S, Alexander A, Kaluve R, Ramesh R, et al. Premenopausal women with breast cancer in the early post-partum period show molecular profiles of invasion and are associated with poor prognosis. Breast Cancer Res Treat. 2023;200(1):139–49.
23. Krueger F, Galore T. A wrapper tool around Cutadapt and FastQC to consistently apply quality and adapter trimming to FastQ files. Babraham Bioinforma. 2012;516.
24. Martin M. Cutadapt removes adapter sequences from high-throughput sequencing reads. EMBnet J. 2011;17(1):10–2.
25. Li H, Durbin R. Fast and accurate short read alignment with Burrows–Wheeler transform. Bioinformatics. 2009;25(14):1754–60.
26. Langmead B, Trapnell C, Pop M, Salzberg SL. Ultrafast and memory-efficient alignment of short DNA sequences to the human genome. Genome Biol. 2009;10(3):1–10.
27. Kim D, Langmead B, Salzberg SL. HISAT: A fast spliced aligner with low memory requirements. Nat Methods. 2015;12(4):357–60.
28. Pirooznia M., Kramer M., Parla J., Goes F.S., Potash J.B., McCombie W.R., et al. Validation and assessment of variant calling pipelines for next-generation sequencing. Human genomics. 2014;8(1):1–10.
29. Cingolani P, Platts A, Wang LL, Coon M, Nguyen T, Wang L, et al. A program for annotating and predicting the effects of single nucleotide polymorphisms, SnpEff: SNPs in the genome of Drosophila melanogaster strain w1118; iso-2; iso-3. Fly (Austin). 2012;6(2):80–92.
30. Wang K, Li M, Hakonarson H. ANNOVAR: Functional annotation of genetic variants from high-throughput sequencing data. Nucleic Acids Res. 2010;38(16):e164–e164.
31. McLaren W, Gil L, Hunt SE, Riat HS, Ritchie GRS, Thormann A, et al. The ensembl variant effect predictor. Genome Biol. 2016;17(1):1–14.
32. Hu H, Huff CD, Moore B, Flygare S, Reese MG, Yandell M. VAAST 2.0: Improved variant classification and disease-gene identification using a conservation-controlled amino acid substitution matrix. Genet Epidemiol. 2013;37(6):622–34.
33. Douville C, Masica DL, Stenson PD, Cooper DN, Gygax DM, Kim R, et al. Assessing the pathogenicity of insertion and deletion variants with the variant effect scoring tool (VEST-indel). Hum Mutat. 2016;37(1):28–35.
34. Rentzsch P, Witten D, Cooper GM, Shendure J, Kircher M. CADD: Predicting the deleteriousness of variants throughout the human genome. Nucleic Acids Res. 2019;47(D1):D886–D894.
35. Neuman JA, Isakov O, Shomron N. Analysis of insertion–deletion from deep-sequencing data: Software evaluation for optimal detection. Brief Bioinform. 2013;14(1):46–55.
36. Betschart RO, Thiéry A, Aguilera-Garcia D, Zoche M, Moch H, Twerenbold R, et al. Comparison of calling pipelines for whole genome sequencing: An empirical study demonstrating the importance of mapping and alignment. Sci Rep. 2022;12(1):21502.
37. Lin Y-L, Chang P-C, Hsu C, Hung M-Z, Chien Y-H, Hwu W-L, et al. Comparison of GATK and DeepVariant by trio sequencing. Sci Rep. 2022;12(1):1809.

5 Implementation of WGCNA for Identifying Regulatory Modules in Biological Networks

Subarna Thakur, Subhajit Ghosh, and Maheswari Dasgupta

INTRODUCTION

Next-generation sequencing (NGS) technology has revolutionised gene expression studies by enabling researchers to analyse the transcriptome in a more comprehensive and high-throughput manner. Unlike traditional methods, such as microarrays, which can only detect known genes, NGS allows for the detection of all RNA transcripts present in a sample, including novel and low-abundance transcripts. NGS has also greatly increased the depth and accuracy of gene expression analysis. Gene expression studies play a crucial role in understanding biological networks by providing insights into the regulation and function of genes within a cell or organism [1]. By measuring the level of mRNA transcripts or protein abundance, gene expression studies can reveal which genes are active or inactive under different conditions or in different cell types, providing clues to their functional roles. One important application of gene expression studies is in the identification of gene regulatory networks [2]. By analysing the co-expression patterns of genes across different samples, researchers can identify gene modules, which are clusters of genes that are co-regulated and may be involved in the same biological processes or pathways. This information can be used to construct gene regulatory networks that provide a systems-level view of the interactions between genes and their products. In bioinformatics, a correlation network is a type of graph that shows the relationship between multiple variables, such as genes or proteins, based on their correlation coefficients [3]. These networks are often used to analyse large-scale biological datasets, such as gene expression data, to identify groups of genes or proteins that are co-regulated or functionally related. Their construction and analysis have been extensively used in several areas of bioinformatics, including gene co-expression analysis, protein–protein interaction (PPI) networks, and metabolic pathway analysis.

WEIGHTED GENE CO-EXPRESSION NETWORK ANALYSIS (WGCNA)

The increasing availability of high-throughput biological data has led to a greater understanding of the complex networks that regulate gene expression in cells. However, the analysis of such large-scale datasets can be challenging, and new methods are needed to identify regulatory modules and understand their functions. WGCNA is a powerful method for identifying gene expression modules and their relationships in biological networks [4]. It constructs co-expression networks based on pairwise correlations between gene expression profiles and then identifies modules of co-expressed genes using a hierarchical clustering algorithm [5]. These modules can then be analysed to identify their biological functions, regulatory mechanisms, and potential drug targets. WGCNA has been implemented in the statistical programming language R and is widely used, maintained, and improved by an active community. Initially, it was developed to analyse transcriptomic profiling experiments; however, the method is equally powerful and illustrative as a way to

DOI: 10.1201/9781003354062-5

analyse metabolomic fingerprinting experiments [6]. A similar approach has also been applied to proteomics datasets to analyse protein abundance correlations and networks [7].

GENERAL PROTOCOL OF WGCNA

WGCNA analyses 'high-dimensional' data across multiple samples by forming a scale-free network and assigning the genes to different modules; from there, candidate genes can be identified that may be associated with experimental conditions. It has been designed to be a correlation-based, unsupervised analysis method that clusters genes based on their expression profiles. The following subsections describe a general step-by-step protocol for conducting a general WGCNA analysis (Figure 5.1).

Step I. Data Preparation

Gene expression data (e.g., microarray or RNA-seq) are obtained from the appropriate experimental conditions, including samples under different treatments or conditions relevant to the research question. The data are then preprocessed by steps like quality control, filtering, normalisation, and log transformation to ensure the data's suitability for WGCNA analysis. In the case of an RNA-seq dataset, processed raw reads are first aligned to a reference followed by gene expression quantification and then data normalisation.

Step II. Construction of a Co-Expression Network

Pairwise correlation or similarity measures (e.g., Pearson correlation, Spearman correlation, or weight mid correlation) between genes using the expression data are calculated. The correlation matrix is then converted into an adjacency matrix using a power function to emphasise strong correlations and down-weight weak correlations. The power value is determined based on scale-free topology criteria. Selecting a suitable soft-thresholding power value (β) is essential for attaining a

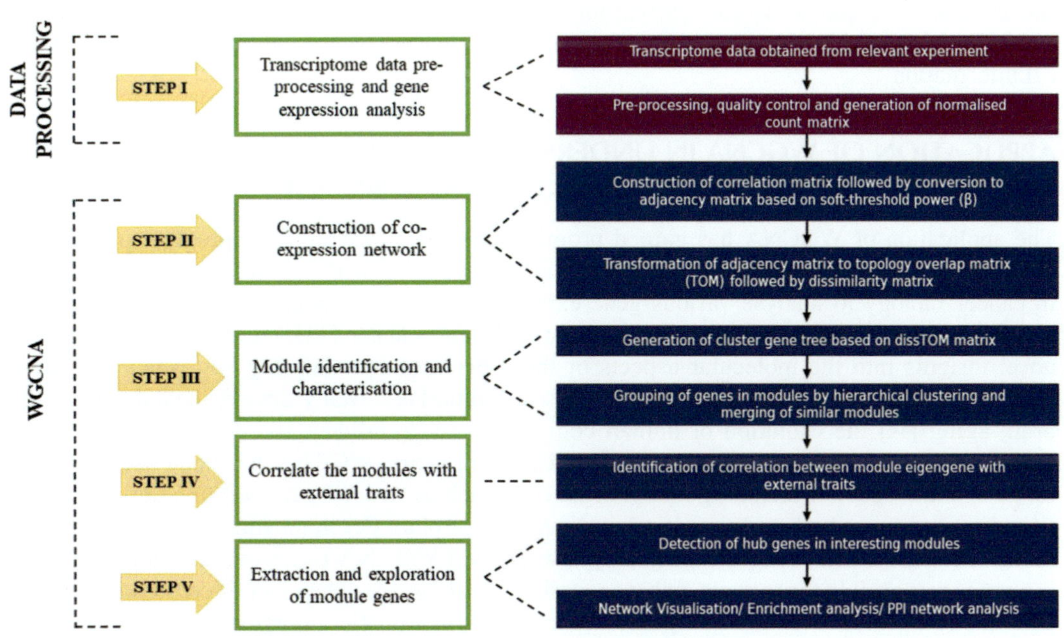

FIGURE 5.1 Overview of the steps involved in WGCNA from transcriptome data.

scale-free network. Network connectivity measures, such as average connectivity and network density, need to be evaluated to ensure the network's optimal properties. The topological overlap matrix (TOM) is constructed from the adjacency matrix. The TOM measure quantifies the interconnectedness of genes in the network and the dissimilarity matrix is then calculated from TOM.

Step III. Module Identification and Characterisation

Clustering algorithms (e.g., hierarchical clustering or dynamic tree cut) are applied to identify modules based on the TOM-based dissimilarity measure. Detection parameters may be adjusted to obtain biologically meaningful modules and then merged or split as needed to optimise size and gene membership. The eigengene for each module can be calculated, which represents the first principal component of the gene expression profiles within the module. It is used as a representative expression profile for the entire module.

Step IV. Correlate the Modules with External Traits

The relationship between the module eigengenes and external traits of interest is then assessed. This can be done using correlation analysis, where the correlation coefficient between the eigengenes and each trait is calculated. Statistical tests, such as linear regression, can also be performed to determine the significance of the module–trait associations. Modules that show significant correlations or associations with the external traits are then further examined.

Step V. Extraction and Exploration of Module Genes

Genes of the module of interest can be extracted, and highly connected genes (hub genes) within the module can be identified using measures such as intramodular connectivity or gene significance. Co-expression networks can be visualised using visualisation tools like Cytoscape [8]. The hub genes of the module of interest can be further subjected to Kyoto Encyclopedia of Genes and Genomes (KEGG) pathway analysis [9], gene set enrichment analysis (GSEA) [10], survival analysis, and even PPI analysis to gain insight into their biological mechanism. Finally, module results can be integrated with other omics data or external resources to explore additional layers of biological information.

APPLICATION OF WGCNA IN UNDERSTANDING REGULATORY MODULES IN THE CONTEXT OF PLANT STRESS RESPONSE

The application of WGCNA has gained prominence in deciphering co-expression patterns and identifying functional modules in complex biological systems. It has been extensively applied in cancer biology to analyse various gene expression profiles of hepatocellular carcinoma [11], stomach adenocarcinoma [12], ovarian cancer [12], lung cancer [13], etc. WGCNA has yielded novel insights into the molecular aspects of these diseases by analysing key genes and identifying potential therapeutic targets and predictive biomarkers. Many developmental aspects like brain aging [14] the evolution of human cognition, and germline testis development [15] have also been studied with the help of WGCNA. In recent years, WGCNA has been used in a wide range of plant species, including model organisms such as *Arabidopsis thaliana* and crop plants like rice, maize, and wheat to study various aspects of plant development, stress responses, and metabolism. De Silva and Colleagues [12] have applied WGCNA to decode the embryo patterning genes responsible for somatic embryogenesis in *Arabidopsis*. They identified highly correlated gene modules to uncover the hub genes associated with somatic embryogenesis that may serve as potential molecular targets. Transcriptional regulation of different floral organs in plants is a very complicated process due to the interlinking of different developmental pathways. A wide range of genes are interconnected in this process, and the findings of the highly enriched modules have been done using WGCNA [16,17]. Plant stress response is another important arena where WGCNA has been applied to understand regulatory modules and gene networks. Heat, drought, salt, cold, heavy metal, and even biotic stress situations have been explored by WGCNA. The method was used to build a co-expression network of differentially expressed genes (DEGs) between the rice genotypes IR64 (resistant to heat stress) and Koshihikari (susceptible to heat stress) [18]. Four modules in both genotypes were identified that were highly correlated with the time points after heat stress in the seedling. Li et al. [19] conducted a comprehensive study on cold stress and vernalisation using WGCNA. Their analysis identified three significant networks involving: (1) ethylene and jasmonic acid response, (2) cold acclimation, and (3) chromatin alteration in response to extended cold exposure. WGCNA was employed to identify modules and genes associated with Cd-accumulative traits in different varieties of rice [20]. Meta-analysis and gene co-expression network analysis were used to identify citrus genes involved in interactions during Huanglongbing disease [21]. In 2023, Nemati et al. combined microarray meta-analysis with WGCNA to pinpoint key gene network modules and hub genes associated with the wheat response to biotic stress [22]. The drought response mechanism in two sunflower inbred lines was explored using time-course transcriptome analysis in conjunction with WGCNA [23]. The mechanism of potassium regulating cotton root adaptation to salt stress was investigated, and the method was used to identify key modules [24]. WGCNA has recently emerged as a powerful way to analyse co-expressed miRNA networks responsive to plant stress. MiRNAs play a pivotal role in responding to unfavourable conditions by regulating the expression of target genes in plants. The connections between miRNAs, their targets, and phenotypic variations were explored in maize roots under drought stress through the application of WGCNA [25]. MiRNA-target regulatory modules during the grain-filling stage in foxtail millet have been investigated using a similar approach [26]. Zhang et al. explored the co-expression network at the isoform level in *Nelumbo nucifera*, emphasising the pivotal regulatory role of miRNAs due to the extensive interactions between miRNAs and hub isoforms [27].

LIMITATIONS OF WGCNA

While WGCNA is a powerful method for identifying co-expression modules and hub genes, it also has several limitations:

Overfitting: WGCNA can be sensitive to overfitting, especially when dealing with small sample sizes or noisy data. It is important to carefully select the soft thresholding power and other parameters to avoid this.

Sample size: Robustness or repeatability is highly dependent on the sample size and quality of the data. Attempting WGCNA on a dataset consisting of fewer than 15 samples is not recommended. Correlations on so few will simply be too noisy for the network to be biologically meaningful.

Module size: WGCNA is designed to identify modules, but their size can vary widely depending on the data and parameters used. Small modules may be difficult to interpret, while large ones may contain genes with diverse functions.

Reproducibility: The results of WGCNA can be sensitive to the choice of parameters and the specific dataset used. Reproducibility can be a challenge, especially when comparing results across different studies.

Interpretation: While WGCNA can identify modules and hub genes, it does not necessarily provide information on the causal relationships between genes or the underlying biological mechanisms. Another problem is that it assumes only a linear or monotonic relationship among genes. Further experimental validation and functional analysis are often necessary to confirm the biological relevance of the identified modules and genes.

Data preprocessing: WGCNA assumes that the data are preprocessed and normalised appropriately. However, different preprocessing methods can lead to different results, and it is important to carefully evaluate the effects of different normalisation methods and quality control measures on the results of WGCNA.

Overall, while WGCNA is a powerful method for identifying co-expression modules and their hub genes, it is important to carefully consider its limitations and ensure that the results are supported by additional experimental validation and functional analysis.

COMPARISON OF WGCNA AND GSEA

WGCNA and GSEA are two popular methods used in gene expression analysis to gain insights into biological processes and identify key genes or gene sets associated with specific phenotypes or conditions. While both approaches have similarities, they differ in their underlying principles and applications. Table 5.1 shows a comparison of WGCNA and GSEA in several key aspects.

Case Study

This section describes the steps involved in WGCNA analysis, utilising a normalised gene expression dataset for plant stress conditions, followed by hub gene identification and network visualisation in the R environment and Cytoscape. The aim is to create co-expression networks, grouping genes with similar expression patterns in clusters and relating these clusters with phenotypic characteristics. This protocol has been revised from available online resources like the tutorial from Peter Langfelder and Steve Horvath [28] (https://web.archive.org/web/20230323144343/horvath.genetics.ucla.edu/html/CoexpressionNetwork/Rpackages/WGCNA/), and subsequent modifications have been made to the code to adapt according to our dataset.

Dataset

We demonstrate our protocol using the raw gene count matrix obtained from gene expression profiles of *Arabidopsis* seedlings treated with or without ethanol under drought conditions [29]. The RNA-seq dataset is publicly available from the GenBank/GEO database under the accession number GSE201379. There are a total of 30 samples present in the dataset. As per the paper, the RNA-Seq analysis was done by removing the low-quality reads and trimming of adapters using Trimmomatic version 0.39 [30] followed by mapping the reads to the reference genome using HISAT2 version 2.2.1 [31]. Finally, aligned reads within gene models were counted using featureCounts version 2.0.1 [32].

Hardware and Software Tools Requirement

This pipeline requires a computer with at least 16 GB RAM, Intel Core i5 (64 bits) or higher processors, with at least 1 TB hard drive storage. The software tools required are RStudio with R-4.3.0, the WGCNA package, a few other requisite R packages, and the latest version of Cytoscape.

Method

Step I. Transcriptome Data Processing and Preparation

'WGCNA' is an R package with built-in functions for WGCNA analysis. Before starting to work with the input dataset, the following packages must be installed in the R environment:

```
# Install the requisite packages.
install.packages("BiocManager")
```

TABLE 5.1
Comparison of WGCNA and GSEA

Variables/Features	WGCNA	GSEA
Principle	WGCNA is a systems biology method that focuses on identifying co-expressed gene modules or networks. It constructs a gene co-expression network and assigns genes into modules based on their expression patterns [28].	GSEA is an enrichment analysis method that assesses whether predefined sets of genes (gene sets) exhibit statistically significant differences between two biological states or conditions [10].
Input data	WGCNA tools typically require gene expression data across multiple samples. They can handle various types of data, including microarray and RNA sequencing data, and do not rely on predefined gene sets.	GSEA tools primarily work with gene expression data from two or more distinct groups or conditions. They require a ranked gene list based on differential expression analysis, typically obtained through methods such as t-tests or linear models. They also rely on predefined gene sets, which can be derived from various sources, including pathway databases or expert-curated collections.
Sample size	WGCNA is preferable with a moderate to large sample size (e.g., 20 to hundreds of samples) [33].	GSEA can be applied to datasets with varying sample sizes. It is commonly used even with smaller sample sizes (e.g., tens of samples) [10,34].
Output	WGCNA tools generate modules or gene networks, where genes within a module exhibit similar expression patterns. They assign each module a unique colour or identifier, and provide information on the relationships between modules, module eigengenes, and phenotypic traits.	GSEA tools produce an enrichment score for each gene set, which represents the degree to which the gene set is overrepresented at the top or bottom of the ranked gene list. They also provide a normalised enrichment score (NES) that accounts for the size and correlation structure of the gene sets [10].
Key statistical test	The WGCNA approach identifies modules using hierarchical clustering and dynamic tree-cutting algorithms [4].	The gene set enrichment analysis is performed using statistical tests like the Kolmogorov–Smirnov test or gene set permutation test [35].
Advantages	WGCNA reduces the complexity of high-dimensional gene expression data by aggregating genes into modules. WGCNA can identify highly connected genes, or hub genes, within each module that play important roles in regulating the module's activity. Hub gene identification helps in biomarker or regulatory gene exploration and study [36].	GSEA employs a non-parametric and permutation-based approach to calculate statistically significant gene sets. This helps in accounting for the complex dependencies and correlations within gene expression data, which is more reliable and accurate [34]. GSEA is useful in comparing gene expression data across multiple studies. Analysis of gene sets across different datasets and identification of common biological pathways helps in the identification of generalised, robust, and reproducible findings.
Limitations	WGCNA involves constructing a correlation matrix, calculating network adjacency, and performing hierarchical clustering, which sometimes requires extensive computational power and time. WGCNA depends on setting a soft threshold to transform the correlation matrix into a network adjacency matrix. Selecting an appropriate threshold is important. The choice can impact the resulting network structure and interpretation [4,37].	One of the major disadvantages of GSEA is that the analysis results can be influenced by the size of the gene sets used. Larger gene sets have a higher statistical power to detect enrichment, while smaller ones may be more susceptible to noise and random variations. GSEA typically treats gene expression data as static snapshots and does not explicitly consider temporal dynamics. Sometimes, temporal changes in gene expression patterns may not be properly captured by GSEA analysis [38,39].

```
BiocManager::install("WGCNA")
BiocManager::install("flashClust")
BiocManager::install("igraph")
BiocManager::install("edgeR")

# Load the libraries.
library(BiocManager)
library(WGCNA)
library(flashClust)
library(edgeR)
```

The raw read count datasets are first normalised into fragments per kilobase of transcript per million of mapped fragments (FPKM) and then loaded in the R environment. The data are then transformed for the appropriate file format for WGCNA analysis using the following code:

```
# Set up the directory.
setwd("/home/Protocol/WGCNA")

# Read the raw data file.
counts <- read.table(
  "Raw_gene_counts.txt",
  header = TRUE,
  row.names = 1
)

# Normalise the raw data into FPKM gene count matrix.
raw_counts <- DGEList(counts = counts)
tmm_normalized <- calcNormFactors(raw_counts)

normalized_counts <- cpm(
  Tmm_normalized,
  normalized.lib.sizes = TRUE
) * 1e6

library_sizes <- colSums(normalized_counts)
gene_lengths <- c(1000)
fpkm <- normalized_counts / (library_sizes / 1e6) / (gene_lengths / 1000)
GeneExpression.csv = ("fpkm")
datExpr = read.csv("GeneExpression.csv")

# Manipulate file so it matches the format WGCNA needs.
row.names(datExpr) = datExpr$ID
datExpr$ID = NULL
datExpr = as.data.frame(t(datExpr))
dim(datExpr)
```

As WGCNA involves constructing a gene co-expression network based on pairwise correlations between gene expression profiles, outliers can significantly affect these correlation measures, potentially leading to biased network structures. The outliers are the samples which do not follow the pattern of most of the samples. To determine and remove the outliers, the following code was used:

```
# Group the data in a dendrogram to check outliers.
sampleTree = hclust(dist(datExpr), method = "average")
par(cex = 0.6)
par(mar = c(0, 4, 2, 0))
```

```
plot(
  sampleTree,
  main = "Sample clustering to detect outliers",
  sub = "",
  xlab = "",
  cex.lab = 1.5,
  cex.axis = 1.5,
  cex.main = 2
)

# Plot a line showing the cutoff value.
abline(h = 200000, col = "red")

# Select clusters below the cutoff line.
clust = cutreeStatic(sampleTree, cutHeight = 200000, minSize = 10)

# Select cluster 1 samples for the analysis.
keepSamples = (clust == 1)
expression = datExpr[keepSamples,]
nGenes = ncol(expression)
nSamples = nrow(expression)

# Create an object called "datTraits" that contains our trait
# data.
datTraits = read.csv("phenodata.csv", row.names = 1)

# Create a data frame analogous to expression data that can
# hold the clinical traits.
datTraits$Sample = NULL
table(rownames(datTraits) == rownames(datExpr))
save(datExpr, datTraits, file = "SamplesAndTraits.RData")
load("SamplesAndTraits.RData")
```

Figure 5.2 shows a dendrogram with clusters of data that are used for checking outliers. The red line in the figure shows the cutoff value for the cluster height to detect outliers. All the 30 samples were below the cutoff and were further included in the co-expression network.

The phenotype data (`phenodata.csv`) are also uploaded here. Correlating expression and phenotype data could reveal the role or effects of each gene on traits. The phenotype data that we considered for this analysis were time series data. The goal is to find genes that correlate with traits through time and differences in gene expression between samples treated with or without ethanol after the onset of drought.

Step II. Creating the Co-Expression Network

Choosing the Soft-Threshold Power for Generating a Scale-Free Topology Network

A soft-threshold power needs to properly be chosen to construct a co-expression network. It is an important parameter that determines the strength of the correlation between genes. It is needed to transform the gene expression data into an adjacency matrix, which represents the connectivity between genes.

```
# Choose the soft-threshold power.
powers = c(c(1:10), seq(from = 10, to = 30, by = 1))

sft = pickSoftThreshold(
  datExpr,
```

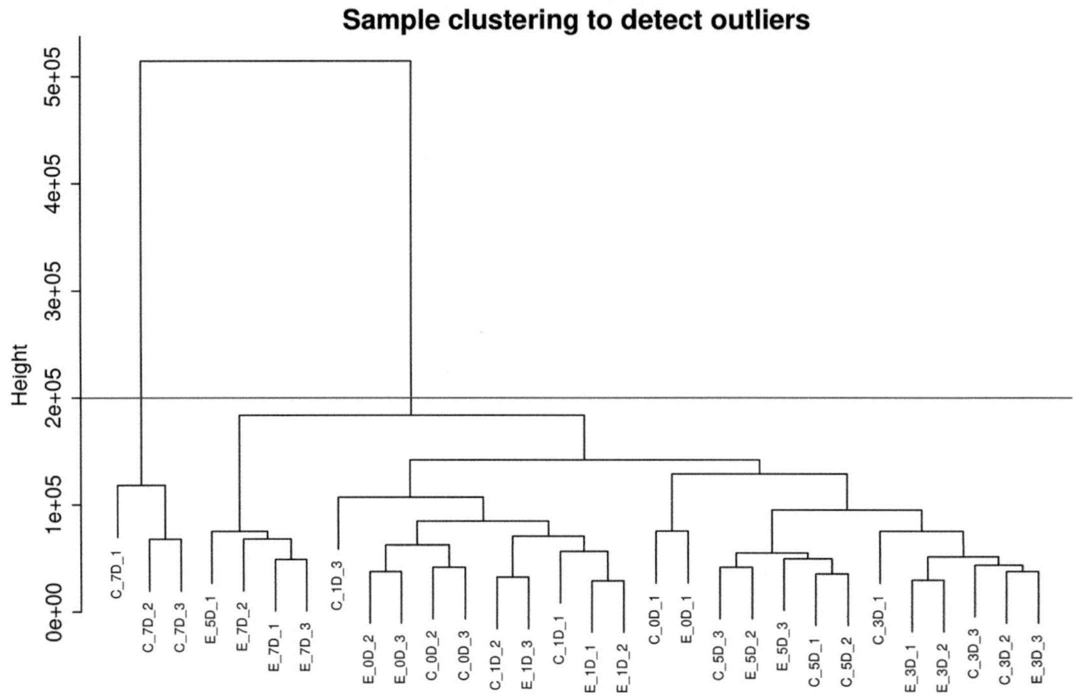

FIGURE 5.2 Clustering the samples for outlier detection. The clustered dendrogram shows outlier clusters based on our data with a cut-off value of 200,000.

```
  powerVector = powers,
  verbose = 5,
  networkType = "signed" # Call network topology analysis func.
)

sizeGrWindow(5, 9)
par(mfrow = c(1, 2))
cex1 = 0.9

plot(
  sft$fitIndices[,1],
  -sign(sft$fitIndices[,3]) * sft$fitIndices[,2],
  xlab = "Soft Threshold (power)",
  ylab = "Scale Free Topology Model Fit, signed R^2",
  type = "n",
  main = paste("Scale independence")
)

text(
  sft$fitIndices[,1],
  -sign(sft$fitIndices[,3]) * sft$fitIndices[,2],
  labels = powers,
  cex = cex1,
  col = "red"
)

abline(h = 0.90, col = "red")
```

```
plot(
  sft$fitIndices[,1],
  sft$fitIndices[,5],
  xlab = "Soft Threshold (power)",
  ylab = "Mean Connectivity",
  type = "n",
  main = paste("Mean connectivity")
)

text(
  sft$fitIndices[,1],
  sft$fitIndices[,5],
  labels = powers,
  cex = cex1,
  col = "red"
)
```

Figure 5.3 shows the plot for analysis of the scale-free fit index for various soft-thresholding powers. From this plot, we chose a power of 8 because it is the lowest power for which the scale-free topology index reaches 0.90 (the red line).

Gene Co-Expression Similarity and Adjacency Identification

Once the soft thresholding power is chosen, the next step is to identify the connection strength between nodes for co-expression module detection. Adjacency measures the strength of pairwise connections between genes based on their co-expression patterns. It quantifies the similarity in expression profiles between genes and determines the level of connectivity between them. The

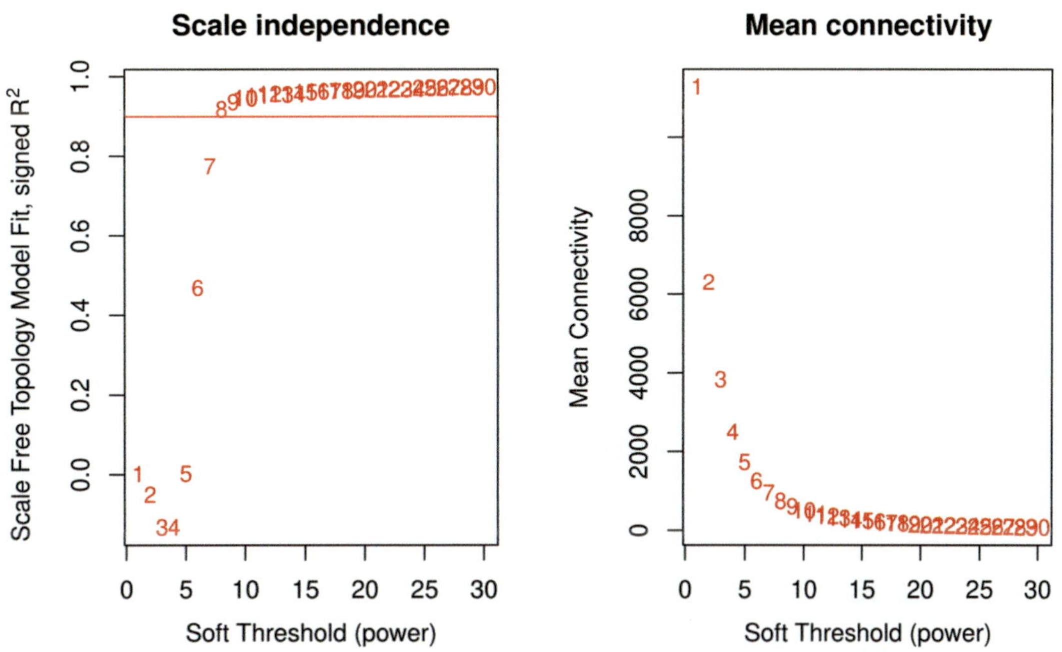

FIGURE 5.3 Analysis for co-expression network topology for different soft-thresholding powers. (a, b) The scale-free fit index (y-axis) is a function of the soft thresholding power (x-axis). (b) The mean connectivity (degree, y-axis) is a function of the soft-thresholding power (x-axis).

adjacency matrix is transformed into a TOM considering the number of neighbours the genes shared. TOM is then transformed into a dissimilarity measure (dissTOM).

```
# Build an adjacency correlation matrix.
enableWGCNAThreads()

softPower = 8

adjacency = adjacency(datExpr, power = softPower, type = "signed")

# Create a topology overlap matrix from the adjacency matrix.
TOM = TOMsimilarity(adjacency, TOMType = "signed")

dissTOM = 1 - TOM
```

Step III. Module Identification and Characterisation

Based on the dissTOM matrix, a clustered gene tree is generated. The dissTOM matrix, which represents the topological overlap between genes, serves as the input for constructing the hierarchical clustered gene tree. The next step involves module detection. The dynamic tree cut method was used for module identification, where the minimum module gene number is set to 30. Modules in WGCNA are defined as groups of genes that exhibit similar expression patterns across the samples under investigation.

```
# Generate a clustered gene tree.
geneTree = flashClust(as.dist(dissTOM), method = "average")

plot(
  geneTree,
  xlab = "",
  sub = "",
  main = "Gene Clustering on TOM-based dissimilarity",
  labels = FALSE,
  hang = 0.04
)

# Choose minimum module size based on dataset.
minModuleSize = 30

# Implement the dynamic tree-cut method to identify modules.
dynamicMods = cutreeDynamic(
  dendro = geneTree,
  distM = dissTOM,
  deepSplit = 2,
  pamRespectsDendro = FALSE,
  minClusterSize = minModuleSize
)

dynamicColors = labels2colors(dynamicMods)

MEList = moduleEigengenes(
  datExpr,
  colors = dynamicColors,
  softPower = softPower
)
```

```
MEs = MEList$eigengenes

MEDiss <- 1 - cor(MEs, use = "pairwise.complete.obs")

# Convert the dissimilarity matrix to a distance matrix and
# hierarchical clustering.
MEDiss_dist <- as.dist(MEDiss)

# Hierarchical clustering using the dissimilarity matrix
hclust_result <- hclust(
  as.dist(cleaned_MEDiss),
  method = "average"
)

METree = hclust_result

save(
  dynamicMods,
  MEList,
  MEs,
  MEDiss,
  METree,
  file = "Network_allSamples_signed_RLDfiltered.RData"
)

# Create a plot of eigengene clusters.
plot(
  METree,
  main = "Clustering of module eigengenes",
  xlab = "",
  sub = ""
)

MEDissThres = 0.0

merge = mergeCloseModules(
  datExpr,
  dynamicColors,
  cutHeight = MEDissThres,
  verbose = 3
)

mergedColors = merge$colors
mergedMEs = merge$newMEs
```

To plot the dendrogram, we used the following code:

```
# Plot the dendrogram.
plotDendroAndColors(
  geneTree,
  cbind(dynamicColors, mergedColors),
  c("Dynamic Tree Cut", "Merged dynamic"),
  dendroLabels = FALSE,
  hang = 0.03,
  addGuide = TRUE,
```

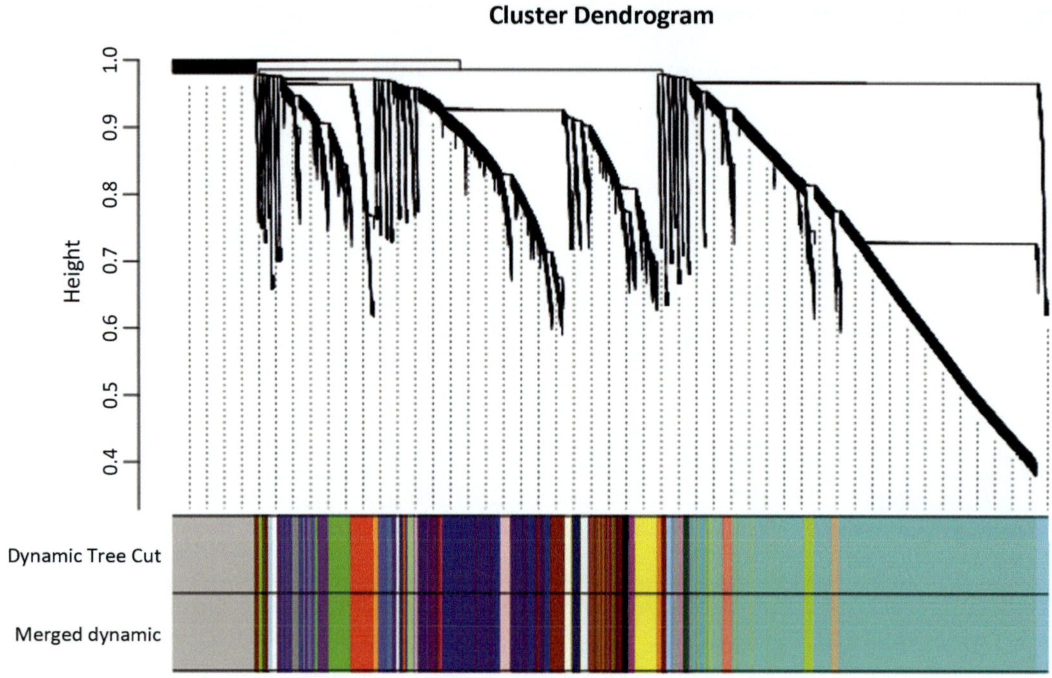

FIGURE 5.4 Cluster dendrogram representing hierarchical clustering of genes based on co-expression patterns. The x-axis represents the individual genes or modules, while the y-axis represents a measure of dissimilarity or similarity between the genes. The top is the gene dendrogram, and the bottom is genes' modules with different colours. Dynamic tree cut corresponds to the original module, and merged dynamic corresponds to the merged module.

```
  guideHang = 0.05
)

moduleColors = mergedColors
colorOrder = c("grey", standardColors(50))
moduleLabels = match(moduleColors, colorOrder) - 1
MEs = mergedMEs
```

A clustering dendrogram of all genes, with dissimilarities based on topological overlap and assigned module colours, is seen in Figure 5.4.

Step IV. Correlating Traits with Modules

Correlation between the module eigengene (the first principal component of gene expression within a module) with external traits is used to identify modules that are significantly associated with specific phenotypes. This association analysis helps uncover potential biomarkers, gene sets, or pathways that are relevant to the phenotype. To correlate phenotypic traits with modules, we used the following code:

```
# Correlate traits with eigengene modules.
nGenes = ncol(datExpr)
nSamples = nrow(datExpr)
MEs0 = moduleEigengenes(datExpr, moduleColors)$eigengenes
MEs = orderMEs(MEs0)
moduleTraitCor = cor(MEs, datTraits, use = "p")
```

```
moduleTraitPvalue = corPvalueStudent(
  moduleTraitCor,
  nSamples
)
```

To plot the module trait relationship heatmap, we used the following code:

```
# Print the correlation heatmap between traits and modules.
textMatrix = paste(
  signif(moduleTraitCor, 2),
  "\t(",
  signif(moduleTraitPvalue, 1),
  ")",
  sep = ""
)

dim(textMatrix) = dim(moduleTraitCor)
par(mar = c(8, 8.5, 5, 3))
```

Figure 5.5 displays the relationships between modules and traits in the form of a heatmap.

```
# Plot the module trait relationship heatmap.
labeledHeatmap(
  Matrix = moduleTraitCor,
  xLabels = names(datTraits),
  yLabels = names(MEs),
  ySymbols = names(MEs),
  colorLabels = TRUE,
  colors = blueWhiteRed(50),
  textMatrix = textMatrix,
  setStdMargins = FALSE,
  cex.text = 0.4,
  cex.axis = 0.5,
  zlim = c(-1),
  main = "Module-trait relationships",
  textAdj = c(0.9, 0.9),
  height = 10,
  pdf(file = "plot112.pdf", width = 10, height = 10)
)

dev.off()
```

Step V. Extraction and Exploration of Module Genes

Within each module, hub genes are highly connected genes that serve as key regulators or drivers of the module. The following code can be employed to extract and explore the hub genes of a module. Visualisation tools such as Cytoscape [8] can be used to visualise these hub genes within the network. The key genes of the modules can be further subjected to gene and pathway enrichment analysis, PPI network analysis, etc.

```
# Identify and plot hub genes from module "tan".
chooseTopHubInEachModule(datExpr, moduleColors)
inModule.tan = moduleColors == "tan"
modTOM.tan = TOM[inModule.tan, inModule.tan]
dim(modTOM.tan)
dimnames(modTOM.tan) = list(modGenes.tan, modGenes.tan)
```

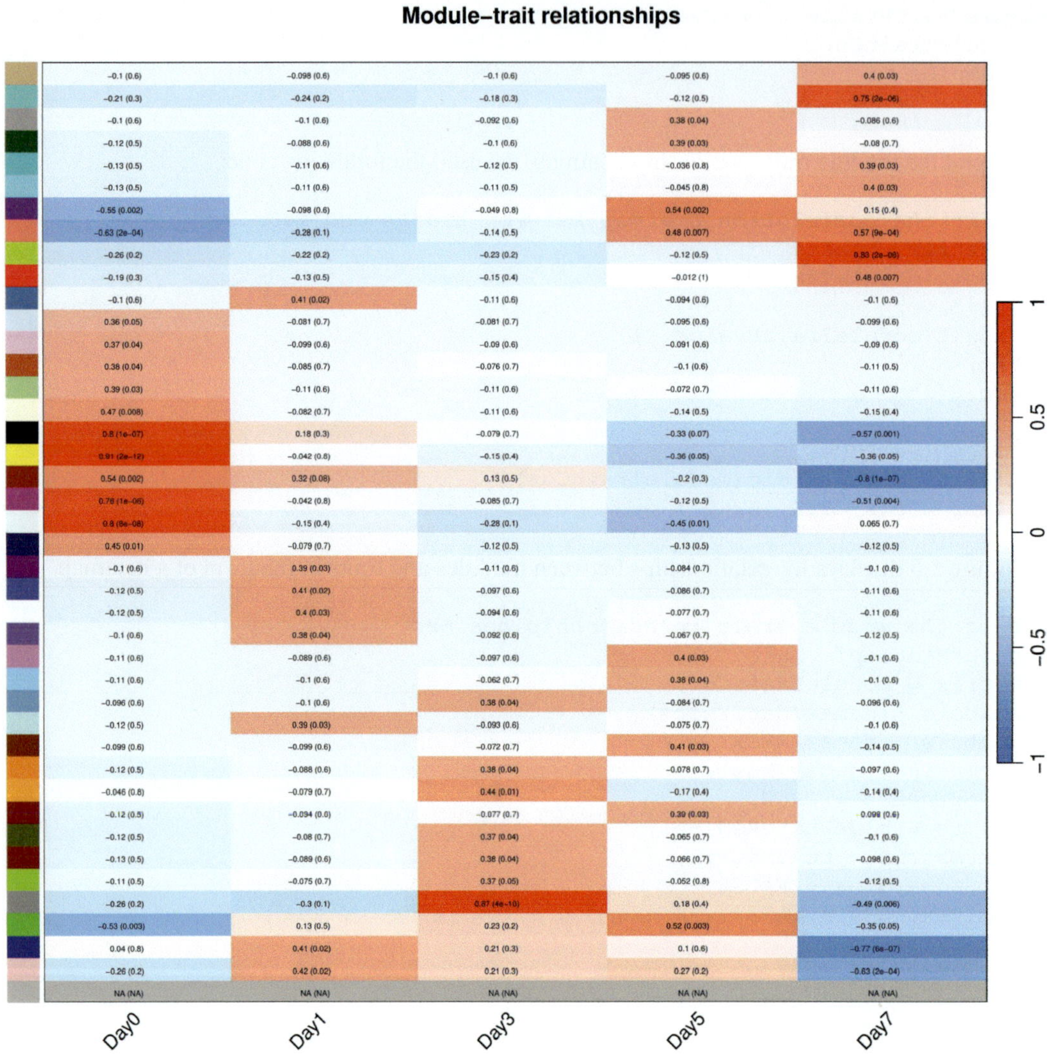

FIGURE 5.5 Module–trait relationship heatmap. The x-axis represents phenotypic traits used in the experiment; the y-axis represents the modules. The correlation coefficient decreases from red to blue. Numbers in the table represent the correlation coefficient between the gene module and the corresponding traits, and the numbers in parentheses represent the P value.

```
# To produce a clearer plot, we can subsample the top hub genes.
nTop = 60
genes = names(datExpr)

IMConn = softConnectivity(
  datExpr[, genes[(moduleColors == "tan")]]
)

top = (rank(-IMConn) <= nTop)
modTOM.tan[top, top]
library(igraph)
```

```
graph <- graph.adjacency(
  modTOM.tan[top, top],
  weighted = TRUE,
  mode = "undirected",
  diag = FALSE
)

pdf("NetworkHubGenes.tan.pdf")
plot(graph)
dev.off()
```

CONCLUSIONS

WGCNA is an important method in systems biology for describing the patterns of gene expression correlation among genes. Clusters of highly correlated genes are identified as modules based on hierarchical clustering combined with user-defined rules for defining discrete modules from the clustering dendrogram. These gene modules may reflect genes that interact in expression responses or are expressed similarly. WGCNA has been widely used in various biological setups, including plant stress conditions, to explore the relationships between different gene sets (modules) or between gene sets and external features.

CODE AVAILABILITY

The R codes used in this chapter can be found in GitHub page: "https://github.com/Subho7jit/WGCNA.git".

REFERENCES

1. Kusonmano K. Gene expression analysis through network biology: Bioinformatics approaches. Adv Biochem Eng Biotechnol [Internet]. 2017 [cited 2023 Dec 27];160:15–32. Available from: https://link.springer.com/chapter/10.1007/10_2016_44
2. Grimes T, Potter SS, Datta S. Integrating gene regulatory pathways into differential network analysis of gene expression data. Sci Rep. 2019;9(1):1–12.
3. Batushansky A, Toubiana D, Fait A. Correlation-based network generation, visualization, and analysis as a powerful tool in biological studies: A case study in cancer cell metabolism. Biomed Res Int. 2016;2016. https://doi.org/10.1155/2016/8313272
4. Langfelder P, Horvath S. WGCNA: An R package for weighted correlation network analysis. BMC Bioinformatics. 2008;9. https://doi.org/10.1186/1471-2105-9-559
5. Zhang B, Horvath S. A general framework for weighted Gene co-expression network analysis. Stat Appl Genet Mol Biol [Internet]. 2005;4(1). Available from: https://doi.org/10.2202/1544-6115.1128
6. DiLeo MV, Bakker Md, Chu EY, Hoekenga OA. An assessment of the relative influences of genetic background, functional diversity at major regulatory genes, and transgenic constructs on the tomato fruit metabolome. Plant Genome. 2014;7(1):1–16. https://doi.org/10.3835/plantgenome2013.06.0021
7. Yang X, Yang S, Li Q, Wuchty S, Zhang Z. Prediction of human-virus protein-protein interactions through a sequence embedding-based machine learning method. Comput Struct Biotechnol J. 2020;18:153–161. https://doi.org/10.1016/j.csbj.2019.12.005
8. Shannon P, Markiel A, Ozier O, Baliga NS, Wang JT, Ramage D, et al. Cytoscape: A software environment for integrated models of biomolecular interaction networks. Genome Res [Internet]. 2003 Nov [cited 2023 Aug 11];13(11):2498. Available from:/pmc/articles/PMC403769/
9. Chen L, Chu C, Lu J, Kong X, Huang T, Cai YD. Gene ontology and KEGG pathway enrichment analysis of a drug target-based classification system. PLoS One. 2015;10(5):1–12.
10. Subramanian A, Tamayo P, Mootha VK, Mukherjee S, Ebert BL, Gillette MA, et al. Gene set enrichment analysis: A knowledge-based approach for interpreting genome-wide expression profiles. Proc Natl Acad Sci U S A [Internet]. 2005 Oct 25 [cited 2023 Aug 12];102(43):15545–50. Available from: https://www.pnas.org/doi/abs/10.1073/pnas.0506580102

11. Zhang J, Baddoo M, Han C, Strong MJ, Cvitanovic J, Moroz K, et al. Gene network analysis reveals a novel 22-gene signature of carbon metabolism in hepatocellular carcinoma. Oncotarget. 2016;7(31):49232–49245. https://doi.org/10.18632/oncotarget.10249.
12. Ally A, Balasundaram M, Carlsen R, Chuah E, Clarke A, Dhalla N, et al. Comprehensive and integrative genomic characterization of hepatocellular carcinoma. Cell [Internet]. 2017 Jun 6 [cited 2023 Apr 29];169(7):1327. Available from:/pmc/articles/PMC5680778/
13. Luo X, Feng L, Xu WB, Bai XJ, Wu MN. Weighted gene co-expression network analysis of hub genes in lung adenocarcinoma. Evol Bioinforma. 2021;17:1–7. https://doi.org/10.1177/11769343211009898
14. Yang H, Junping P, Yirong X, Xiangnan M, Jiahui W, Qin G, Huanmin L. 2018. Gene expression analysis reveals novel gene signatures between young and old adults in the human prefrontal cortex. Front Aging Neurosci. 2018;10. https://doi.org/10.3389/fnagi.2018.00259.
15. Guo J, Sosa E, Chitiashvili T, Nie X, Rojas EJ, Oliver E, et al. Single-cell analysis of the developing human testis reveals somatic niche cell specification and fetal germline.
16. Yang H, Nukunya K, Ding Q, Thompson BE. Tissue-specific transcriptomics reveal functional differences in floral development. Plant Physiol. 2022;188(2):1158–73. https://doi.org/10.1093/plphys/kiab557
17. Zhang Y, Chen X, Cao Y, Yang Z. C8b in complement and coagulation cascades signaling pathway is a predictor for survival in HBV-related hepatocellular carcinoma patients. Cancer Manag Res. 2021;13:3503–15. https://doi.org/10.2147/CMAR.S302917
18. Wang Y, Wang Y, Liu X, Zhou J, Deng H, Zhang G, et al. WGCNA analysis identifies the hub genes related to heat stress in seedling of rice (Oryza sativa L.). Genes (Basel). 2022;13(6): 1–14. https://doi.org/10.3390/genes13061020
19. Li N, Euring D, Cha JY, Lin Z, Lu M, Huang LJ, Kim WY. Plant Hormone-mediated regulation of heat tolerance in response to global climate change. Front Plant Sci. 2021 Feb 11;11: 627969. doi: 10.3389/fpls.2020.627969
20. Wang Q, Zeng X, Song Q, Sun Y, Feng Y, Lai Y. Identification of key genes and modules in response to cadmium stress in different rice varieties and stem nodes by weighted gene co-expression network analysis. Sci Rep. 2020;10(1):1–13.
21. Rawat N, Kiran SP, Du D, Gmitter FG, Deng Z. Comprehensive meta-analysis, co-expression, and miRNA nested network analysis identifies gene candidates in citrus against huanglongbing disease. BMC Plant Biol [Internet]. 2015;15(1). https://doi.org/10.1186/s12870-015-0568-4
22. Nemati M, Zare N, Hedayat-Evrigh N, Asghari R. Identification of key Gene network modules and hub genes associated with wheat response to biotic stress using combined microarray meta-analysis and WGCN analysis. Mol Biotechnol [Internet]. 2023 Mar 1 [cited 2023 Dec 28];65(3):453–65. Available from: https://link.springer.com/article/10.1007/s12033-022-00541-w
23. Wu Y, Wang Y, Shi H, Hu H, Yi L, Hou J. Time-course transcriptome and WGCNA analysis revealed the drought response mechanism of two sunflower inbred lines. PLoS One [Internet]. 2022 Apr 1 [cited 2023 Dec 28];17(4):e0265447. Available from: https://journals.plos.org/plosone/article?id=10.1371/journal.pone.0265447
24. Ju F, Sun L, Xiong C, Wang Z, Yu H, Pang J, et al. Weighted gene co-expression network analysis revealed the key pathways and hub genes of potassium regulating cotton root adaptation to salt stress. Front Plant Sci. 2023;14:1–15. https://doi.org/10.3389/fpls.2023.1132877
25. Tang Q, Lv H, Li Q, Zhang X, Li L, Xu J, et al. Characteristics of microRNAs and target genes in maize root under drought stress. Int J Mol Sci. 2022;23(9):1–13. https://doi.org/10.3390/ijms23094968
26. Deng Y, Zhang H, Wang H, Xing G, Lei B, Kuang Z, et al. The construction and exploration of a comprehensive MicroRNA centered regulatory network in foxtail millet (Setaria italica L.). Front Plant Sci. 2022;13.
27. Zhang Y, Rahmani RS, Yang X, Chen J, Shi T. Integrative expression network analysis of microRNA and gene isoforms in sacred lotus. BMC Genomics. 2020;21(1):1–15. https://doi.org/10.3389/fpls.2022.848474
28. Langfelder P, Horvath S. WGCNA: An R package for weighted correlation network analysis. BMC Bioinformatics [Internet]. 2008 Dec 29 [cited 2023 Aug 2];9(1):1–13. Available from: https://bmcbioinformatics.biomedcentral.com/articles/10.1186/1471-2105-9-559
29. Bashir K, Todaka D, Rasheed S, Matsui A, Ahmad Z, Sako K, et al. Ethanol-mediated novel survival strategy against drought stress in plants. Plant Cell Physiol. 2022;63(9):1181–1192. https://doi.org/10.1093/pcp/pcac114
30. Bolger AM, Lohse M, Usadel B. Trimmomatic: A flexible trimmer for Illumina sequence data. Bioinformatics. 2014;30(15):2114–2120. https://doi.org/10.1093/bioinformatics/btu170

31. Kim D, Paggi JM, Park C, Bennett C, Salzberg SL. Graph-based genome alignment and genotyping with HISAT2 and HISAT-genotype. Nat Biotechnol. 2019 378 [Internet]. 2019 Aug 2 [cited 2023 Dec 28];37(8):907–15. Available from: https://www.nature.com/articles/s41587-019-0201-4
32. Liao Y, Smyth GK, Shi W. FeatureCounts: An efficient general purpose program for assigning sequence reads to genomic features. Bioinformatics. 2014;30(7):923–30.
33. Allen JD, Xie Y, Chen M, Girard L, Xiao G. Comparing statistical methods for constructing large scale gene networks. PLoS One. 2012;7(1):17–9.
34. Maleki F, Ovens K, McQuillan I, Kusalik AJ. Size matters: How sample size affects the reproducibility and specificity of gene set analysis. Hum Genomics. 2019;13(Suppl 1):42.
35. Irizarry RA, Wang C, Zhou Y, Speed TP. Gene set enrichment analysis made simple. Stat Methods Med Res. 2009;18(6):565–75.
36. Zhang C, Luo G, Lin J, Zhao Z, Luo M, Li H. Identification of significant modules and hub genes involved in hepatic encephalopathy using WGCNA. Eur J Med Res [Internet]. 2022;27(1):1–15. Available from: https://doi.org/10.1186/s40001-022-00898-3
37. Rahmani B, Zimmermann MT, Grill DE, Kennedy RB, Oberg AL, White BC, et al. Recursive indirect-paths modularity (RIP-M) for detecting community structure in RNA-seq co-expression networks. Front Genet. 2016;7:1–15. https://doi.org/10.3389/fgene.2016.00080
38. Simillion C, Liechti R, Lischer HEL, Ioannidis V, Bruggmann R. Avoiding the pitfalls of gene set enrichment analysis with SetRank. BMC Bioinformatics. 2017;18(1):1–14.
39. Oh S, Song S, Dasgupta N, Grabowski G. The analytical landscape of static and temporal dynamics in transcriptome data. Front Genet. 2014;5(FEB):1–12.

6 Meta-Analysis of RNA-seq and Microarray Data

Pavithra Madamarandawala

INTRODUCTION

RNA plays a pivotal role in cellular function, orchestrating processes from the translation of genetic information to the intricate regulation of gene expression. A set of RNA transcripts (i.e., a transcriptome) functioning in an organism at a particular time may elicit a response towards internal/external stimuli and/or perform regular functions such as cellular differentiation and development. Several years after the discovery of RNA, scientists understood that studying the real-time transcriptome can give an understanding of the cellular behaviour under different experimental conditions, spurring a surge in RNA biology research. Over time, gene expression techniques evolved from Northern blotting, first described in 1977, to qPCR, then microarray technology, and more recently to high-throughput sequencing methods such as RNA sequencing (RNA-seq) [1]. Prior to the introduction of RNA-seq, microarray technologies were widely used for gene expression analysis (mRNA). The technology is based on hybridisation of the target nucleic acid sample (fluorescently labelled cDNA) to a group of oligonucleotide probes on a microarray slide. Expression patterns are identified by scanning the microarray slide for fluorescent emission [2]. Microarray technologies were used in a wide range of studies to identify gene expression patterns: cancer [3], antibiotic treatment studies [4,5], response to plant pathogens [6], etc. However, with time, the method of gene expression analysis transitioned from array-based technologies into RNA-seq, owing to its high-quality, quantitative nature and direct approach. Rather than relying on additional techniques, such as gel electrophoresis (e.g., Northern blotting) and hybridisation (e.g., microarrays), RNA-seq directly sequences RNA in each sample: the input RNA is fragmented and ligated to adapter molecules to generate libraries, which are then amplified and sequenced in parallel. The entire process is conducted employing various quality-control check points and repetition of sequencing to ensure the highest quality and resolution [7]. Raw data are in the form of reads, and the basic steps of bioinformatics analysis include (i) assessment of quality control, (ii) preprocessing of reads (adapter removal, duplicate removal, etc.), (iii) alignment to a reference genome (or *de novo* method if no reference genome is available), (iv) quantification of transcripts, and (v) differential gene expression analysis. RNA-seq is contributing to the advancement of many fields today, including food and agriculture (in the development of drought-tolerant crops [8] and the fermentation industry [9]) and medicine (in monitoring treatment, cancer [10], disease associated functional genomics studies, diabetes [11], biomarker studies, tuberculosis [12], kidney fibrosis [13], forensics [14], etc.).

IMPORTANCE OF META-ANALYSIS

The versatility, declining cost, and increasing global accessibility to these technologies have increased the use of gene expression analysis. To increase transparency, many journals instruct authors to submit the sequencing data to a public data repository such as the Gene Expression Omnibus (GEO)/Sequence Read Archive (SRA) of the National Center for Biotechnology Information (NCBI), ArrayExpress of the European Bioinformatics Institute (EBI), and the Encyclopedia of DNA Elements (ENCODE). Thus, these repositories store millions of array-based and next-generation sequencing (NGS) datasets (including those generated via RNA-seq)

DOI: 10.1201/9781003354062-6

of different organisms, tissues/cell lines, diseases/experimental conditions, etc. [15]. Figures 6.1(A) and 6.2 show, respectively, how the number of RNA-seq records submitted to the GEO and number of array-based studies submitted to ArrayExpress increased over the years (human-specific RNA-seq records are separately shown). Each GEO series shown may bear a varying number of samples ranging from 1 to 7501. The study which deposited the highest number (7,501) used *Mus musculus* samples to identify architectural variations in two different epithelial tissue types [16].

In general context, the higher the number of samples, the more generalisable and statistically stronger the findings are. However, the cost, lack of resources, low accessibility to high-quality samples, and ethical considerations in acquiring clinical samples can limit the number of samples included in an experiment. As the costs go down, some researchers tend to re-sequence more biological replicates to reinvestigate biological questions which are poorly understood [17]. As of late 2022, the average cost of RNA-seq per mRNA sample is around $36.9 to $173, which could be still unaffordable to many researchers [18]. Thus, many RNA biologists follow the general rule of using a minimum of three biological replicates per RNA-seq experiment, as is seen with the majority of GEO records (49%) in Figure 6.1(B) [19].

The goal of meta-analysis is to integrate multiple datasets bearing the same null hypothesis to bring about a common result [15]. When the number of biological replicates of an experiment is low, the current trend is to integrate one's own data with data from public repositories and present in meta-analytic approaches: (i) meta-analysis of RNA-seq data, (ii) meta-analysis of microarray data, or (iii) integrated meta-analysis of microarray and RNA-seq data. Under all these circumstances, meta-analysis of gene expression data is useful for the following reasons:

- It increases the statistical power.
- Its output is reproducible.
- It has an unbiased approach and is therefore generalisable to a majority of datasets.
- It is feasible, as the previous datasets are publicly available in repositories.
- It has generated reliable results [20–23].
- It allows for a better comparison between datasets, which is otherwise not practical due to differences between them [15,17].

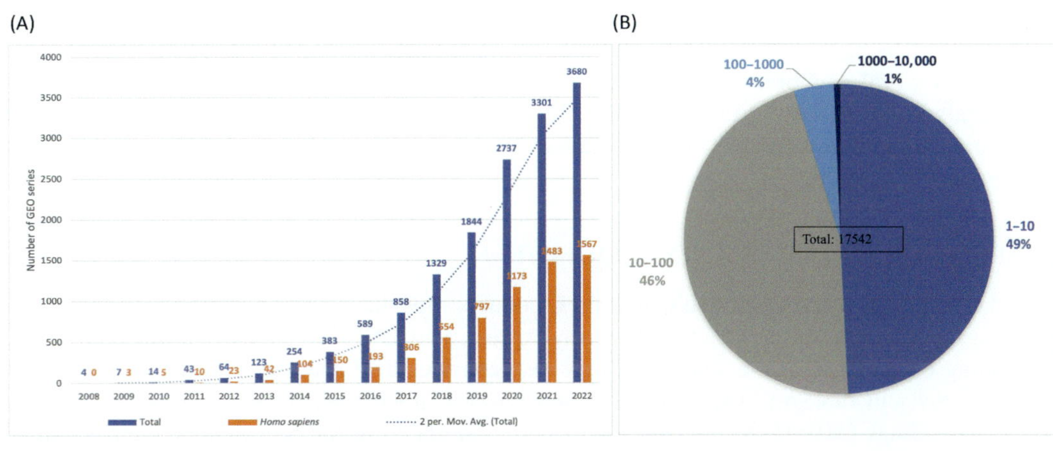

FIGURE 6.1 (A) The number of RNA-seq series (submitted records of an experiment) documented in the GEO database from 2008 to 2022 (total: 17,542, including 2462 in the year of 2023 alone). (B) The number of samples sequenced under each GEO series submitted. Source: www.ncbi.nlm.nih.gov/geo/, Search criteria: `("sequence analysis, rna" [MeSH Terms] OR RNA SEQUENCING[All Fields]) AND ("Expression profiling by high throughput sequencing"[Filter])`.

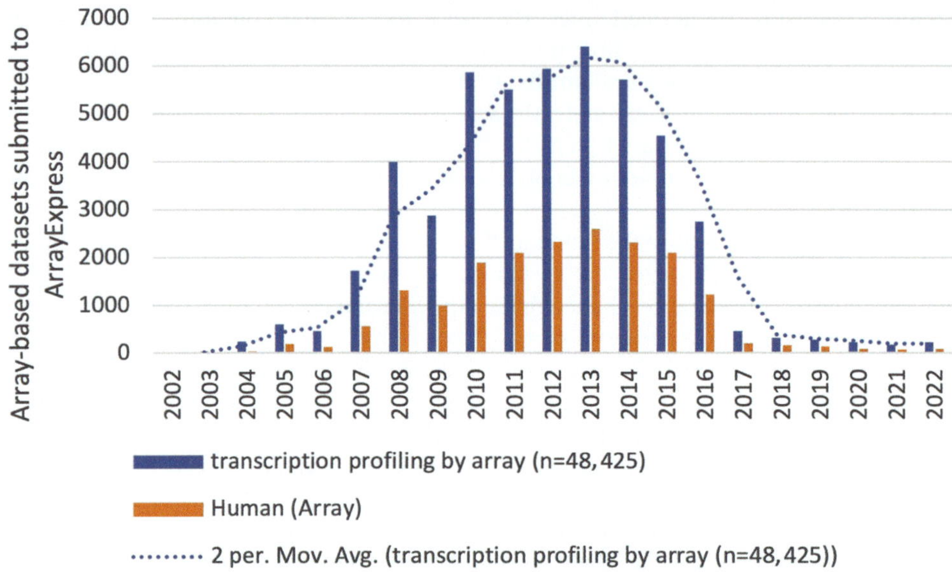

FIGURE 6.2 Array-based gene expression studies submitted to ArrayExpress during the years 2002–2022.

GENE EXPRESSION DATA META-ANALYSIS WORKFLOW

Generally, the goal of a meta-analysis of transcriptome data is to generate more reliable differential expression results. Tools for unified meta-analysis, such as Galaxy (usegalaxy.org, galaxyproject.eu) based metaRNAseq (meta-analysis of RNA-seq data), metaMA (meta-analysis of microarray data), and online tools such as NetworkAnalyst (www.networkanalyst.ca), were introduced [15]; however, some of these tools may be discontinued or not regularly updated. Important considerations in a meta-analysis are (i) selection of appropriate datasets, (ii) incorporating the same analysis tools and parameters to all samples to maintain uniformity from beginning to end, (iii) use of the same gene annotation ID algorithm for all samples, and (iv) performing a batch correction (if required). If an integrated meta-analysis of microarray and RNA-seq data is performed, the most common approach is to perform individual analyses for each platform using appropriate tools followed by integrating the end results using either p-value combination, rank combination, or effect size combination methods [15,24]. The choice of the meta-analysis method should be made based on the dataset characteristics (different or similar experimental conditions/platforms). A detailed description of each method is reviewed in Toro-Dominguez et al 2020 [15].

SEARCH FOR DATASETS AND ACQUISITION

The datasets may come from one's own labs, collaborating labs, and/or downloaded from public data repositories. While GEO/SRA, ArrayExpress, and ENCODE are the go-to repositories, depending on the field of the study, one may also consider field-specific repositories such as the examples given in Table 6.1. A search for datasets is mostly done manually using keywords, Medical Subject Headings (MESH) terms, and filter options available in the repositories. General guidelines for dataset search in repositories can be found on their respective websites and in detailed publications [25,26]. Data can be downloaded directly, via SRA Toolkit, or imported directly into analysis platforms like R, via packages such as GEOquery [27] and ArrayExpress [28]. Given the interconnectedness of repositories, it is crucial to screen for and remove duplicate studies prior to analysis. This screening process should ideally be conducted independently by at least two

TABLE 6.1
Public Gene Expression Data Repositories

Usage	Repository	URL
General	GEO/SRA	www.ncbi.nlm.nih.gov/geo/ www.ncbi.nlm.nih.gov/sra
	ENCODE	www.encodeproject.org
	ArrayExpress	www.ebi.ac.uk/biostudies/arrayexpress
	Expression Atlas: Curated and/or analysed datasets which were submitted/integrated to ArrayExpress. May be linked to subcategories as a single-cell expression atlas, FlyAtlas (*Drosophila melanogaster*)	www.ebi.ac.uk/gxa/home flyatlas.org/atlas.cgi www.ebi.ac.uk/gxa/sc/home
Animal genomics	Functional Annotation of Animal Genomes (FAANG) [29]	data.faang.org/home
	Farm Animal Genotype-Tissue Expression (FarmGTEx)/ pigGTEx/CattleGTEx [30]	www.farmgtex.org
Human tissues	Genotype-Tissue Expression (GTEx) [31]	gtexportal.org/home/
Human and mouse RNA-seq data	Recount3 [32]	rna.recount.bio
Cancer	Cancer Genome Atlas (TCGA)	www.cancer.gov/ccg/research/ genome-sequencing/tcga
Single-cell RNA (scRNA)-seq data	scRNAseq (a Bioconductor package)	bioconductor.org/packages/release/data/ experiment/html/scRNAseq.html
	Single cell portal [33]	singlecell.broadinstitute.org/single_cell
	CZ cell x gene discover [34]	cellxgene.cziscience.com

Note: All URLs were last accessed on 21 January 2025.

researchers to minimise human error and ensure thoroughness, particularly in systematic reviews and meta-analyses.

DATASET INCLUSION/EXCLUSION CRITERIA

In any meta-analysis study, selection of appropriate datasets is very important. Datasets may be different from one another based on (i) technical considerations (RNA type: total-RNA, small-RNA, single-cell RNA (scRNA), mRNA [7], RNA extraction method, platform used, bioinformatics analysis method) and/or (ii) biological considerations (patient samples, variations in the cohort such as age, cells/tissues from which RNA was extracted, disease status/stage, treatment given, geographical location, co-morbidities, animal samples, species/breed, bacterial samples, drug resistance, growth phase, etc.) [27,28]. Generally, all or some of the previously mentioned dataset descriptions are also available in the repositories; however, it is possible that some information might be incomplete, as was seen by a study on public meta-data evaluation [35]. If a particular criterion of consideration is missing from a sample, and cannot be obtained by other means such as communicating with the authors, it is wise to exclude the sample from analysis. A dataset is considered eligible based on the aim of the study. For example, if the aim is to investigate the differential gene expression in patients infected with drug-resistant tuberculosis bacteria in comparison to those infected with drug-susceptible tuberculosis, one should include datasets of samples collected prior to drug administration, as drug susceptibility can vary with the drugs [12]. If the aim of the study is broad – for example, if one aims to identify differentially expressed genes (DEGs) in all stages of breast cancer – the study should include datasets of all stages of breast cancer. However, if the aim is to identify early biomarkers, one might not find it useful to include gene expression data extracted from later stages of cancer. It is advisable to build up dataset inclusion/exclusion criteria to select the datasets without bias.

ASSESSMENT OF QUALITY CONTROL AND PREPROCESSING OF RAW DATA

Each raw data file selected for meta-analysis should be examined for its quality and preprocessed prior to its inclusion in the meta-analysis. While the data repositories may have downloadable preprocessed data, it is highly recommended to use raw data for a meta-analysis to ensure that all the datasets are treated the same way [36]. Tools used for preprocessing the RNA-seq data are different from those used for microarray data.

RNA-seq Datasets

Quality control and preprocessing (adapter trimming, duplicate removal, analysis of overrepresented reads and low-quality bases, etc.) can be done using RNA-seq analysis tools such as FastQC (for quality-control assessment of Illumina reads) or fastp (an all-in-one quality-control and preprocessing tool), MultiQC or NGSQC (applicable to any NGS platform) [37–40]. It is necessary to keep all parameters, such as Phred value and trim length, constant for all raw data files. If all the raw data were derived from the same organism, alignment should be done to the same reference genome. The common tools used for alignment are STAR, HISAT2, and TopHat. The quality of aligned reads should also be assessed; RSeQC, and RNA-QC chain are two examples of tools used for the purpose. For example, RSeQC screens the aligned data for optimal transcript integrity, read distribution, and gene body coverage and eliminates any unsuitable samples from the analysis [41]. The mapped reads should then be counted using an appropriate tool, such as featureCounts or htseq-count. If different platforms were used, gene IDs in each dataset might be different; therefore, an annotation approach should be used to identify the genes. The most common approach is biomaRt, an R interface linked to ENSEMBL (bioconductor.riken.jp/packages/3.4/bioc/vignettes/biomaRt/inst/doc/biomaRt.html). If the RNA-seq datasets are to be integrated with microarray data, before batch correction, the data might need to be log transformed.

Microarray Datasets

Microarray datasets included in a meta-analysis might come from either Agilent chip or Affymetrix chip or both, and the analysis tools might differ. General analysis workflow includes (i) quality-control analysis to remove duplicates, and background correction by removing probes with non-specific hybridisation and low-intensity values (tools: principal component analysis or PCA plots, heatmaps, relative log expression plot, oligo tool in R); (ii) data normalisation between arrays (robust multiarray averaging method, or RMA); (iii) log transformation; (iv) gene annotation, with the most common approach being biomaRt; (v) screening for duplicate gene IDs and replacement with their average value; and (vi) removal of normalisation control probes [42].

BATCH CORRECTION

Batch effects are variations within the batches of data which may arise due to technical variations, differences in handling sequences in different facilities, reagents or batches used, and/or unbalanced experimental design [43]. These effects need to be corrected prior to differential gene expression analysis as they can mask the real DEGs and/or introduce false DEGs solely due to the batch effect(s). Since meta-analysis considers data from diverse backgrounds, batch correction is an essential step in meta-analysis to reduce heterogeneousness between batches [43]. One can examine if the datasets selected for meta-analysis truly exhibit a batch effect using unsupervised clustering or PCA, which generates a plot that clusters samples based on their similarity.

The most commonly used tools for batch correction are ComBat-seq and limma. ComBat-seq employs an empirical Bayes (eBayes) approach, while limma uses a two-way ANOVA method [44]. Although originally developed for microarray data, limma has become popular for batch effect removal in various types of expression data. To use limma, researchers must input a numeric matrix or log-transformed expression values along with corresponding batch information. [45]. Both methods work well for bulk RNA-seq data analysis; however, if the batches are small, ComBat-seq should be the go-to batch correction tool, as overcorrection is unlikely to happen with it. ComBat-seq is a surrogate variable analysis (SVA)-based R package widely used for batch effect correction in RNA-seq count matrices [46]. An untransformed raw count data matrix is generally used, as the input and the batches should be defined. The main two advantages of ComBat-seq are (i) its output is directly compatible with differential gene expression packages such as DESeq2 and edgeR and (ii) it effectively removes batch effects while preserving the true biological signal in the data. One should also ensure the success of batch correction either via unsupervised clustering or PCA plots. PCA plots are also useful in removal of outliers. In addition to the two common tools, there are tools available such as scBatch [47], POIBM [48], Harmony, LIGER, and Seurat3 for scRNA-seq batch effect correction [49].

META-ANALYSIS TO IDENTIFY DEGs AND ADDITIONAL CONSIDERATIONS

For meta-analysis of RNA-seq data, count matrices (or, if required, batch-corrected count matrices) can be directly used in differential gene expression analysis tools DESeq2 and edgeR [50,51]. For meta-analysis of microarray data, normalised and log-transformed microarray data (and, if required, batch-corrected data) can be used directly for statistical analysis of DEGs using the linear regression model and eBayes adjustment in R [42].

For integrated meta-analysis of RNA-seq and microarray data, there is no defined protocol and hence different studies report different approaches. A study on identifying candidate genes for salt tolerance conducted differential gene expression analysis separately for microarray and RNA-seq datasets and screened for common DEGs [52]. Another study, which aimed to identify DEGs and pathways associated with age-related macular degeneration, conducted differential gene expression analysis on both microarray and RNA-seq datasets using limma. The resulting p-values were then combined using Fisher's method [24]. Another study related to pancreatic cancer used a WGCNA R package to build up a co-expression network analysis, which builds a weighted correlation analysis of the gene expression data [21,53]. A web-based tool called Omics Compendia Commons (OMiCC) was developed by a group of scientists in the United States for meta-analysis of gene expression data which would perform the entire workflow of meta-analysis, from downloading data from repositories to visualisation of the results. However, the tool has been tested only for microarray data [54,55]. In addition, the online tool NetworkAnalyst (www.networkanalyst.ca) is a recently developed web-based tool which can also perform integrated meta-analysis of both microarray and RNA-seq data. Multiple gene expression tables can be used as input for meta-analysis, and integration is performed based on p-value, effect size, rank order combination, vote counting, and direct merging approaches. However, the tool currently supports the analysis of data from five organisms: human, mouse, *Drosophila melanogaster, Caenorhabditis elegans*, and *Saccharomyces cerevisiae*, and the server response is known to be slow. A detailed protocol is given in Xia et al. 2015 [56]. All this work follows a three-step process: (i) conducting individual analyses for each platform using platform-specific tools, (ii) obtaining results from these analyses, and (iii) integrating these results using either p-value combination, rank combination, or effect size combination methods.

While the outputs (gene lists) of meta-analysis can be used for downstream analysis such as enrichment and pathway analysis, protein–protein interaction analysis, etc., one should also be aware that several meta-analysis studies have reported disagreement in the direction of differential expression of some DEGs [57,58]. Also, RNA-seq can identify more DEGs compared to microarray platforms [59,60]. These factors underscore the need for developing meta-analysis approaches that can effectively address such discrepancies and technological differences when integrating data from multiple sources.

CONCLUSIONS

An increasing number of studies use meta-analysis of gene expression data: either RNA-seq, microarray, or an integrated meta-analysis of both. While a meta-analysis of gene expression data has the potential to produce reliable and reproducible biological information of a higher statistical power, the meta-analysis should be done carefully in order not to overshadow the best biological interpretation of the question investigated. Most of the general tools used in the individual analysis of gene expression data are applicable in meta-analysis; however, a standard protocol is yet to be defined.

REFERENCES

1. Singh KP, Miaskowski C, Dhruva AA, Flowers E, Kober KM. Mechanisms and measurement of changes in gene expression. Biol Res Nurs. 2018;**20**(4):369–82.
2. Govindarajan R, Duraiyan J, Kaliyappan K, Palanisamy M. Microarray and its applications. J Pharm Bioallied Sci. 2012;**4**(Suppl 2):S310–2.
3. Virtanen C, Woodgett J. Clinical uses of microarrays in cancer research. Methods Mol Med. 2008;**141**:87–113.
4. Brazas MD, Hancock REW. Using microarray gene signatures to elucidate mechanisms of antibiotic action and resistance. Drug Discov Today. 2005 Sep 15;**10**(18):1245–52.
5. Microbiology Society. Microarray analysis of the transcriptional responses of Clostridium difficile to environmental and antibiotic stress [Internet]. Available from: www.microbiologyresearch.org/content/journal/jmm/10.1099/jmm.0.47657-0
6. Moy P, Qutob D, Chapman BP, Atkinson I, Gijzen M. Patterns of gene expression upon infection of soybean plants by *Phytophthora sojae*. Mol Plant-Microbe Interactions. 2004;**17**(10):1051–62.
7. Hrdlickova R, Toloue M, Tian B. RNA-seq methods for transcriptome analysis. WIREs RNA. 2017;**8**(1):e1364.
8. Saidi A, Hajibarat Z. Application of next generation sequencing, GWAS, RNA seq, WGRS, for genetic improvement of potato (Solanum tuberosum L.) Under drought stress. Biocatal Agric Biotechnol. 2020;**29**:101801.
9. Wang HK, Ng YK, Koh E, Yao L, Chien AS, Lin HX, et al. RNA-seq reveals transcriptomic interactions of *Bacillus* subtilis Natto and *Bifidobacterium animalis* subsp. *lactis* in whole soybean solid-state co-fermentation. Food Microbiol. 2015 Oct 1;**51**:25–32.
10. Wang J, Dean DC, Hornicek FJ, Shi H, Duan Z. RNA sequencing (RNA-seq) and its application in ovarian cancer. Gynecol Oncol. 2019;**152**(1):194–201.
11. Yi H, Peng R, Zhang LY, Sun Y, Peng HM, Liu HD, et al. LincRNA-Gm4419 knockdown ameliorates NF-κB/NLRP3 inflammasome-mediated inflammation in diabetic nephropathy. Cell Death Dis. 2017 Feb;**8**(2):e2583.
12. Madamarandawala P, Rajapakse S, Gunasena B, Madegedara D, Magana-Arachchi D. A host blood transcriptional signature differentiates multi-drug/rifampin-resistant tuberculosis (MDR/RR-TB) from drug susceptible tuberculosis: A pilot study. Mol Biol Rep. 2023;**50**(4):3935–43.
13. Craciun FL, Bijol V, Ajay AK, Rao P, Kumar RK, Hutchinson J, et al. RNA sequencing identifies novel translational biomarkers of kidney fibrosis. J Am Soc Nephrol JASN. 2016 Jun;**27**(6):1702–13.
14. Hanson E, Ingold S, Haas C, Ballantyne J. Messenger RNA biomarker signatures for forensic body fluid identification revealed by targeted RNA sequencing. Forensic Sci Int Genet. 2018;**34**:206–21.
15. Toro-Domínguez D, Villatoro-García JA, Martorell-Marugán J, Román-Montoya Y, Alarcón-Riquelme ME, Carmona-Sáez P. A survey of gene expression meta-analysis: Methods and applications. Brief Bioinform. 2021;**22**(2):1694–705.
16. Andrews N, Serviss JT, Geyer N, Andersson AB, Dzwonkowska E, Šutevski I, et al. An unsupervised method for physical cell interaction profiling of complex tissues. Nat Methods. 2021 Aug;**18**(8):912–20.
17. Rau A, Marot G, Jaffrézic F. Differential meta-analysis of RNA-seq data from multiple studies. BMC Bioinformatics. 2014 Mar 29;**15**(1):91.
18. Alithea Genomics, Budgeting for an mRNA-seq project? How much does RNA Seq cost? [Internet]. Available from: alitheagenomics.com/blog/budgeting-for-an-mrna-seq-project-here-are-the-main-cost-drivers-to-keep-an-eye-on. Accessed on 10 February 2024.
19. Todd EV, Black MA, Gemmell NJ. The power and promise of RNA-seq in ecology and evolution. Mol Ecol. 2016;**25**(6):1224–41.

20. Benny J, Pisciotta A, Caruso T, Martinelli F. Identification of key genes and its chromosome regions linked to drought responses in leaves across different crops through meta-analysis of RNA-seq data. BMC Plant Biol. 2019 May 10;**19**(1):194.
21. Nisar M, Paracha RZ, Arshad I, Adil S, Zeb S, Hanif R, et al. Integrated analysis of microarray and RNA-seq data for the identification of hub genes and networks involved in the pancreatic cancer. Front Genet. 2021 Jun 23;**12**:663787.
22. Tamura K, Bono H. Meta-analysis of RNA sequencing data of arabidopsis and rice under hypoxia. Life. 2022;12(7):1079.
23. Bairakdar MD, Tewari A, Truttmann MC. A meta-analysis of RNA-seq studies to identify novel genes that regulate aging. Exp. Gerontolt. 2023;173:12107.
24. Dhirachaikulpanich D, Li X, Porter LF, Paraoan L. Integrated microarray and RNAseq transcriptomic analysis of retinal pigment Epithelium/Choroid in age-related macular degeneration. Front Cell Dev Biol [Internet]. 2020;**8**. Available from: www.frontiersin.org/articles/10.3389/fcell.2020.00808
25. Wang Z, Lachmann A, Ma'ayan A. Mining data and metadata from the gene expression omnibus. Biophys Rev. 2019;**11**(1):103–10.
26. Consortium TEP. A user's guide to the encyclopedia of DNA elements (ENCODE). PLOS Biol. 2011 Apr 19;**9**(4):e1001046.
27. Davis S, Meltzer PS. GEOquery: A bridge between the gene expression omnibus (GEO) and bioconductor. Bioinforma Oxf Engl. 2007Jul 15;**23**(14):1846–7.
28. Kauffmann A, Rayner TF, Parkinson H, Kapushesky M, Lukk M, Brazma A, et al. Importing array express datasets into R/bioconductor. Bioinformatics. 2009 Aug 15;**25**(16):2092–4.
29. Harrison PW, Sokolov A, Nayak A, Fan J, Zerbino D, Cochrane G, et al. The FAANG data portal: Global, open-access, "FAIR", and richly validated genotype to phenotype data for high-quality functional annotation of animal genomes. Front Genet [Internet]. 2021;**12**. Available from: www.frontiersin.org/articles/10.3389/fgene.2021.639238
30. Liu, S., Gao, Y., Canela-Xandri, O. et al. A multi-tissue atlas of regulatory variants in cattle. Nat Genet 54, 1438–1447 (2022). https://doi.org/10.1038/s41588-022-01153-5
31. Lonsdale J, Thomas J, Salvatore M, Phillips R, Lo E, Shad S, et al. The genotype-tissue expression (GTEx) project. Nat Genet. 2013 Jun;**45**(6):580–5.
32. Wilks C, Zheng SC, Chen FY, Charles R, Solomon B, Ling JP, et al. recount3: Summaries and queries for large-scale RNA-seq expression and splicing. Genome Biol. 2021 Nov 29;**22**(1):323.
33. Tarhan L, Bistline J, Chang J, Galloway B, Hanna E, Weitz E. Single cell portal: An interactive home for single-cell genomics data. bioRxiv. 2023 Jul 17. Available from: 2023.07.13.548886
34. Cellxgene Data Portal [Internet]. Cellxgene Data Portal. Available from: cellxgene.cziscience.com
35. Rajesh A, Chang Y, Abedalthagafi MS, Wong-Beringer A, Love MI, Mangul S. Improving the completeness of public metadata accompanying omics studies. Genome Biol. 2021 Apr 15;**22**(1):106.
36. Frontiers | Recent developments and future directions in meta-analysis of differential gene expression in livestock RNA-Seq [Internet]. Available from: www.frontiersin.org/articles /10.3389/fgene.2022.983043/full
37. Patel RK, Jain M. NGS QC toolkit: A toolkit for quality control of next generation sequencing data. PLoS ONE. 2012;7(2):e30619.
38. Ewels P, Magnusson M, Lundin S, Käller M. MultiQC: Summarize analysis results for multiple tools and samples in a single report. Bioinformatics. 2016;32(19):3047–8.
39. Chen S, Zhou Y, Chen Y, Gu J. Fastp: An ultra-fast all-in-one FASTQ preprocessor. Bioinformatics. 2018;34(17):i884–90.
40. Andrews S. FastQC: A quality control tool for high throughput sequence data. 2010 [Internet]. Available from: http://www.bioinformatics.babraham.ac.uk/projects/fastqc/
41. Wang L, Wang S, Li W. RSeQC: Quality control of RNA-seq experiments. Bioinformatics. 2012 Aug 15;**28**(16):2184–5.
42. Berrar DP, Dubitzky W, Granzow M, editors. A Practical Approach to Microarray Data Analysis [Internet]. Boston: Kluwer Academic Publishers; 2003. Available from: link.springer.com/10.1007/b101875
43. Sprang M, Andrade-Navarro MA, Fontaine JF. Batch effect detection and correction in RNA-seq data using machine-learning-based automated assessment of quality. BMC Bioinformatics. 2022 Jul 14;**23**(6):279.
44. Zhang Y. zhangyuqing/ComBat-seq [Internet]. 2023. Available from: github.com/zhangyuqing/ComBat-seq
45. removeBatchEffect: Remove Batch Effect in limma: Linear Models for Microarray Data [Internet]. Available from: rdrr.io/bioc/limma/man/removeBatchEffect.html
46. Zhang Y, Parmigiani G, Johnson WE. ComBat-seq: Batch effect adjustment for RNA-seq count data. NAR Genomics Bioinforma. 2020 Sep 21;**2**(3):lqaa078.

47. Fei T, Yu T. scBatch: Batch-effect correction of RNA-seq data through sample distance matrix adjustment. Bioinformatics. 2020 May 15;**36**(10):3115–23.
48. Holmström S, Hautaniemi S, Häkkinen A. POIBM: Batch correction of heterogeneous RNA-seq datasets through latent sample matching. Bioinformatics. 2022 Apr 28;**38**(9):2474–80.
49. Tran HTN, Ang KS, Chevrier M, Zhang X, Lee NYS, Goh M, et al. A benchmark of batch-effect correction methods for single-cell RNA sequencing data. Genome Biol. 2020 Jan 16;**21**(1):12.
50. Robinson MD, McCarthy DJ, Smyth GK. edgeR: A bioconductor package for differential expression analysis of digital gene expression data. Bioinformatics. 2010;**26**(1):139–40.
51. Genome Biology. Moderated estimation of fold change and dispersion for RNA-seq data with DESeq2 Full Text [Internet]. Available from: genomebiology.biomedcentral.com/articles/10.1186/s13059-014-0550-8
52. Mirdar Mansuri R, Shobbar ZS, Babaeian Jelodar N, Ghaffari M, Mohammadi SM, Daryani P. Salt tolerance involved candidate genes in rice: An integrative meta-analysis approach. BMC Plant Biol. 2020;**20**(1):452.
53. Langfelder P, Horvath S. WGCNA: An R package for weighted correlation network Analysis. BMC Bioinformatics. 2008 Dec 29;**9**(1):559.
54. Liu CC, Guo Y, Vrindten KL, Lau WW, Sparks R, Tsang JS. OMiCC: An expanded and enhanced platform for meta-analysis of public gene expression data. STAR Protoc. 2022 Sep 16;**3**(3):101474.
55. Haynes WA, Vallania F, Liu C, Bongen E, Tomczak A, Andres-Terrè M, Lofgren S, Tam A, Deisseroth CA, Li MD, Sweeney TE, Khatri P. Empowering multi-cohort gene expression analysis to increase reproducibility, Pac Symp Biocomput. 2017;22:144–53.
56. Xia J, Gill EE, Hancock REW. NetworkAnalyst for statistical, visual and network-based meta-analysis of gene expression data. Nat Protoc. 2015;**10**(6):823–44.
57. Keel BN, Zarek CM, Keele JW, Kuehn LA, Snelling WM, Oliver WT, et al. RNA-seq meta-analysis identifies genes in skeletal muscle associated with gain and intake across a multi-season study of crossbred beef steers. BMC Genomics. 2018 Jun 4;**19**(1):430.
58. MDPI. Animals [free full text] Genes involved in feed efficiency identified in a meta-analysis of rumen tissue from two populations of beef steers [Internet]. Available from: www.mdpi.com/2076-2615/12/12/1514
59. Rao MS, Van Vleet TR, Ciurlionis R, Buck WR, Mittelstadt SW, Blomme EAG, et al. Comparison of RNA-seq and microarray gene expression platforms for the toxicogenomic evaluation of liver from short-term rat toxicity studies. Front Genet. 2019 Jan 22;**9**:636.
60. Ditz B, Boekhoudt JG, Aliee H, Theis FJ, Nawijn M, Brandsma CA, et al. Comparison of genome-wide gene expression profiling by RNA sequencing versus microarray in bronchial biopsies of COPD patients before and after inhaled corticosteroid treatment: Does it provide new insights? ERJ Open Res [Internet]. 2021 Apr 1;**7**(2). Available from: openres.ersjournals.com/content/7/2/00104-2021

7 Best Practices in Single-Cell RNA-seq Data Analysis

Mathias Galati, Xinhui Wang, Muhammad Shoaib, Rajesh Rawal, Irina Balaur, Shaman Narayanasamy, and Venkata Satagopam

INTRODUCTION

Single-cell (SC) research is rapidly ascending to the forefront of biomedicine and biotechnology, due to its ability to elucidate cell-level variation and heterogeneity, particularly in the context of human diseases. Consequently, considerable advances have been made in SC research, spanning from experimental methodologies, technological breakthroughs, data analytics, and outcomes [1]. Single-cell RNA sequencing (scRNA-seq) provides unprecedented depth in exploring cellular heterogeneity [2], unveiling intricate variability within seemingly homogeneous cell populations. Moreover, it facilitates understanding cell trajectories through pseudo-time analysis, which allows the reconstruction of cellular development or differentiation pathways [3]. Furthermore, scRNA-seq can uncover intricate disease mechanisms. Researchers can dissect signalling networks between different cell types, offering novel insights into disease progression and potential therapeutic targets [4]. In oncology, it offers a higher-resolution picture of the intricate and diverse cell populations within tumours, aiding in the identification of unique transcriptional signatures that could be associated with therapy response or disease prognosis [5], a critical asset in the development of personalised medicine strategies.

Historically, the progress of SC studies was limited by the cost, labour, and time-intensive nature of low-throughput SC isolation methods [6]. Today, high-throughput techniques have mitigated those challenges [7]. The success of SC technologies can be largely attributed to advancement in two main techniques: (i) cellular separation and (ii) barcoding [8,9]. Figure 7.1 illustrates an SC protocol that utilises those strategies. Briefly, high-throughput cell isolation methods such as fluorescence-activated cell sorting (FACS) in SC suspensions, plate-based segregation, and robotic droplet microfluidic devices [10] work in tandem with state-of-the-art barcoding to label cells derived from specific samples, subjects, experimental conditions, tissue, or cell types. It is

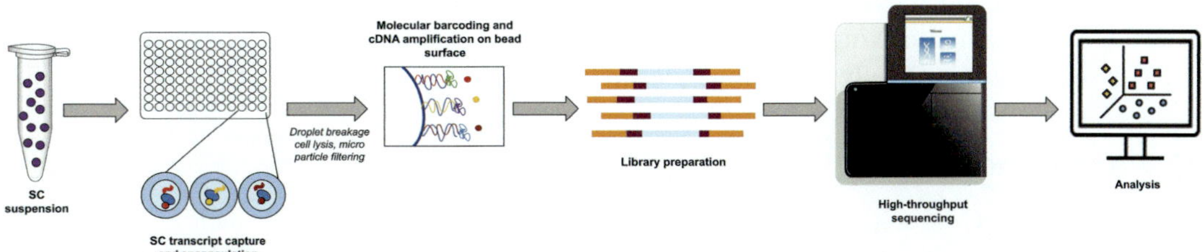

FIGURE 7.1 SC transcriptomics workflow. Tissue samples are preprocessed and SC suspension is collected in vials. Each cell is wrapped with a bead inside a nanoscale droplet (each bead contains a unique molecular identifier [UMI] barcode attached with a transcript). Cell lysis, droplet breakage, and microparticle filtering occurs within each droplet. Transcripts together with the relevant barcodes and adapter bind to the beads followed by amplification of complementary DNA (cDNA). The cDNA then undergoes library preparation procedure for sequencing, which in turn is followed by data analysis. (Graphical elements were obtained from Bioicons https://bioicons.com/ [Accessed 16 October 2024].)

DOI: 10.1201/9781003354062-7

important to note that barcoding can be performed on various biomolecules (e.g., DNA, RNA, proteins, etc.) and staggered or nested to reflect multiple aspects, e.g., separating subjects and cell types [11]. Upon barcoding, the cells can then be lysed to extract the biomolecules of interest (e.g., DNA, RNA, proteins, etc.). Barcoded biomolecules can then be pooled into a single library for an associated high-throughput run, e.g., next-generation sequencing (NGS), thus significantly reducing the overall cost [12]. The process of cellular separation to pooled library preparation can be collectively referred to as multiplexing. To that end, there are various multiplexing strategies [13–16] and platforms [17–19] currently available.

The cell transcriptome, i.e., the collection of all RNA molecules within a cell, exhibits considerable diversity and dynamics relative to the genome and epigenome while offering superior coverage relative to readouts from spectrometry-based proteomics. These hallmark characteristics highlight the versatility of the transcriptome as an effective tool for SC studies. As such, SC transcriptomics enables us to investigate cell-wide gene expression by quantifying mRNA expression at the SC level [20], thereby producing detailed gene expression profiles. These reflect a snapshot of a cell's transcriptomic activity at a particular time point. Such profiles can be applied in various contexts, including but not limited to exploring intracellular dynamics [21], alternative splicing [22], and cell typing [23].

In this chapter, we focus on SC transcriptomics analyses generated using scRNA-seq technologies and the various strategies and tools for the analyses, spanning from large-scale preprocessing of NGS reads (primary analysis) to downstream secondary analysis, such as differential expression. We aim to educate and guide readers who are looking to explore high-throughput SC data for analysis (Figure 7.2).

PRIMARY ANALYSIS

The primary analysis involves the large-scale processing of NGS reads derived from scRNA-seq runs. It covers demultiplexing, quality control (QC), and the alignment and quantification of the data.

Demultiplexing

Demultiplexing [24,25] is a process applied to the raw scRNA-seq data to separate individual cells and distinct samples within a multiplexed library using barcode information (see Figure 7.1 and the 'Introduction'). Briefly, demultiplexing can be carried out based on barcodes (i.e., artificially introduced nucleotides that tag specific samples) [14,15,26–30] or single-nucleotide polymorphisms (SNPs, which are naturally occurring) [31–34]. Popular barcode-based demultiplexing tools were found to be of comparable performance [25], with HTODemux [14] edging out the others for barcode-based demultiplexing. It is important to note that this process can be skipped if a given scRNA-seq dataset is already in a demultiplexed form.

Quality Control

Standard NGS QC of scRNA-seq reads is typically performed using tools such as FastQC [35] to ensure sufficient quality for downstream processes. Additional scRNA-seq-specific QC steps involve, for instance, setting a threshold for UMIs per cell to exclude low-quality cells. UMIs are 'barcodes' that tag individual molecules during library preparation [36]. Tools such as EmptyDrop identify droplets with ambient or 'free-floating' RNA, thus reducing the false discovery rates [26]. Genes from cells that are either dying, dead, or below the detection threshold can be removed to further improve reliability [37]. Last, but not least, the doublets or multiplets, i.e., artefacts of two or more cells encapsulated in a single droplet and sequenced as if they were a single cell (see Figure 7.1) can be removed using tools such as Scrublet [38] and DoubletFinder [39]. Finally, Hong

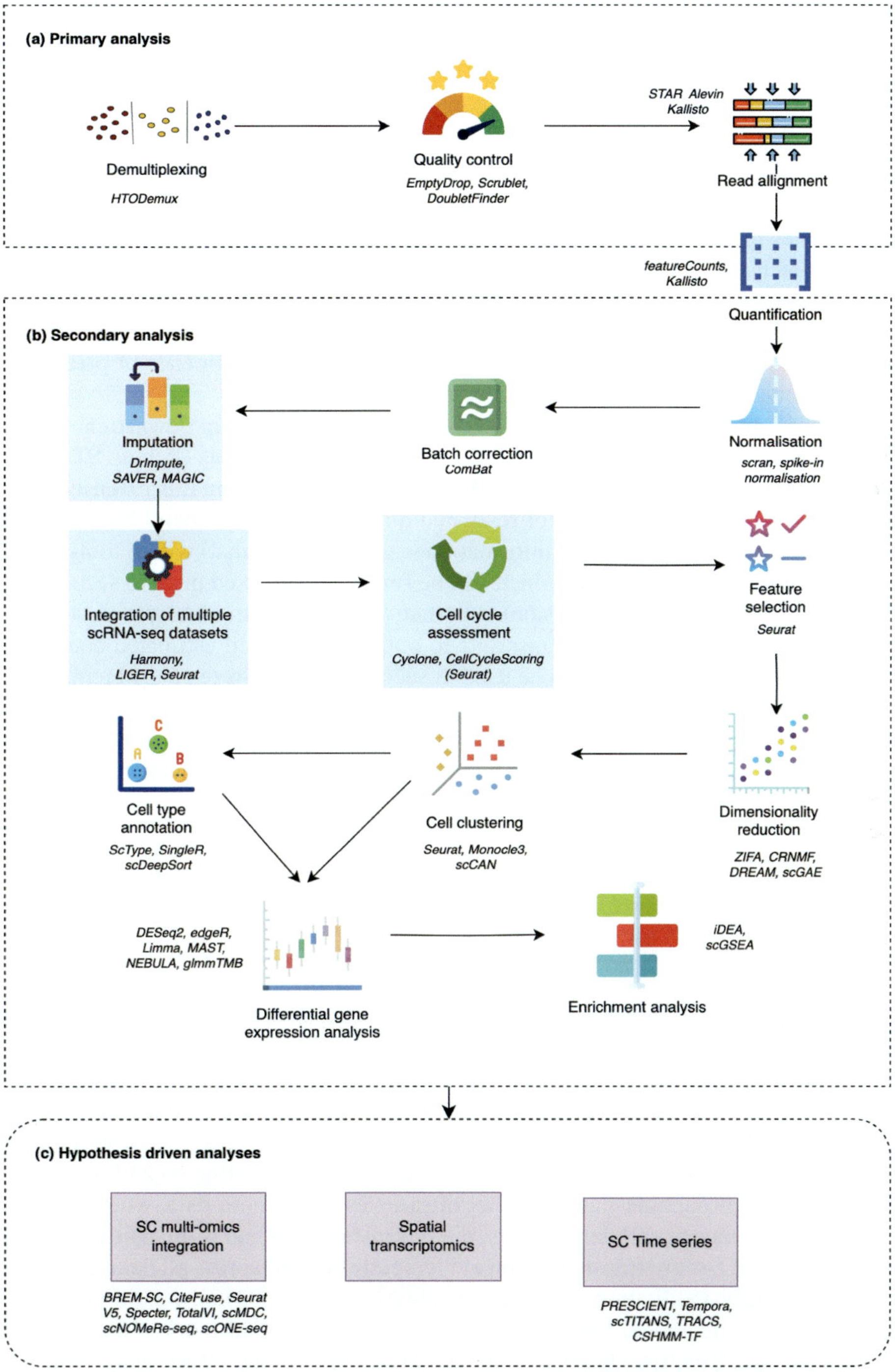

FIGURE 7.2 Overview of the scRNA-seq data processing and analysis workflow. (a) Primary analysis steps include processing of the scRNA-seq reads such as demultiplexing, quality control, and read alignment. (b) Secondary analysis steps that are used to explore and interpret scRNA-seq data, spanning from data normalisation up to functional interpretation. Blue boxes highlight optional steps. (c) Downstream hypothesis-driven analysis could include more advanced multi-omics integration, temporal or spatially resolved analyses. Examples of tools for each step are provided as italic text. (Illustrative icons were obtained from Flaticons https://www.flaticon.com/ and Bioicons https://bioicons.com/ [Accessed 16 October 2024].)

et al. provide an updated and extensive review of various QC tools for scRNA-seq to guide researchers working on such data [40].

Read Alignment and Quantification

NGS reads that pass the QC process then proceed to the read alignment or mapping step. This involves aligning the reads to a reference genome, e.g., a human genome reference. Briefly, alignment programmes require the NGS reads and an indexed reference genome. This process is typically resource-intensive and highly dependent on the size of the reference genome and the depth of sequencing. STAR [41] is a common tool used for transcriptomic analysis that relies on a classical alignment algorithm [42]. However, the resource intensive (i.e., CPU, RAM, and storage) nature of classical read alignment (e.g., in tools like STAR) has led to the prominence of pseudo-alignment tools such as Alevin [43] and Kallisto [44] in scRNA-seq data.

The quantification of reads from scRNA-seq data is a critical step in downstream analysis, and it can be approached in two main ways. First, reads aligned with tools such as STAR result in a sequence/binary alignment map in SAM/BAM file format, which contains extensive information such as read mapping position, quality of read, and quality of the alignment. The SAM/BAM formats are, in turn, used with annotation information (e.g., in GTF format) within tools such as featureCounts to quantify gene expression, which can be further normalised prior to secondary analysis and processing (e.g., differential expression). Alternatively, pseudo-alignment tools such as Kallisto can directly 'align' reads to a set of transcripts to output normalised estimated count tables (e.g., transcripts per million [TPM]) that can be used in secondary analysis (e.g., differential expression) with minimal additional processing. Pseudo-alignment tools may be preferred for scRNA-seq data processing as they are less resource intensive, require less intermediate processing, and are more in line with the goals of scRNA-seq primary analysis, which is gene quantification.

SECONDARY ANALYSIS

Normalisation, Batch Effect Correction, and Imputation

ScRNA-seq requires normalisation procedures to account for technical variabilities, specifically the fact that different cells or sequencing experiments may yield different numbers of reads. After initial quantification, which produces raw count matrices, tools such as scran [45] estimate size factors that are subsequently used to normalise gene expression values. Spike-in normalisation [46] is another strategy that can be used when spike-in controls are added to the samples, providing a reference for correcting technical noise. Moreover, different normalisation strategies are used for varying scales of data and different downstream analyses. For instance, SCnorm is applied to low-throughput (low number of samples), high-depth sequencing data, as it is robust in dealing with high variability [47]. Conversely, for high-throughput (large number of samples), low-depth data, where the emphasis is on capturing a wide range of cellular diversity, the sctransform method is often utilised [48]. Finally, log-normalisation and z-score transformation are typically applied across all datasets to standardise gene expression values, facilitating comparative analysis across cells and conditions.

Diverse sources of technical variability may introduce so-called batch effects in scRNA-seq data, potentially obscuring true biological signals of interest. Batch effects may stem from many sources, for instance, different or distinct extractions, experiments, sequencing runs, and/or reagent lots. To tackle these challenges, numerous batch effect correction tools have been developed [49]. Among them, ComBat [50] is often favoured due to its robustness and simplicity. However, correction methodologies such as this were originally developed for bulk transcriptomics, and thus assume equal cell composition across batches, which is often inaccurate in the context of scRNA-seq data. Hence, batch correction relying on mutual nearest neighbours (MNNs) may lead to more accurate and precise results [51]. Alternatively, the task of integrating multiple scRNA-seq datasets is addressing similar challenges, thus the methodologies and computational strategies often overlap. For more details, please refer to the subsequent section on 'Integration of Multiple scRNA-seq Datasets'.

Dropout events, wherein a gene appears to be unexpressed due to technical limitations, pose significant challenges in scRNA-seq data interpretation. As a result, several imputation methods have been developed to handle this issue, typically replacing dropout events (usually represented as zeros) with estimated values to reduce noise and facilitate downstream data analysis. DrImpute [52] and SAVER [53], for example, employ statistical models to predict missing values based on observed gene expression patterns. Alternatively, methods such as MAGIC [54] use diffusion-based algorithms to smooth gene expression data across similar cells, helping to mitigate noise. While these methods can reduce data noise and improve interpretability, it is important to note potential limitations, namely, the over-reliance on imputation could inadvertently introduce artificial structures or patterns, leading to overfitting. Therefore, these techniques should be employed judiciously, always considering the context of the biological question.

Integration of Multiple scRNA-Seq Datasets

The integration of datasets (also referred as P-integration) in scRNA-seq is a critical task in the analysis of SC genomics data, as it allows for the comparison and consolidation of data from multiple batches, experimental conditions, or technologies (such as scATAC-seq). Various computational tools have been developed to facilitate such integration, including Harmony [55], LIGER [56], and Seurat [57]. Harmony, an algorithm and R package, creates a harmonised embedding of cells where they cluster by cell type rather than by the batch or dataset of origin, and is known for its speed and efficiency. Linked Inference of Genomic Experimental Relationships (LIGER), on the other hand, identifies shared and dataset-specific (or condition-specific) cell types using integrative non-negative matrix factorisation, thereby uncovering shared and unique features across datasets. Finally, Seurat, an R package, offers various methods for data integration, including canonical correlation analysis (CCA)-based integration and an integration method based on MNNs. All three methods have proven effective for batch correction in scRNA-seq data and have been recommended based on benchmarking studies, such as the one by Tran et al. [49], each with their unique strengths and use cases.

Cell Cycle Assignment

Cell cycle stages considerably influence gene expression profiles. Specifically, different phases of the cell cycle will result in distinct transcriptional signatures, thus generating an inherent variability in the expression of genes. This variability, if unaccounted for, can introduce confounding factors into downstream analysis, especially when the objective is to compare cell types or states, which may also be represented by disparate proportions of cells in each phase of the cell cycle [47].

To that end, scRNA-seq data analysis should incorporate the assignment of cell cycle stage information. For instance, during an SC differential gene expression analysis, a population of cells predominantly in the S or G2/M phases (phases marked by DNA replication and thus high expression of replication-related genes), compared to a population largely in other stages, may lead to misleading interpretations. In this event, replication-related genes may seem differentially expressed due to the disproportion in cell cycle stages rather than any true inherent biological differences between the cell populations.

By implementing cell cycle assignment, such as through tools like Cyclone [58] and CellCycleScoring [59] from the Seurat package, cells can be categorised into their respective cell cycle phases. This step provides an opportunity to account for this intrinsic variability. Post-assignment, general linear models (GLM) can be employed to 'regress out' cell cycle associated differences, thereby neutralising their impact on downstream analysis [60]. Some studies may require explicit modelling of cell cycle effects or conduct separate intracell cycle analyses. This process not only minimises the potential confounding influence of these effects but also permits focus on biologically relevant variations over those from different cell cycle stages, enabling accurate and biologically meaningful interpretations.

Feature Selection

The scRNA-seq dataset often contains a vast number of genes, but not all of these genes or cell types may be adequately represented during a given experiment. This underlines the importance of feature selection, a step in the analysis process that focuses on retaining only strong biological signals compared to technical noise. Genes (also referred to as 'features' in this context) that are either too noisy or present in a small number of samples are often eliminated based on parameters like the mean and variance of gene expression or a high frequency of zero values [48,57].

When dealing with multiple datasets, the feature selection process is first performed on an individual dataset. Following this, features are prioritised across multiple experiments based on the number of datasets where they were independently identified as highly variable. This integrative approach allows for a more robust identification of relevant features, ultimately enhancing the potential for meaningful biological insights.

It is important to note that the feature selection process in scRNA-seq is crucial for reducing data complexity and improving computational efficiency. However, it also requires careful consideration, as it can influence downstream analyses and interpretation of the results. Methods available range from univariate filters (such as t-statistics), requiring predefined cell type labels, to less restrictive approaches that filter highly variable genes (HVGs), implemented in frameworks like Seurat. Several other approaches for analysing scRNA-seq data are emerging. These include multivariate methods such as gene pair filtering and COMET (a deviance statistic-based gene filtering technique). Additionally, researchers are developing wrapper and embedded methods that build upon classic analytical techniques [61].

Dimensionality Reduction

The processed scRNA-seq data is typically organised into an expression matrix, where each column represents a single cell, and each row represents a specific gene or transcript. Each element in this matrix shows the expression level of a particular gene in a specific cell. The size of this matrix can be substantial, with the number of cells ranging from thousands to millions, while the number of genes depends on the reference transcriptome (e.g., about 30,000 for humans). This results in a large, high-dimensional dataset, which presents significant challenges for effective visualisation, interpretation, and further analysis. To address these challenges, researchers use dimensionality reduction (DR) techniques. These DR methods allow the data to be represented and visualised in two dimensions, typically as a graph or plot, making it easier to identify patterns and relationships within the complex scRNA-seq data.

DR is an integral step in scRNA-seq analysis, as it facilitates the identification of clusters in the data that are presumed to represent similar cell types. Traditional DR methods, such as principal component analysis (PCA) [62], t-Distributed Stochastic Neighbour Embedding (t-SNE) [63], and especially uniform manifold approximation and projection (UMAP) [64], have been commonly employed in scRNA-seq data analysis, as they generally work well. However, in some cases, those conventional methods do not account for characteristics specific to scRNA-seq data, such as the prevalence of zero values (dropouts), which are false-zero gene expression values in some cells due to technical limitations.

To that end, methods such as Zero Inflated Factor Analysis (ZIFA) [65], Constrained Robust Non-negative Matrix Factorisation (CRNMF) [66], and an improved Deep Variational Autoencoder model (DREAM) [67] are being developed to mitigate SC-specific DR issues. These methods claim to be capable of preserving long-distance relationships in a latent space such as single-cell Graph Autoencoder (scGAE) [68].

To conclude, the choice of suitable methods depends on the specifics of the data and the research question at hand, while future rapid advancements in SC technologies will continue introducing novel techniques for DR.

Cell Clustering

An essential part of scRNA-seq analysis is cell clustering, a process that groups cells based on their features, facilitating the discovery of new cell (sub)types and/or marker genes. However, it can be challenging due to the noisy and high-dimensional nature of the data. Seurat clustering functions [57] and Monocle [38] are the most widely used methods among several developed approaches to address this problem, each offering distinct advantages, drawbacks, and areas of application. A novel approach is scCAN [69], which utilises non-negative kernel autoencoders, stacked variational autoencoders, and graph-based techniques. Authors have demonstrated that scCAN outperforms existing methods in accuracy and scalability.

Nevertheless, while multiple methods exist for clustering scRNA-seq data, choosing the right approach is dependent on the nature and specific needs of the analysis. The advancement of computational methods continues to enhance the accuracy and efficiency of scRNA-seq clustering. As a result, it is advisable to consult recent benchmark studies, such as the article by Yu et al. [70], to make an informed decision.

Cell Type Annotation

Cell type annotation involves associating clusters (generated from the cell clustering step described in the previous section) with different gene expression profiles to distinct cell types. Given the time-consuming and subjective nature of manual annotation, the development of standardised and automated tools for cell type identification has been a significant advancement in the field [71].

Automated cell type annotation methodologies can be classified into three primary categories: (i) marker gene database-based annotation such as ScType [23], (ii) correlation-based methods such as SingleR [72], and (iii) supervised classification such as scDeepSort [73] or scClassify [74]. The first approach utilises specific marker genes to identify cell types, while correlation-based methods compare gene expression profiles of cells with reference datasets to infer cell types. Supervised classification employs machine learning algorithms trained on labelled datasets to classify cells into known cell types [75].

Despite the availability of these automation methods, manual annotation remains necessary in cases of conflicting or absent cell labels, or for novel and critically discussed cell type subclasses. This underscores the importance of refining cell type annotation methods. Useful guidelines are discussed in a review by Clarke et al. [76].

Differential Gene Expression Analysis

Differentially expressed genes (DEGs) refer to genes that exhibit significant variations in their expression levels across two or more biological states or conditions. The identification of these DEGs is fundamental to the analysis of gene expression data, as they are often implicated in the biological processes or the diseases being studied.

In the context of scRNA-seq, the analysis of DEGs is even more important, as it enables researchers to identify differences in gene expression at the level of individual cells. Such granular insights can help in understanding the diversity within cell populations and the specific roles of different cell types.

Classical methods for differential gene expression in scRNA-seq include DESeq2 [77] and edgeR [78], which are based on negative binomial models. In contrast, limma [79] utilises a linear model for data processing. Originally designed for bulk RNA-seq data, these techniques have also found applicability in scRNA-seq analyses despite not being specifically tailored for the task.

The Model-based Analysis of Single-cell Transcriptomics (MAST) approach is interesting because it offers a two-part generalised linear model that simultaneously models the rate of expression over the background of various transcripts and the positive expression mean. By leveraging a hurdle model and by introducing the concept of cellular detection rate, they provide a mechanism for advanced, multiparametric modelling of SC transcriptomics data that can account for both technical and biological factors affecting gene expression and thus potentially enabling more accurate interpretation of experimental results [80].

According to a recent study from Gagnon et al. [81], the NEBULA [82] and glmmTMB [83] methods were found to outperform the commonly used methods in the literature. Nevertheless, selecting an appropriate method depends on the dataset characteristics and the research questions available. Researchers are advised to consult original papers and software documentation to understand the optimal use of these methodologies.

Enrichment Analysis (EA)

EA in scRNA-seq is a computational method used to identify groups or classes of significantly overrepresented or underrepresented features in the data. The goal is to determine if the genes from the previous DEG analysis, whether comparing conditions (e.g., healthy vs. diseased) or cell types, are enriched in certain biological pathways, molecular functions, cellular components, or other defined gene sets (such as MSigDB [84], Gene Ontology [85], etc.), thus providing a deeper understanding of the underlying biological processes being activated or repressed in the given dataset.

To perform an EA, the list of genes is compared to predefined gene sets or databases to see if there is an overlap greater than what is expected by chance, using statistical methods such as Fisher's exact test. There are common tools and software available for EA, such as g:Profiler [86], DAVID [87], and GSEA [88,89]. Some of the most recent and tailored tools for scRNA-seq data are iDEA [90] and scGSEA [91]. iDEA is a statistical method that uses summary statistics from gene expression data, including measures of gene expression changes (such as fold change or effect size estimates) and their associated standard errors. What distinguishes iDEA is its ability to combine differential expression analysis with gene set enrichment analysis within a single, unified statistical framework. scGSEA combines latent data representations and gene set enrichment scores to detect coordinated gene activity at single-cell resolution.

Overall, EA is a form of hypothesis-generating procedure, and while it provides useful biological insights, it does not prove causal relationships, and so the results need to be interpreted with caution and ideally validated with further experiments.

SC Analytic Frameworks

scRNA-seq data analysis is multifaceted, with various factors influencing the success of a given study. We highly recommend using established analytical methodologies (such as those highlighted in this chapter) as a sturdy foundation for analysis and later moving to more tailored approaches that may be better suited to the given experimental design and/or data. To that end, there are several specialised frameworks to facilitate scRNA-seq data analyses that comprise a suite of preselected tools conveniently packaged together and maintained by experts in the field [92]. These frameworks offer a smoother onboarding experience for those who want to delve into scRNA-seq data analysis. They can also be divided by focus: whether it is on primary or secondary analysis.

Primary analysis frameworks typically consist of large-scale processing tools (see the section 'Primary Analysis') that are strung together using workflow automation tools such as Snakemake [93] or Nextflow [94,95]. These workflow languages enhance reproducibility, automation, and scalability, potentially saving considerable time, while being kept up to date by a community of experts.

Secondary analysis frameworks include various secondary analysis steps (see the section 'Secondary Analysis') and can be further divided by platform, such as whether they are R or Python based. Seurat, a package built on R [96], is arguably the most widely used implementation for scRNA-seq secondary analysis due to its flexibility, high-quality implementation, and frequent updates. And for those familiar with R, platforms such as Bioconductor, Seurat, and tidy transcriptomics are highly recommended for their excellent documentation and very active user communities that can provide support and answer questions. Similarly, tidyseurat [97] (within the tidytranscriptomics suite [98]) offers a collection of powerful and cohesive packages for data manipulation, exploration, and visualisation under the R tidyverse framework [98]. This unified framework enables consistent syntax and data structures for efficient, intuitive, and reproducible scRNA-seq analysis. Furthermore, packages such as clustree [99] should be considered to facilitate visualisations and decision-making, while SCpubr [100] could aid in generating publication-quality figures. Similarly, SCEDAR [101] and scanpy [102] offer equivalent secondary analysis capabilities in Python, with the latter further offering an R implementation.

Researchers should consult available resources when starting their scRNA-seq analysis journey; high-quality guides are available for Python [103,104] and R Bioconductor [105]. More importantly, the landscape of scRNA-seq data analysis is highly dynamic, therefore researchers should always refer to the most recent analysis procedures, guidelines, and benchmarks.

BEST PRACTICES FOR SC DATA ANALYSIS

It is necessary for researchers to approach scRNA-seq data analysis with a dynamic mindset, considering the rapid advances in the field. Depending on the time of reading, one may encounter certain tools highlighted within this chapter that have become either outdated or significantly improved. Thus, we aim to address crucial aspects of SC transcriptomics data analysis from two angles: (i) the selection of tools and (ii) the reproducibility of analysis pipelines.

When selecting analytical tools, two factors should be considered: the goals of the study and the quality and implementation of the tools. Always consider the objectives of your analysis and the characteristics of the generated SC data when selecting tools. Given the vast range of available tools, it is generally prudent to select those with a proven track record of generating high-quality output, demonstrating stable implementation, providing regular support and updates, and ensuring interoperability with other tools through standard output formats. Another factor of paramount importance is a collaborative multidisciplinary mindset when designing scRNA-seq experiments, so that you include input from biologists, statisticians, and bioinformaticians to ensure a biologically, statistically, and methodologically robust downstream analysis. Moreover, remember that the specifics of your project – such as the biological question, sample type, and sequencing platform – will often determine the most suitable tools and analysis strategies.

The reproducibility of analysis pipelines applied within an scRNA-seq study should adhere to general open-science best practices [106,107]. Researchers should have a long-term view on being able to replicate a given study even years after publication. Therefore, code for a given analysis protocol should be deposited into repositories such as GitHub, while raw NGS data (e.g., scRNA-seq data) should be uploaded into public archives like the National Center for Biotechnology Information's (NCBI) Sequence Read Archive (SRA) or the European Bioinformatics Institute's (EBI) European Nucleotide Archive (ENA) with detailed accompanying metadata, while gene expression data should be provided on NCBI's Gene Expression Omnibus (GEO) platform. If applicable, non-standard data formats could be uploaded on Zenodo [108]. Finally, containerisation (e.g., using Docker or Singularity) of specific software versions could further facilitate reproducibility.

Recently, the sharing of raw and/or processed NGS data from human subjects has raised regulatory and ethical concerns [109], hence researchers must adhere to policies such as the General Data Protection Regulation (GDPR) in the European Union [110] or the Health Insurance Portability and Accountability Act (HIPAA) in the United States, among other equivalent regional or international data protection policies. Overall, striking a balance between protecting the subjects of data collection and maintaining research transparency is critical. The Findability, Accessibility, Interoperability, and Reuse (FAIR) principles and resources serve as an excellent guide for data discovery, use, and sharing [111–115].

In general, the availability of code, data, and software alongside detailed methodological descriptions or documentation (e.g., in an article) should provide sufficient material for reproducing a given study, which in turn could lead to downstream integration into other independent studies (e.g., meta-analysis; see the section 'Integration of Multiple scRNA-seq Datasets'), scaling up existing studies, and knowledge transfer.

ADVANCED SCRNA-SEQ ANALYSES

Following the scRNA-seq data analysis approach outlined in this chapter, it is imperative to underscore the numerous avenues for further analyses, which include the following:

Cell-to-cell communication analysis can unravel the intricate interactions that occur within and between cellular communities [116]. LIANA is a computational tool worth mentioning here, as it integrates multiple ligand-receptor databases and offers various statistical methods to infer and analyse cell-cell interactions from scRNA-seq data [4].

Gene regulatory network inference is a process that models the complex regulation of gene expression in individual cells. Tools such as SCENIC perform this inference, offering insights into the fundamental mechanisms underpinning cellular diversity and the dynamics of biological processes [117].

Genotyping analysis can supplement scRNA-seq data by relating genetic variations to cellular-level expression, providing insights into their phenotypic consequences. As such, various tools can be utilised to perform copy number variation (CNV) [118], single nucleotide variation (SNV) [119], and expression quantitative trait locus (eQTL) [120] analysis.

RNA velocity is an approach that employs scRNA-seq to approximate the derivative of gene expression over time, thereby elucidating the dynamism of gene expression within individual cells. Through the quantification of RNA expression 'velocity', researchers can predict cellular fate and comprehend developmental trajectories with tools such as scVelo [121].

INTEGRATED APPROACHES IN SC ANALYSIS: TIME-SERIES, SPATIAL, AND MULTI-OMICS

Unveiling biological complexity requires the intersection of the various dimensions of SC analysis, namely time-series, spatial, and multi-omics approaches.

The time-series aspect of SC analysis unravels dynamic changes in cell states and gene expression, elucidating the evolution of cellular processes [122]. Current capabilities extend to inferring cell trajectory [123], predicting cell fates over varying time points and populations [124], assigning transcription factors to activation points [125], and mitigating cell asynchrony [126].

Spatial transcriptomics can be used to complement high-resolution scRNA-seq information to mitigate the inability of SC methods to retain spatial information [127,128]. This union reveals cell-specific changes within their native tissue context, aiding in the exploration of disease biomarkers [127]. Challenges such as data availability and computational complexity are hurdles on the path towards full integration [128,129].

Lastly, SC multi-omics technologies offer the opportunity to analyse multiple types of biomolecules from individual cells, providing an extensive understanding of cellular processes and heterogeneity [130–132]. As such, scRNA-seq derived gene expression can be integrated with genomic, DNA methylation, chromatin accessibility, and/or protein abundance information using a multitude of cutting-edge analytical methods and tools [130,133–137], despite existing technical and computational challenges [138].

Together, these integrated approaches in SC analysis offer powerful tools for comprehensive insights into cellular dynamics, spatial organisation, and multilayered molecular signatures.

THE EXPANDING RESOURCES AND APPLICATIONS OF SC TECHNOLOGIES

scRNA-seq has been applied in various research areas and clinical applications as it enables the high-resolution analysis of individual cells, providing deep details about the cellular heterogeneity and gene expression of complex biological systems and disease mechanisms [139,140].

Precision medicine is one of the most promising applications of scRNA-seq. By elucidating the intricacies of cellular dynamics, it helps identify therapeutic targets and develop innovative disease diagnoses and treatments [140]. In addition, scRNA-seq has enabled the culmination of major resources, including the Human Cell Atlas (HCA) [141], which aims to develop comprehensive reference maps for all human cells, detailing their features, location, and biological functions [129,142]; and the Brain Research through Advancing Innovative Neurotechnologies (BRAIN) Initiative's Cell Census Network (BICCN) [143], which aims for the equivalent with a focus on brain cells. These initiatives have catalysed the development of robust SC analytical tools [144], further enriching our understanding of cellular complexity and heterogeneity across different species and different scales.

From a clinical perspective, scRNA-seq can improve diagnostics, therapeutic choices, and disease monitoring in various fields of medicine. The potential impact of these technologies on precision medicine could see them becoming a key component of the diagnostic process in the future [145].

The application of scRNA-seq in cancer research has been particularly noteworthy with its ability to study individual cells in a tumour, providing insights into cancer biology, with the potential to enhance diagnosis, prognosis, and therapeutics specific to the tumour heterogeneity [146,147].

Moreover, scRNA-seq holds significant promise for drug discovery and development by enabling detailed descriptions of cellular heterogeneity and the related disease mechanisms, and it can help with the identification of new therapeutic targets and understanding cellular responses to treatments [139].

Finally, scRNA-seq facilitates the study of rare cell (sub)types that might have a significant impact on health or disease. These cells are often missed in bulk sequencing techniques but can be effectively studied with scRNA-seq [148]. Also, the integration of scRNA-seq with spatial transcriptomics provides a holistic understanding of the cellular environment, maintaining crucial information about the spatial organisation of the tissue.

In summary, scRNA-seq has a wide range of applications and holds immense promise for advancing our understanding of complex biological systems and disease mechanisms, potentially transforming various fields of research and clinical practice.

CONFLICT OF INTEREST

Authors declare no conflict of interest.

AUTHOR CONTRIBUTIONS

VS conceived and supervised the effort. MG, XW, MS, RR, and IB reviewed the literature. MG and SN compiled the first draft. MS, RR, and IB prepared the figures. All authors wrote and reviewed the final manuscript.

ACKNOWLEDGEMENTS

The authors acknowledge the funding agencies and grants: Luxembourg National Research Fund (FNR) PRIDE grant i2TRON to MG; Eurostars E! 113726 grant to SN; FNR/ERA PerMed (INTER/ERAPerMed/19/13589768) grant to IB; Innovative Medicines Initiative 2 Joint Undertaking, ImmUniverse IMI2-RI-853995 to XW; FNR NCER-PD grant FNR/NCER13/BM/11264123 to RR, and EU Horizon 2020 grant no. 101016072 to MS.

REFERENCES

1. Stuart T & Satija R. Integrative single-cell analysis. *Nat Rev Genet.* 2019;**20**:257–72.
2. Choi YH & Kim JK. Dissecting cellular heterogeneity using single-cell RNA sequencing. *Mol Cells.* 2019;**42**:189–99.
3. Wang L et al. Current progress and potential opportunities to infer single-cell developmental trajectory and cell fate. *Curr Opin Syst Biol.* 2021;**26**:1–11.
4. Dimitrov D et al. Comparison of methods and resources for cell-cell communication inference from single-cell RNA-seq data. *Nat Commun.* 2022;**13**:3224.
5. Noureen N, Ye Z, Chen Y, Wang X & Zheng S. Signature-scoring methods developed for bulk samples are not adequate for cancer single-cell RNA sequencing data. *eLife.* 2022;**11**:e71994.
6. Hu P, Zhang W, Xin H & Deng G. Single cell isolation and analysis. *Front Cell Dev Biol.* 2016;**4**:4–116. doi: 10.3389/fcell.2016.00116
7. Lähnemann D et al. Eleven grand challenges in single-cell data science. *Genome Biol.* 2020;**21**:31.
8. Cao J et al. The single-cell transcriptional landscape of mammalian organogenesis. *Nature* 2019;**566**: 496–502.
9. Zilionis R et al. Single-cell barcoding and sequencing using droplet microfluidics. *Nat Protoc.* 2017; **12**:44–73.
10. Prakadan SM, Shalek AK & Weitz DA. Scaling by shrinking: Empowering single-cell 'omics' with microfluidic devices. *Nat Rev Genet.* 2017;**18**:345–61.
11. Rosenberg AB et al. Single-cell profiling of the developing mouse brain and spinal cord with split-pool barcoding. *Science.* 2018;**360**:176–82.
12. Svensson V, Vento-Tormo R & Teichmann SA. Exponential scaling of single-cell RNA-seq in the past decade. *Nat Protoc.* 2018;**13**:599–604.
13. Gehring J, Hwee Park J, Chen S, Thomson M & Pachter L. Highly multiplexed single-cell RNA-seq by DNA oligonucleotide tagging of cellular proteins. *Nat Biotechnol.* 2020;**38**:35–38.
14. Stoeckius M et al. Cell hashing with barcoded antibodies enables multiplexing and doublet detection for single cell genomics. *Genome Biol.* 2018;**19**:224.
15. McGinnis CS et al. MULTI-seq: Sample multiplexing for single-cell RNA sequencing using lipid-tagged indices. *Nat Methods.* 2019;**16**:619–26.
16. Single Cell Gene Expression Flex. 10x Genomics https://www.10xgenomics.com/products/single-cell-gene-expression-flex.
17. Jariani A et al. A new protocol for single-cell RNA-seq reveals stochastic gene expression during lag phase in budding yeast. *eLife.* 2020;**9**:e55320.
18. Kim J & Marignani PA. Single-Cell RNA Sequencing Analysis Using Fluidigm C1 Platform for Characterization of Heterogeneous Transcriptomes. in Cancer Cell Biology: Methods and Protocols (ed. Christian, S. L.) 261–278 (Springer US, 2022). doi:10.1007/978-1-0716-2376-3_19
19. Wen L et al. Single-cell technologies: From research to application. *The Innovation.* 2022;**3**:100342.
20. Haque A, Engel J, Teichmann SA & Lönnberg T. A practical guide to single-cell RNA-sequencing for biomedical research and clinical applications. *Genome Med.* 2017;**9**:75.
21. Schwabe D, Formichetti S, Junker JP, Falcke M & Rajewsky N. The transcriptome dynamics of single cells during the cell cycle. *Mol Syst Biol.* 2020;**16**:e9946.
22. Arzalluz-Luque Á & Conesa A. Single-cell RNAseq for the study of isoforms—how is that possible? *Genome Biol.* 2018;**19**:110.
23. Ianevski A, Giri AK & Aittokallio T. Fully-automated and ultra-fast cell-type identification using specific marker combinations from single-cell transcriptomic data. *Nat Commun.* 2022;**13**:1246.
24. Zhang Y et al. Sample-multiplexing approaches for single-cell sequencing. *Cell Mol Life Sci.* 2022; **79**:466.
25. Howitt G et al. Benchmarking single-cell hashtag oligo demultiplexing methods. 2023. Preprint at doi: 10.1101/2022.12.20.521313
26. Lun ATL et al. EmptyDrops: Distinguishing cells from empty droplets in droplet-based single-cell RNA sequencing data. *Genome Biol.* 2019;**20**:63.
27. Xin H et al. GMM-Demux: Sample demultiplexing, multiplet detection, experiment planning, and novel cell-type verification in single cell sequencing. *Genome Biol.* 2020;**21**:188.
28. Tuddenham JF et al. A cross-disease human microglial framework identifies disease-enriched subsets and tool compounds for microglial polarization. 2022. Preprint at doi: 10.1101/2022.06.04.494709
29. Boggy GJ et al. BFF and cellhashR: Analysis tools for accurate demultiplexing of cell hashing data. *Bioinformatics* 2022;**38**:2791–801.

30. Bernstein NJ, Fong NL, Lam I, Roy MA, Hendrickson DG, Kelley DR. Solo: Doublet identification in single-cell RNA-Seq via semi-supervised deep learning. *Cell Systems*. 2020;**11**(1): 95–101.e5. doi: 10.1016/j.cels.2020.05.010
31. Kang HM et al. Multiplexed droplet single-cell RNA-sequencing using natural genetic variation. *Nat Biotechnol*. 2018;**36**:89–94.
32. Neavin D et al. Demuxafy: Improvement in droplet assignment by integrating multiple single-cell demultiplexing and doublet detection methods. *Genome Biology*. doi: 10.1186/s13059-024-03224-8
33. Heaton H et al. Souporcell: Robust clustering of single-cell RNA-seq data by genotype without reference genotypes. *Nat Methods*. 2020;**17**:615–20.
34. Xu J et al. Genotype-free demultiplexing of pooled single-cell RNA-seq. *Genome Biol*. 2019;**20**:290.
35. Babraham bioinformatics - FastQC A quality control tool for high throughput sequence data. https://www.bioinformatics.babraham.ac.uk/projects/fastqc/
36. Chen W et al. UMI-count modeling and differential expression analysis for single-cell RNA sequencing. *Genome Biol*. 2018;**19**:70.
37. Ilicic T et al. Classification of low quality cells from single-cell RNA-seq data. *Genome Biol*. 2016;**17**:29.
38. Wolock SL, Lopez R & Klein AM. Scrublet: Computational identification of cell doublets in single-cell transcriptomic data. *Cell Syst*. 2019;**8**:281–91.
39. McGinnis CS, Murrow LM & Gartner ZJ. DoubletFinder: Doublet detection in single-cell RNA sequencing data using artificial nearest neighbors. *Cell Syst*. 2019;**8**:329–37.
40. Hong R et al. Comprehensive generation, visualization, and reporting of quality control metrics for single-cell RNA sequencing data. *Nat Commun*. 2022;**13**:1688.
41. Dobin A & Gingeras TR. Mapping RNA-seq reads with STAR. *Curr Protoc Bioinforma*. 2015;**51**: 11.14.1–11.14.19.
42. Malhotra A, Das S & Rai SN. Analysis of single-cell RNA-sequencing data: A step-by-step guide. *BioMedInformatics*. 2022;**2**:43–61.
43. Srivastava A, Malik L, Smith T, Sudbery I & Patro R. Alevin efficiently estimates accurate gene abundances from dscRNA-seq data. *Genome Biol*. 2019;**20**:65.
44. Melsted P et al. Modular, efficient and constant-memory single-cell RNA-seq preprocessing. *Nat Biotechnol*. 2021;**39**:813–18.
45. Lun ATL, McCarthy DJ & Marioni JC. A step-by-step workflow for low-level analysis of single-cell RNA-seq data with Bioconductor. 2016. Preprint at doi: 10.12688/f1000research.9501.2
46. DeBerardine M. BRGenomics for analyzing high-resolution genomics data in R. *Bioinformatics*. 2023; **39**(6):btad331. doi: 10.1093/bioinformatics/btad331
47. Andrews TS, Kiselev VY, McCarthy D & Hemberg M. Tutorial: Guidelines for the computational analysis of single-cell RNA sequencing data. *Nat Protoc*. 2021;**16**:1–9.
48. Hafemeister C & Satija R. Normalization and variance stabilization of single-cell RNA-seq data using regularized negative binomial regression. *Genome Biol*. 2019;**20**:296.
49. Tran HTN et al. A benchmark of batch-effect correction methods for single-cell RNA sequencing data. *Genome Biol*. 2020;**21**:12.
50. Zhang Y, Parmigiani G & Johnson WE. ComBat-seq: Batch effect adjustment for RNA-seq count data. *NAR Genom Bioinform*. 2020;**2**(3):lqaa078. doi: 10.1093/nargab/lqaa078
51. Haghverdi L, Lun ATL, Morgan MD & Marioni JC. Batch effects in single-cell RNA-sequencing data are corrected by matching mutual nearest neighbors. *Nat Biotechnol*. 2018;**36**:421–27.
52. Gong W, Kwak IY, Pota P, Koyano-Nakagawa N & Garry DJ. DrImpute: Imputing dropout events in single cell RNA sequencing data. *BMC Bioinformatics*. 2018;**19**:220.
53. Huang M et al. SAVER: Gene expression recovery for single-cell RNA sequencing. *Nat Methods*. 2018;**15**:539–42.
54. van Dijk D et al. Recovering gene interactions from single-cell data using data diffusion. *Cell*. 2018; **174**:716–29.e27.
55. Korsunsky I et al. Fast, sensitive and accurate integration of single-cell data with harmony. *Nat Methods*. 2019;**16**:1289–96.
56. Welch JD et al. Single-cell multi-omic integration compares and contrasts features of brain cell identity. *Cell*. 2019;**177**:1873–87.e17.
57. Stuart T et al. Comprehensive integration of single-cell data. *Cell*. 2019;**177**:1888–902.e21.
58. Scialdone A et al. Computational assignment of cell-cycle stage from single-cell transcriptome data. *Methods*. 2015;**85**:54–61.
59. Butler A, Hoffman P, Smibert P, Papalexi E & Satija R. Integrating single-cell transcriptomic data across different conditions, technologies, and species. *Nat Biotechnol*. 2018;**36**:411–20.

60. Piper M, Mistry M, Liu J, Gammerdinger W & Khetani R. hbctraining/scRNA-seq_online: scRNA-seq lessons from HCBC (first release). Zenodo. 2022. doi:10.5281/zenodo.5826256
61. Yang P, Huang H & Liu C. Feature selection revisited in the single-cell era. *Genome Biol.* 2021;**22**:321.
62. Pearson KL III. On lines and planes of closest fit to systems of points in space Lond. *Edinb Dublin Philos Mag J Sci.* 1901;**2**:559–72.
63. vanMaaten L & Hinton G. Visualizing data using t-SNE. *J Mach Learn Res.* 2008;**9**:2579–605.
64. McInnes L, Healy J & Melville J. UMAP: Uniform manifold approximation and projection for dimension reduction. 2020. Preprint at doi: 10.48550/arXiv.1802.03426
65. Pierson E & Yau C. ZIFA: Dimensionality reduction for zero-inflated single-cell gene expression analysis. *Genome Biol.* 2015;**16**:241.
66. Zhang S, Yang L, Yang J, Lin Z & Ng MK. Dimensionality reduction for single cell RNA sequencing data using constrained robust non-negative matrix factorization. NAR Genom Bioinform. 2020;**2**(3):lqaa064. doi: 10.1093/nargab/lqaa064
67. Jiang J et al. Dimensionality reduction and visualization of single-cell RNA-seq data with an improved deep variational autoencoder. *Brief Bioinform.* 2023;**24**:bbad152.
68. Luo Z, Xu C, Zhang Z & Jin W. A topology-preserving dimensionality reduction method for single-cell RNA-seq data using graph autoencoder. *Sci Rep.* 2021;**11**:20028.
69. Tran B, Tran D, Nguyen H, Ro S & Nguyen T. scCAN: Single-cell clustering using autoencoder and network fusion. *Sci Rep.* 2022;**12**:10267.
70. Yu L, Cao Y, Yang JYH & Yang P. Benchmarking clustering algorithms on estimating the number of cell types from single-cell RNA-sequencing data. *Genome Biol.* 2022;**23**:49.
71. Sreenivasan VKA, Henck J & Spielmann M. Single-cell sequencing: Promises and challenges for human genetics. *Med Genet.* 2022;**34**:261–73.
72. Aran D et al. Reference-based analysis of lung single-cell sequencing reveals a transitional profibrotic macrophage. *Nat Immunol.* 2019;**20**:163–72.
73. Shao X et al. scDeepSort: A pre-trained cell-type annotation method for single-cell transcriptomics using deep learning with a weighted graph neural network. *Nucleic Acids Res.* 2021;**49**:e122.
74. Lin Y et al. scClassify: Sample size estimation and multiscale classification of cells using single and multiple reference. *Mol Syst Biol.* 2020;**16**(6). doi: 10.15252/msb.20199389
75. Pasquini G, Rojo Arias JE, Schäfer P & Busskamp V. Automated methods for cell type annotation on scRNA-seq data. *Comput Struct Biotechnol J.* 2021;**19**:961–9.
76. Clarke ZA et al. Tutorial: Guidelines for annotating single-cell transcriptomic maps using automated and manual methods. *Nat Protoc.* 2021;**16**:2749–64.
77. Love MI, Huber W & Anders S. Moderated estimation of fold change and dispersion for RNA-seq data with DESeq2. *Genome Biol.* 2014;**15**:550.
78. Robinson MD, McCarthy DJ & Smyth GK. edgeR: A bioconductor package for differential expression analysis of digital gene expression data. *Bioinformatics.* 2010;**26**(1):139–40. doi: 10.1093/bioinformatics/btp616
79. Ritchie ME et al. Limma powers differential expression analyses for RNA-sequencing and microarray studies. *Nucleic Acids Res.* 2015;**43**:e47.
80. Finak G et al. MAST: A flexible statistical framework for assessing transcriptional changes and characterizing heterogeneity in single-cell RNA sequencing data. *Genome Biol.* 2015;**16**:278.
81. Gagnon J et al. Recommendations of scRNA-seq differential Gene expression analysis based on comprehensive benchmarking. *Life.* 2022;**12**:850.
82. He L et al. NEBULA is a fast negative binomial mixed model for differential or co-expression analysis of large-scale multi-subject single-cell data. *Commun Biol.* 2021;**4**:1–17.
83. Brooks ME et al. glmmTMB balances speed and flexibility among packages for zero-inflated generalized linear mixed modeling. *R J.* 2017;**9**:378–400.
84. Liberzon A et al. The molecular signatures database (MSigDB) hallmark gene set collection. *Cell Syst.* 2015;**1**:417–25.
85. The Gene Ontology Consortium. The gene ontology knowledgebase in 2023. *Genetics.* 2023;**224**:iyad031.
86. Raudvere U et al. g:Profiler: A web server for functional enrichment analysis and conversions of gene lists (2019 update). *Nucleic Acids Res.* 2019;**47**:W191–8.
87. Sherman BT et al. DAVID: A web server for functional enrichment analysis and functional annotation of gene lists (2021 update). *Nucleic Acids Res.* 2022;**50**:W216–21.
88. Subramanian A et al. Gene set enrichment analysis: A knowledge-based approach for interpreting genome-wide expression profiles. *PNAS.* 2005;**102**(43):15545–50. doi: 10.1073/pnas.0506580102
89. Mootha VK et al. PGC-1α-responsive genes involved in oxidative phosphorylation are coordinately downregulated in human diabetes. *Nat Genet.* 2003;**34**:267–73.

90. Ma Y et al. Integrative differential expression and gene set enrichment analysis using summary statistics for scRNA-seq studies. *Nat Commun*. 2020;**11**:1585.
91. Franchini M, Pellecchia S, Viscido G & Gambardella G. Single-cell gene set enrichment analysis and transfer learning for functional annotation of scRNA-seq data. *NAR Genomics Bioinforma*. 2023;**5**:lqad024.
92. Zappia L & Theis FJ. Over 1000 tools reveal trends in the single-cell RNA-seq analysis landscape. *Genome Biol*. 2021;**22**:301.
93. Snakemake workflow: single-cell-rna-seq. 2023.
94. Peltzer A et al. nf-core/scrnaseq: nf-core/scrnaseq v2.3.2 Sepia Samarium Salmon. 2023. doi:10.5281/ZENODO.3568187
95. Khozoie C et al. scFlow: A scalable and reproducible analysis pipeline for single-cell RNA sequencing data. *Authorea*. August 19, 2021. doi: 10.22541/au.162912533.38489960/v1
96. https://www.r-project.org/
97. Mangiola S, Doyle MA & Papenfuss AT. Interfacing Seurat with the R tidy universe. *Bioinformatics*. 2021;**37**:4100–7.
98. Mangiola S, Molania R, Dong R, Doyle MA & Papenfuss AT. Tidybulk: An R tidy framework for modular transcriptomic data analysis. *Genome Biol*. 2021;**22**:42.
99. Zappia L & Oshlack A. Clustering trees: A visualization for evaluating clusterings at multiple resolutions. *GigaScience*. 2018;**7**:giy083.
100. https://www.biorxiv.org/content/10.1101/2022.02.28.482303v1
101. Zhang Y, Kim MS, Reichenberger ER, Stear B & Taylor DM. Scedar: A scalable python package for single-cell RNA-seq exploratory data analysis. *PLOS Comput Biol*. 2020;**16**:e1007794.
102. Wolf FA, Angerer P & Theis FJ. SCANPY: Large-scale single-cell gene expression data analysis. *Genome Biol*. 2018;**19**:15.
103. Luecken MD & Theis FJ. Current best practices in single-cell RNA-seq analysis: A tutorial. *Mol Syst Biol*. 2019;**15**:e8746.
104. https://www.sc-best-practices.org/introduction/prior_art.html.
105. https://satijalab.org/seurat/articles/pbmc3k_tutorial.html.
106. Sandve GK, Nekrutenko A, Taylor J & Hovig E. Ten simple rules for reproducible computational research. *PLOS Comput Biol*. 2013;**9**:e1003285.
107. Niarakis A et al. Addressing barriers in comprehensiveness, accessibility, reusability, interoperability and reproducibility of computational models in systems biology. *Brief Bioinform*. 2022;**23**:bbac212.
108. Martínez Arbas S et al. Challenges, strategies, and perspectives for reference-independent longitudinal multi-omic microbiome studies. *Front Genet*. 2021;**12**:666244.
109. https://www.frontiersin.org/articles/10.3389/fgene.2020.00303/full.
110. Dove ES. The EU general data protection regulation: Implications for international scientific research in the digital era. *J Law Med Ethics*. 2018;**46**:1013–30.
111. Wilkinson MD et al. The FAIR guiding principles for scientific data management and stewardship. *Sci Data*. 2016;**3**:160018.
112. https://www.nature.com/articles/s41597-023-02166-3.
113. Welter D et al. FAIR in action — A flexible framework to guide FAIRification. *Sci Data*. 2023;**10**:291.
114. https://fairplus.github.io/the-fair-cookbook/content/home.html.
115. https://eosc-portal.eu/.
116. Liu Z, Sun D & Wang C. Evaluation of cell-cell interaction methods by integrating single-cell RNA sequencing data with spatial information. *Genome Biol*. 2022;**23**:218.
117. Aibar S et al. SCENIC: Single-cell regulatory network inference and clustering. *Nat Methods*. 2017;**14**: 1083–86.
118. https://github.com/broadinstitute/infercnv/blob/master/inst/CITATION, accessed on 16/10/2024, accessed on 16 October 2024,
119. Bahonar S & Montazeri H. Somatic Single-Nucleotide Variant Calling from Single-Cell DNA Sequencing Data Using SCAN-SNV in Variant Calling: Methods and Protocols (eds. Ng C & Piscuoglio S) 267–277 (Springer US, 2022). doi:10.1007/978-1-0716-2293-3_17
120. Hu Y, Xi X, Yang Q & Zhang X. SCeQTL: An R package for identifying eQTL from single-cell parallel sequencing data. *BMC Bioinformatics*. 2020;**21**:184.
121. Bergen V, Lange M, Peidli S, Wolf FA & Theis FJ. Generalizing RNA velocity to transient cell states through dynamical modeling. *Nat Biotechnol*. 2020;**38**:1408–14.
122. Ding J, Sharon N & Bar-Joseph Z. Temporal modelling using single-cell transcriptomics. *Nat Rev Genet*. 2022;**23**:355–68.

123. Tran TN & Bader GD. Tempora: Cell trajectory inference using time-series single-cell RNA sequencing data. *PLOS Comput. Biol.* 2020;**16**:e1008205.
124. Yeo GHT, Saksena SD & Gifford DK. Generative modeling of single-cell time series with PRESCIENT enables prediction of cell trajectories with interventions. *Nat Commun.* 2021;**12**:3222.
125. Lin C, Ding J & Bar-Joseph Z. Inferring TF activation order in time series scRNA-seq studies. *PLOS Comput Biol.* 2020;**16**:e1007644.
126. Shao L et al. Identify differential genes and cell subclusters from time-series scRNA-seq data using scTITANS. *Comput Struct Biotechnol J.* 2021;**19**:4132–41.
127. Longo SK, Guo MG, Ji AL & Khavari PA. Integrating single-cell and spatial transcriptomics to elucidate intercellular tissue dynamics. *Nat Rev Genet.* 2021;**22**:627–44.
128. Moffitt JR, Lundberg E & Heyn H. The emerging landscape of spatial profiling technologies. *Nat Rev Genet.* 2022;**23**:741–59.
129. Elmentaite R, Domínguez Conde C, Yang L & Teichmann SA. Single-cell atlases: Shared and tissue-specific cell types across human organs. *Nat Rev Genet.* 2022;**23**:395–410.
130. Lee J, Hyeon DY & Hwang D. Single-cell multiomics: Technologies and data analysis methods. *Exp Mol Med.* 2020;**52**:1428–42.
131. Baysoy A, Bai Z, Satija R & Fan R. The technological landscape and applications of single-cell multi-omics. *Nat Rev Mol Cell Biol.* 2023;1–19 doi:10.1038/s41580-023-00615-w
132. Vandereyken K, Sifrim A, Thienpont B & Voet T. Methods and applications for single-cell and spatial multi-omics. *Nat Rev Genet.* 2023;1–22 doi:10.1038/s41576-023-00580-2
133. Wang X et al. BREM-SC: A Bayesian random effects mixture model for joint clustering single cell multi-omics data. *Nucleic Acids Res.* 2020;**48**:5814–24.
134. Kim HJ, Lin Y, Geddes TA, Yang JYH & Yang P. CiteFuse enables multi-modal analysis of CITE-seq data. *Bioinformatics.* 2020;**36**:4137–43.
135. Do VH, Ringeling FR & Canzar S. Linear-time cluster ensembles of large-scale single-cell RNA-seq and multimodal data. *Genome Res.* 2021;**31**:677–88.
136. Gayoso A et al. Joint probabilistic modeling of single-cell multi-omic data with totalVI. *Nat Methods.* 2021;**18**:272–82.
137. Lin X, Tian T, Wei Z & Hakonarson H. Clustering of single-cell multi-omics data with a multimodal deep learning method. *Nat Commun.* 2022;**13**:7705.
138. Marx V. How single-cell multi-omics builds relationships. *Nat Methods.* 2022;**19**:142–6.
139. Van de Sande B et al. Applications of single-cell RNA sequencing in drug discovery and development. *Nat Rev Drug Discov.* 2023;**22**:496–520.
140. Jovic D et al. Single-cell RNA sequencing technologies and applications: A brief overview. *Clin Transl Med.* 2022;**12**:e694.
141. Regev A et al. The human cell atlas. *eLife.* 2017;**6**:e27041.
142. Rood JE, Maartens A, Hupalowska A, Teichmann SA & Regev A. Impact of the human cell atlas on medicine. *Nat Med.* 2022;**28**:2486–96.
143. BRAIN Initiative Cell Census Network (BICCN). A multimodal cell census and atlas of the mammalian primary motor cortex. *Nature.* 2021;**598**:86–102.
144. Papatheodorou I et al. Expression atlas update: From tissues to single cells. *Nucleic Acids Res.* 2020;**48**:D77–D83.
145. Lim J et al. Transitioning single-cell genomics into the clinic. *Nat Rev Genet.* 2023;1–12. doi:10.1038/s41576-023-00613-w
146. Sun G et al. Single-cell RNA sequencing in cancer: Applications, advances, and emerging challenges. *Mol Ther Oncolytics.* 2021;**21**:183–206.
147. Li L et al. What are the applications of single-cell RNA sequencing in cancer research: A systematic review. *J Exp Clin Cancer Res.* 2021;**40**:163.
148. Ke M, Elshenawy B, Sheldon H, Arora A & Buffa FM. Single cell RNA-sequencing: A powerful yet still challenging technology to study cellular heterogeneity. *BioEssays.* 2022;**44**:2200084.

8 Integration of Spatial Transcriptomics and Single-Cell RNA-seq

Vaishnavi Srinivasan

INTRODUCTION

Over the years, Single-Cell (SC) analysis methods have proliferated, prompting many researchers to shift from traditional bulk sequencing to more nuanced approaches that examine individual cells [1]. SC sequencing is a process of analysing individual cells to understand cellular structure and function at a much finer level of detail than traditional methods allow. To illustrate the difference between SC and bulk sequencing technologies, we can draw an analogy to music. Bulk sequencing can be thought of as listening to an entire orchestra playing a composition; it provides an understanding of the music as a whole. In contrast, SC technology is akin to listening to each individual instrument, allowing us to discern their unique contributions to the overall performance [2]. SC technologies have been integrated into various omics platforms and have been proven to be invaluable in genome and transcriptome sequencing [3]. Despite its limitations, SC technologies have opened up new avenues of research, such as the study of rare cell types, heterogeneous samples, and cell-to-cell variability within samples [1].

Spatial transcriptomics (ST), on the other hand, is an innovative discipline that permits the investigation of tissue-specific gene expression patterns. Although traditional RNA sequencing might help provide gene expression data, it lacks spatial context, which is now understood to be extremely important. ST integrates molecular biology and computational analyses to capture the spatial organisation of gene expression patterns [4]. Consider a puzzle that represents a complex tissue or cellular environment. In bulk RNA sequencing (RNA-seq) technologies, the puzzle pieces would be disassembled and individually analysed. While this can provide a little insight, it becomes easy to miss vital information about the bigger picture and know where each piece belongs. ST helps retain information about 'the bigger picture' by examining gene expression within the context of tissue architecture [5]. It involves analysing individual tissue sections whilst preserving spatial resolution. Incorporating ST therefore allows researchers to study processes such as cell differentiation, intercellular communication, and the influence of the tissue microenvironment [6].

SC Technologies in Transcriptomics: scRNA-seq

Over the years, RNA-seq technology has been used extensively in numerous fields of study, including alternative splicing studies. Traditional bulk RNA-seq methods, however, come with their own set of limitations. They provide averaged gene expression profiles across all cells in a sample, obscuring cell-to-cell variability. Additionally, bulk methods often have reduced ability to distinguish between transcript isoforms and can struggle to detect rare transcripts or subtle gene expression changes [7,8]. SC transcriptomics, particularly methods like single-cell RNA sequencing (scRNA-seq), is a powerful approach that allows researchers to examine gene expression in individual cells, enabling the investigation of cellular behaviour and heterogeneity at high resolution. Bolstered by technological advancements, scRNA-seq has become extraordinarily useful for identifying rare cell populations, discovering novel biomarkers, and researching diseases such as cancer [9]. These

DOI: 10.1201/9781003354062-8

advancements span improved cell isolation techniques, refined library preparation protocols, and sophisticated computational analysis tools.

Experimental Workflow

Cell Preparation

1. *Sample preparation*: The initial step in SC analysis is to carefully gather and process the tissue or cells per experimental requirements.
2. *Cell isolation and sorting*: Next, the tissues are dissociated into individual cells using enzymatic digestion to enable cell isolation and sorting. Cell isolation can be performed on whole cells, cells with specific marker proteins, or cell-specific nuclei, depending on the experiment [10]. The objective of cell sorting is to obtain highly purified cells without removing any viable cells [11]. Magnetic-activated cell sorting (MACS) and fluorescence-activated cell sorting (FACS) are two standard cell-sorting techniques used for this purpose [10].

Library Preparation

3. *Cell capture*: Subsequently, each cell from the isolate is captured using one of two techniques: either the microwell technique or the droplet-based method, depending on the experimental protocol. The microwell technique uses microwell plates, while the droplet-based method uses microfluidics to cover the cells. For example, in the droplet-based method, each droplet contains the cell and a bead that contains primers with a unique barcode and the lysis buffer [11].
4. *Cell lysis and messenger RNA (mRNA) capture*: After individual cells are lysed to release their contents, mRNA molecules are selectively captured using the primers. This process isolates the mRNA from other cellular components and prepares it for subsequent steps in library construction [12]. The unique primers contain three main parts: (i) a cellular barcode that can group all of the transcripts within a cell, (ii) a unique molecule identifier that can uniquely identify each transcript after amplification, and (iii) a poly-A tail that attaches to the mRNA sequencing. Barcoding is crucial in all SC technology, as it helps with cell differentiation [11].

Reverse Transcription, Amplification, and Sequencing

5. *Reverse transcription, amplification, and sequencing*: The mRNA is then reverse transcribed and amplified using either polymerase chain reaction (PCR) or *in vitro* transcription. Sample-specific indexes are used to tag the resulting complementary DNA (cDNA), allowing multiple libraries to be pooled together and sequenced. These sample-specific indexes help identify and segregate the cDNAs from the pool (demultiplexing). Lastly, standard sequencing platforms such as Illumina are employed to sequence the pool of cDNAs [11].

Computational Workflow

Data Preprocessing

1. *Demultiplexing*: Demultiplexing is the process of sorting pooled sequencing reads back to their original samples. This is performed using bioinformatics tools like 10x Genomics Cell Ranger. Subsequently, duplicate reads created during amplification are removed through a deduplication step. This process utilises unique molecular identifiers (UMIs) to identify and eliminate these amplification artifacts, ensuring a more accurate representation of the original transcripts [11].
2. *Trimming*: Tools such as Trimmomatic can be used to trim the adapters and remove bad reads.
3. *Alignment*: After the raw reads are obtained, they are aligned to a reference genome or transcriptome. This process provides valuable information, including the origin of the

reads, and enables comparison between the samples. Different bioinformatics tools, such as STAR aligner or Cell Ranger, can be used for alignment. Additionally, mapping helps to filter out inconsistent reads and incompatible barcodes and UMIs [11].

4. *Quantifying gene expression*: After preprocessing, the gene expression values can be quantified using tools like STARsolo or Cell Ranger, which can take the reads (typically in a FASTQ format) and generate count matrices containing the gene expression levels [11].

Normalisation and Dimensionality Reduction

5. *Normalisation*: This is a crucial step to eliminate gene expression bias and technical noise. It helps to correct for differences in sequencing depth between cells, reduce amplification biases, and minimise batch effects [11,13]. A common normalisation method involves dividing the raw UMI count by the total detected RNA count, then multiplying this by a scaling factor; adding a pseudo-count, and then applying a log transformation [14]. The most common normalisation method used is counts per million (CPM) [15] to compensate for sparse data caused by dropouts.
6. *Dimensionality reduction (DR)*: This is another important step that further eliminates data noise before the clustering process [16]. By reducing data complexity during the visualisation process, DR can improve the analysis. There are three types of DR methods: tSNE, UMAP, and PCA. Principal component analysis (PCA) is a linear method that helps with DR in large datasets. It is an easy process and can be efficiently implemented. The non-linear methods include t-distributed stochastic neighbour embedding (t-SNE), which plots points into two-dimensional (2D) or three-dimensional (3D) space, and uniform manifold approximation and projection (UMAP), which works on the Riemannian geometry [17] and algebraic topology to reduce dimensionality [18].

Clustering and Differential Expression

7. *Clustering*: scRNA-seq technologies use machine learning (ML) methods such as hierarchical clustering and *k*-means to group the cells based on their gene expression similarities [12]. This process helps in understanding the identities of different cell types or states and how they behave. The resulting cell clusters often represent distinct cell types or functional states within the sample.
8. *Differential expression*: Comparing gene expression levels across samples or different conditions can be done through differential expression analysis. This method can pinpoint overexpressed genes in specific diseases or genetic conditions. Popular techniques in scRNA-seq include Seurat, scanpy, and model-based analysis of SC transcriptomics (MAST) [18].

Challenges in SC Transcriptomics

In the field of scRNA-seq, there are various challenges that need to be addressed. Firstly, the data are highly dimensional with more sets of genes, which requires the reduction of resolution before it can be viewed. Secondly, there is a high variability in cells due to issues with RNA capture, such as low transcript capture, resulting in transcript variations with low read counts [19]. Technical noises, such as low quality of reads and amplification bias, can also lead to Type I error inflation in differential expression analysis [20,21]. This means an increased rate of false positives, where genes are incorrectly identified as differentially expressed between conditions or cell types when no true biological difference exists. Additionally, the process of analysing individual cells through scRNA-seq technologies does not preserve the spatial orientation of cells, which is important for intricate gene analysis. Lastly, data normalisation is a challenging task in scRNA-seq, as there is no universally applicable strategy that properly accounts for variability in sequencing depth and cell size [19,22].

TABLE 8.1
Overview of the Different Types of ST

Imaging-Based Technology	Capture/Sequencing-Based Technology
In Situ Hybridisation (ISH) This technology uses fluorescence probes to target specific mRNA molecules. Repeated hybridisation generates a gene-specific barcode [24] which shows the expression patterns within the native tissue environment. The most common methods for *in situ* hybridisation are FISH, merFISH, and smFISH.	*Laser Capture Microdissection* One of the most commonly used types of microdissection is laser capture microdissection (LCM). This technology uses microarrays to profile the transcriptome of specific tissues while retaining the location of the mRNA targets. LCMs are advantageous as they do not compromise the integrity of the tissue. Other types of microdissection techniques include Tomo-seq and Geo-seq [24].
In Situ Sequencing (ISS) This technology sequences RNA molecules directly within tissues, which combines NGS information with spatial data. This is done by using shorter probes to profile fewer bases in the amplified transcript [24]. Some examples of ISS techniques include STARmap and Barista-seq.	*Array-Based Capture* With this technology, polyadenylated mRNA from a tissue section is captured using a specialised array slide containing spatially barcoded, non-gene-specific probes. The captured mRNA is then converted to cDNA, maintaining spatial information, and subsequently sequenced to provide a spatially-resolved transcriptome [24,25].

Types of ST

ST technologies can be categorised into two primary types: imaging-based and capture/sequencing-based. The former uses imaging probes for *in situ* transcript amplification [23], while the latter uses next-generation sequencing (NGS) to sequence transcripts that contain positional data [23]. Table 8.1 explains the different types of ST.

The process for carrying out ST depends on the technology employed and the specific objectives of the experiment. Techniques that rely on imaging-based sequencing are favoured when the goal is to obtain high-resolution output images for a limited number of genes. In contrast, capture-based methods are the priority when imaging more genes [24]. As illustrated in Figure 8.1, the general workflow entails collecting tissue data for imaging, followed by processing the data using one of the available technologies to produce a preprocessed image. For example, in imaging-based methods, the preprocessing phase involves transforming raw images into gene-spot or gene expression matrices [24], aligning and barcoding signals corresponding to genes. Afterward, the data is analysed using standardised spatial analysis toolkits such as Seurat and scanpy [26]. The choice of the toolkit used significantly impacts the inference drawn from the data.

Integration of ST and scRNA-seq

Data integration is a method of consolidating data from various sources into a single location for easier analysis. It helps to combine different sets of data that were collected in different ways, such as with ST and scRNA-seq, and put them together in one place. This makes it easier to analyse and comprehend them [27].

Current ST techniques, while not yet achieving SC resolution with precise positioning in tissues, offer valuable insights into the distribution of specific gene sets. This capability enables the identification of cell subpopulations, such as immune cells, within tissue neighbourhoods [28]. By combining ST with scRNA-seq, researchers can investigate interactions between individual cells in their natural environment. Computational methods can accurately map cell types by combining spatial and SC transcriptomic data. These integrated approaches are particularly useful for studying cell subpopulation interactions during development, homeostasis, and disease processes [28,29].

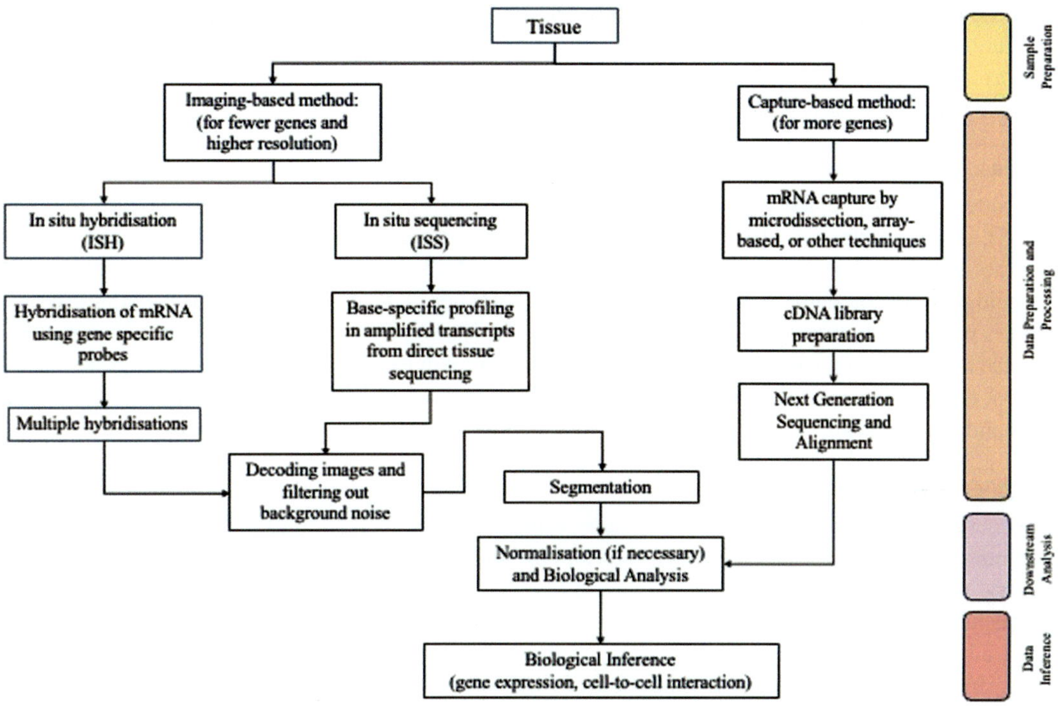

FIGURE 8.1 Overall procedure for different methods in ST: sample preparation, data preparation and pre-processing, downstream analysis, and data inference.

The following are the methods to integrate ST and scRNA-seq:

Cell type mapping: Some tools that can be used to integrate SC data and ST include spISeudoMap and cell2location, which use clustering algorithms and Bayesian models to resolve fine-grained cell types in ST data [30].

Convolution and deconvolution: In ST, convolution can mean constructing gene profiles from raw transcript data, while deconvolution involves deconstructing gene expression to identify signals from specific cells. This can help localise cell subpopulations and proportions, which can be mapped using scRNA-seq. Understanding cellular heterogeneity is important for research, and approaches like conditional autoregressive-based deconvolution and Cell-type Assignment by Regression and Deconvolution (CARD) [31] can aid in this process [28].

Applications of ST and scRNA-seq

ST and scRNA-seq are valuable tools in clinical studies and applications. Some important uses of these technologies include identifying cell types and their relationship to disease models, detecting gene expression changes in specific tissues, and developing personalised medicine approaches. It is important to select appropriate algorithms and methods to effectively utilise these techniques and achieve accurate results.

Cancer Biology and Tumour Microenvironment Characterisation

In cancer research, both ST and scRNA-seq play an important role in understanding tumour heterogeneity and the tumour microenvironment (TME). The TME includes the cells and environment surrounding the cancer cells, which could be the adjacent non-cancerous cells. Tumour cells can interact with these adjacent cells, which could eventually play a part in the invasion and metastasis of these tumour cells. Both ST and scRNA-seq can help us look at individual cell images in high resolution to understand this cell-to-cell interaction in the TME and also tumour progression. It is important to select appropriate algorithms and methods to effectively utilise these techniques and achieve accurate results [32]. (See also the Case Study section below.)

Disease Identification

To better understand the underlying mechanisms of diseases like amyotrophic lateral sclerosis (ALS), it is important to use advanced techniques like integrating ST and scRNA-seq. By utilising the spatial information of gene expression from scRNA-seq, researchers can locate the specific pathological mechanisms of the disease and gain a deeper understanding of what is being affected. This is particularly important for diseases like ALS, where motor neuron degeneration can result in muscle denervation atrophy [33]. With ST, researchers can examine changes in ALS gene expression at the cellular level, identifying subpopulations associated with each stage of the disease process and investigating the molecular mechanisms that cause and sustain the disease [34].

Regenerative Medicine

Regenerative medicine utilises tissue engineering and stem cell therapy to transform non-viable and non-functional cells into functional ones, potentially restoring organs to near-normal function. ST and scRNA-seq can both be employed in regenerative medicine to analyse and identify hidden subpopulations of cells [35].

In addition, there are other practical applications for these technologies. For example, gene mapping, drug response, and the identification of rare cell types are areas where these techniques are commonly utilised. By leveraging ST and scRNA-seq, researchers can gain valuable insights that can help in the development of effective treatments and therapies for a wide range of diseases and conditions [35].

Case Study: PDAC

When studying cancer and its microenvironment, it is important to consider the limitations of both ST and scRNA-seq. However, by integrating these two modalities, researchers can gain a better understanding of the conversation between the microenvironment and the tumour. This approach has been used successfully in case studies and is a promising way to improve the analysis of heterogeneous diseases like cancer. One such workflow example is a case study of the TME in pancreatic ductal adenocarcinoma (PDAC).

The data in this case study were processed using a range of methods, including KNN smoothing to remove any data noise, and hierarchical clustering was performed to identify the cell types and subpopulations and determine the similarity between the cells, which allowed for a more nuanced analysis of the cells. Additionally, a statistical analysis was conducted to further refine the results. A map of the expressed tissues was created using ST cryosection. Raw sequencing data were processed with a custom pipeline, which included the normalisation and hierarchical clustering of ST spots. A multimodal intersection analysis (MIA) was then employed to evaluate the overlap between spatially-resolved gene expression data and known cell type-specific marker genes [36].

Through this case study, it was discovered that certain cell types were prevalent in tumour tissues by comparing the genes associated with the area to cell type-specific genes found through scRNA-seq data. The MIA data for the primary PDAC tumours from multiple patient samples were used to locate distinct cell types and subpopulations in the TME, which helped to understand the relationship between the cells in the cancer subregion.

Although the integration of ST and scRNA-seq has aided in understanding certain subpopulations, this method does have limitations. Improving ST technology could help increase both the sensitivity and spatial resolution across the entire tissue sample. This might involve refining the array design to capture more transcripts from smaller tissue areas. Tissue optimisation may also be necessary to minimise issues with transcript diffusion [36].

Limitations in Method Integration

A key limitation of ST data is that individual spots cannot be directly interpreted as SCs. To address this, multiple computational methods are employed to accurately map cell types by integrating ST with SC transcriptomics data. As ST generates increasingly complex datasets, the challenge of untangling cell types and their relationships to disease models grows. Consequently, it is crucial to carefully select and evolve analytical tools to effectively interpret this data, ensuring a comprehensive understanding of cellular heterogeneity in health and disease contexts [28]. This is especially important as ST methods that can spatially resolve tissue at SC resolution, with scRNA-seq depth and whole-transcriptome coverage, are still in development. Along with developing enhanced methods, choosing the right algorithms is paramount for meeting this challenge [37].

CONCLUSIONS

Both ST and SC technologies present researchers with several technical hurdles. These include managing batch effects, optimising library preparation efficiency, and navigating the complexities of multi-cell analysis. Transitioning from genomics to transcriptomics using SC technologies requires careful attention to these hurdles and the implementation of standard operating procedures (SOPs). ST and scRNA-seq each offer unique insights, and their integration is crucial for comprehensive transcriptomic analysis. While current research is addressing how these methods can be combined, further advancements are needed. Continued collaboration among researchers, technological innovations, and deeper understanding of data integration will be key to making these approaches more accessible and widely applicable in biomedical research.

REFERENCES

1. Tang, X., Y. Huang, J. Lei et al. 'The Single-Cell Sequencing: New Developments and Medical Applications'. *Cell Bioscience* 9 (2019): 53. https://doi.org/10.1186/s13578-019-0314-y
2. Wang, Xuran, Jihwan Park, Katalin Susztak, Nancy R. Zang, Mingyao Li. 'Bulk Tissue Cell Type Deconvolution with Multi-Subject Single-Cell Expression Reference'. *Nature Communications* 10 (2019): 380. https://www.nature.com/articles/s41467-018-08023-x
3. Nawy, Tal. 'Single-Cell Sequencing'. *Nature Methods* 11, no. 1 (1 January 2014): 18. https://doi.org/10.1038/nmeth.2771
4. Ståhl, Patrik L., Fredrik Salmén, Sanja Vickovic, Anna Lundmark, José Fernández Navarro, Jens Magnusson, Stefania Giacomello, et al. 'Visualization and Analysis of Gene Expression in Tissue Sections by Spatial Transcriptomics'. *Science (New York, N.Y.)* 353, no. 6294 (1 July 2016): 78–82. https://doi.org/10.1126/science.aaf2403
5. Popescu, Medeea. 'Spatial transcriptomics methods and multi-omic integration'. *Biodock* (2022). https://blog.biodock.ai/spatial-transcriptomics-methods/
6. Fang, Shuangsang, Bichao Chen, Yong Zhang, Haixi Sun, Longqi Liu, Shiping Liu, Yuxiang Li, and Xun Xu. 'Computational Approaches and Challenges in Spatial Transcriptomics'. *Genomics, Proteomics & Bioinformatics* 21, no. 1 (1 February 2023): 24–47. https://doi.org/10.1016/j.gpb.2022.10.001
7. Hegenbarth, Jana-Charlotte, Giuliana Lezzoche, Leon J. De Windt, Monika Stoll, and Monika Stoll. 'Perspectives on Bulk-Tissue RNA Sequencing and Single-Cell RNA Sequencing for Cardiac Transcriptomics'. *Frontiers in Molecular Medicine* (February 2022). https://doi.org/10.3389/fmmed.2022.839338
8. Hwang, Byungjin, Ji Hyun Lee, and Duhee Bang. 'Single-Cell RNA Sequencing Technologies and Bioinformatics Pipelines'. *Experimental & Molecular Medicine* 50, no. 8 (2018): 1–14. https://doi.org/10.1038/s12276-018-0071-8
9. Chen, Geng, Baitang Ning, and Tieliu Shi. 'Single-Cell RNA-Seq Technologies and Related Computational Data Analysis'. *Frontiers in Genetics* 10 (5 April 2019): 317. https://doi.org/10.3389/fgene.2019.00317
10. Valihrach, L., P. Androvic, and M. Kubista. 'Platforms for Single-Cell Collection and Analysis'. *International Journal of Molecular Sciences* 19, no. 3 (2018): 807. https://doi.org/10.3390/ijms19030807

11. Slovin, Shaked, Annamaria Carissimo, Francesco Panariello, Antonio Grimaldi, Valentina Bouché, Gennaro Gambardella, and Davide Cacchiarelli. 'Single-Cell RNA Sequencing Analysis: A Step-by-Step Overview'. *Methods in Molecular Biology (Clifton, N.J.)* 2284 (2021): 343–65. https://doi.org/10.1007/978-1-0716-1307-8_19
12. Haque, A., J. Engel, S. A. Teichmann, and T. Lönnberg. 'A practical guide to single-cell RNA-sequencing for biomedical research and clinical applications'. *Genome Medicine* 9, no. 1 (18 August 2017): 75. https://doi.org/10.1186/s13073-017-0467-4.
13. Lytal, Nicholas, Di Ran, and Lingling An. 'Normalization Methods on Single-Cell RNA-Seq Data: An Empirical Survey'. *Frontiers in Genetics* 11 (7 February 2020): 41. https://doi.org/10.3389/fgene.2020.00041
14. Hafemeister, C., and R. Sajita. 'Normalization and Variance Stabilization of Single-Cell RNA-seq Data Using Regularized Negative Binomial Regression'. *Genome Biology* 20, no. 1 (2019): 296. https://doi.org/10.1186/s13059-019-1874-1
15. Townes, F. William, and Rafael A. Irizarry. 'Quantile Normalization of Single-Cell RNA-Seq Read Counts Without Unique Molecular Identifiers'. *Genome Biology* 21, no. 1 (2020): 160. https://doi.org/10.1186/s13059-020-02078-0
16. Linderman, George C. 'Dimensionality Reduction of Single-Cell RNA-Seq Data'. *Methods in Molecular Biology (Clifton, N.J.)* 2284 (2021): 331–42. https://doi.org/10.1007/978-1-0716-1307-8_18
17. Eisenhart, L. P., and O. Veblen. 'The Riemann Geometry and its Generalization'. *Proceedings of the National Academy of Sciences* 8, no. 2 (1922): 19–23. https://doi.org/10.1073/pnas.8.2.19
18. Mu, Quanhua, Yiyun Chen, and Jiguang Wang. 'Deciphering Brain Complexity Using Single-Cell Sequencing'. *Genomics, Proteomics & Bioinformatics* 17, no. 4 (2019): 344–66. https://doi.org/10.1016/j.gpb.2018.07.007
19. Adil, Asif, Vijay Kumar, Arif Tasleem Jan, and Mohammed Asger. 'Single-Cell Transcriptomics: Current Methods and Challenges in Data Acquisition and Analysis'. *Frontiers in Neuroscience* 15 (22 April 2021): 591122. https://doi.org/10.3389/fnins.2021.591122
20. Brennecke, Philip, Simon Anders, Jong Kyoung Kim, Aleksandra A. Kołodziejczyk, Xiuwei Zhang, Valentina Proserpio, Bianka Baying, et al. 'Accounting for Technical Noise in Single-Cell RNA-Seq Experiments'. *Nature Methods* 10, no. 11 (22 September 2013): 1093–95. https://doi.org/10.1038/nmeth.2645
21. Jovic, Dragomirka, Xue Liang, Hua Zeng, Lin Lin, Fengping Xu, and Yonglun Luo. 'Single-Cell RNA Sequencing Technologies and Applications: A Brief Overview'. *Clinical and Translational Medicine* 12, no. 3 (2022): e694. https://doi.org/10.1002/ctm2.694
22. Sandberg, Rickard. 'Entering the Era of Single-Cell Transcriptomics in Biology and Medicine'. *Nature Methods* 11, no. 1 (1 January 2014): 22–24. https://doi.org/10.1038/nmeth.2764
23. Rao, Anjali, Dalia Barkley, Gustavo S. França, and Itai Yanai. 'Exploring Tissue Architecture Using Spatial Transcriptomics'. *Nature* 596, no. 7871 (11 August 2021): 211–20. https://doi.org/10.1038/s41586-021-03634-9
24. Williams, Cameron G., Hyun Jae Lee, Takahiro Asatsuma, Vento-Tormo Roser, and Haque Ashraful. 'An Introduction to Spatial Transcriptomics for Biomedical Research'. *Genome Medicine* 14, no. 1 (27 June 2022): 68. https://doi.org/10.1186/s13073-022-01075-1
25. Lee, Jongwon, Minsu Yoo, and Jungmin Choi. 'Recent Advances in Spatially Resolved Transcriptomics: Challenges and Opportunities'. *BMB Reports* 55, no. 3 (2022): 113–24. https://doi.org/10.5483/BMBRep.2022.55.3.014
26. Moses, Lambda, and Lior Pachter. 'Museum of Spatial Transcriptomics'. *Nature Methods* 19, no. 5 (2022): 534–46. https://doi.org/10.1038/s41592-022-01409-2
27. Ghazanfar, Shila, Carolina Guibentif, and John C. Marioni. 'Stabilized Mosaic Single-Cell Data Integration Using Unshared Features'. *Nature Biotechnology* (25 May 2023). https://doi.org/10.1038/s41587-023-01766-z
28. Longo, Sophia K., Margaret G. Guo, Andrew L. Ji, and Paul A. Khavari. 'Integrating Single-Cell and Spatial Transcriptomics to Elucidate Intercellular Tissue Dynamics'. *Nature Reviews. Genetics* 22, no. 10 (18 June 2021): 627–44. https://doi.org/10.1038/s41576-021-00370-8
29. Wu, Hongwei, Fanna Liu, Yu Shangguan, Yane Yang, Wei Shi, Wenlong Hu, Zhipeng Zeng, et al. 'Integrating Spatial Transcriptomics With Single-Cell Transcriptomics Reveals a Spatiotemporal Gene Landscape of the Human Developing Kidney'. *Cell & Bioscience* 12, no. 1 (3 June 2022): 80. https://doi.org/10.1186/s13578-022-00801-x
30. Zhang, Yun, Brian Aevermann, Rohan Gala, and Richard H. Scheuermann. 'Cell Type Matching in Single-Cell RNA-Sequencing Data Using FR-Match'. *bioRxiv* (1 January 2021). https://doi.org/10.1101/2021.10.17.464718

31. Ma, Ying, and Xiang Zhou. 'Spatially Informed Cell-Type Deconvolution for Spatial Transcriptomics', *Nature Biotechnology*, 40, no. 9 (1 September 2022): 1359. https://doi.org/10.1038/s41587-022-01273-7
32. Hunter, Miranda V., Reuben Moncada, Joshua M. Weiss, Itai Yanai, and Richard M. White. 'Spatially Resolved Transcriptomics Reveals the Architecture of the Tumor-Microenvironment Interface'. *Nature Communications* 12, no. 1 (1 November 2021): 6278. https://doi.org/10.1038/s41467-021-26614-z
33. Lépine, Sarah, Maria José Castellanos-Montiel, and Thomas Martin Durcan. 'TDP-43 Dysregulation and Neuromuscular Junction Disruption in Amyotrophic Lateral Sclerosis'. *Translational Neurodegeneration* 11, no. 1 (27 December 2022): 56. https://doi.org/10.1186/s40035-022-00331-z
34. Ya, Dongshan, Yingmei Zhang, Qi Cui, Yanlin Jiang, Jiaxin Yang, Ning Tian, Wenjing Xiang, Xiaohui Lin, Qinghua Li, and Rujia Liao. 'Application of Spatial Transcriptome Technologies to Neurological Diseases'. *Frontiers in Cell and Developmental Biology* 11 (3 March 2023): 1142923. https://doi.org/10.3389/fcell.2023.1142923
35. Yan, Ruojin, Chunmei Fan, Zi Yin, Tingzhang Wang, and Xiao Chen. 'Potential Applications of Deep Learning in Single-Cell RNA Sequencing Analysis for Cell Therapy and Regenerative Medicine'. *Stem Cells* 39, no. 5 (2021): 511–21. https://doi.org/10.1002/stem.3336
36. Moncada, Reuben, Dalia Barkley, Florian Wagner, Marta Chiodin, Joseph C. Devlin, Maayan Baron, Cristina H. Hajdu, Diane M. Simeone, and Itai Yanai. 'Author Correction: Integrating Microarray-Based Spatial Transcriptomics and Single-Cell RNA-Seq Reveals Tissue Architecture in Pancreatic Ductal Adenocarcinomas'. *Nature Biotechnology* 38, no. 12 (December 2020): 1476. https://doi.org/10.1038/s41587-020-00776-5
37. Wan, Xiaomeng, Jiashun Xiao, Sindy Sing Ting Tam, Mingxuan Cai, Ryohichi Sugimura, Yang Wang, Xiang Wan, Zhixiang Lin, Angela Ruohao Wu, and Can Yang. 'Integrating spatial and single-cell transcriptomics data using deep generative models with SpatialScope'. *Nature Communications* 14, no. 1 (November 2023): 7848. https://doi.org/10.1038/s41467-023-43629-w

9 Metagenomic Analysis Pipelines for Microbiome Studies

QIIME and mothur

Bharati Pandey, Romi Wahengbam, Moirangthem Goutam Singh, Amartya Nandi, Sneha Murmu, Sachin Kumar, and Rina Yumnam

INTRODUCTION

Microbes are integral to Earth's biosphere, playing a crucial role in ecological processes across various habitats and scales, even within the human body. Despite their fundamental importance to Earth's existence and human well-being, there remains much to uncover about the intricacies of microbial interactions and their relationship with the environment [1]. The term 'microbiome' refers to the assembly of microorganisms (including fungi, bacteria, and viruses) present in a specific setting. In the context of humans, this expression is commonly employed to portray the microorganisms inhabiting a specific area of the body, such as the skin or gastrointestinal tract [2]. For example, an individual's microbiome could impact their vulnerability to infectious diseases and play a role in chronic gastrointestinal conditions like Crohn's disease and irritable bowel syndrome. Certain groups of microorganisms dictate an individual's response to drug therapies. Furthermore, the maternal microbiome might influence the well-being of a mother's offspring [3].

Comprising an array of prokaryotic and eukaryotic entities such as bacteria, archaea, viruses, fungi, and protozoans, the microbiome has attracted interest for its combined functional significance in regulating host nutrition, metabolism, physiology, and immunology [4]. The significance of these distinct taxa can vary based on the macroorganism they are linked to. Within animals, the bacterial microbiome holds greater numerical superiority and wields more influence over the health and wellness of its host compared to the fungal microbiome [5]. The converse holds true for plants. Fungi, encompassing those residing in leaves, shoots, and roots like mycorrhizae, are the dominant symbiotic entities in terms of functionality [6,7].

With the maturation of modern microbiology and the advancement of NGS technologies, there has been a growing emphasis on investigating intricate microbial communities that engage with the host and impact health and disease processes [8]. There are several fundamental methods available to study the microbiome, primarily driven by massively parallel high-throughput technologies. These include (i) marker gene analysis, for profiling microbiota community structure and composition; (ii) shotgun metagenomics, for assembling microbial genomes from environmental samples and determining the functional potential of microbiota; (iii) metatranscriptomics, for investigating the functional activity of the microbiota; (iv) metabolomics, for profiling small-molecule metabolites of the microbes, hosts, and environment; and (v) metaproteomics, for determining the proteins and enzymes involved in metabolic pathways

DOI: 10.1201/9781003354062-9

to describe the functional activities of microbes [9]. These studies have paved the way for the field of metagenomics, which involves the study of all genetic material directly obtained from environmental or living samples [10]. The term 'metagenomics' refers to the direct genetic analysis of genomes present within an environmental sample without the need for culturing the microorganisms [11].

In any metagenomics project, the initial and pivotal stage is sample processing. It is imperative that the extracted DNA accurately represent all the cells in the sample, and it is essential to procure ample quantities of high-quality nucleic acids to facilitate the creation and sequencing of libraries in the subsequent steps. The process necessitates distinct protocols tailored to each sample type, with a range of sturdy DNA extraction techniques at one's disposal [12–14]. Consequently, bioinformatics software tools play a vital role in analysing these DNA samples, though this aspect of microbiome research can be challenging for researchers. With established workflows, targeted metagenomic data from environmental samples can be efficiently retrieved and comprehensively analysed using user-friendly tools. The resultant assembly of metagenomic genomes further facilitates the characterisation of microbial clades that were previously unidentified. Various such pipelines have been developed, among which are three popular microbial community analysis pipelines, viz., quantitative insights into microbial ecology (QIIME) [15], metagenomics rapid annotation using subsystem technology (MG-RAST) [16,17], and mothur [18]. These pipelines offer comprehensive suites of tools for analysing metagenomic data and are discussed in this chapter.

NGS AND METAGENOMICS

The term 'metagenomics' was first used in 1998 [19] and refers to an approach that uses genome sequencing or tests of functional features to analyse complex and diverse ('meta') populations of bacteria without the use of a culture. A metagenomics study aims to directly characterise the composition and function of microbial communities in various habitats, including but not limited to soil, water, plants, fermented food, the human gut, etc. [13,20–22]. In contrast to standard microbial genomic sequencing initiatives, researchers can investigate their complicated relationships and biochemical activities (Figure 9.1). The rapid development of NGS methodologies [23] has augmented the current metagenomics effort by offering lower-cost experimental instruments free from the time-consuming and labour-intensive cloning procedure inherent in traditional capillary-based approaches. Researchers can sequence thousands of species concurrently, thanks to NGS. NGS-based metagenomic sequencing can detect members of the microbial community that are of very low abundance and may be missed or are too expensive to identify using other approaches. This is because the approach can combine several samples in a single sequencing run and produce high sequence coverage per sample (https://sapac.illumina.com/areas-of-interest/microbiology/microbial-sequencing-methods/shotgun-metagenomic-sequencing.html, last accessed on 19 November 2024). Metagenomics makes it possible to examine every microbial organism, including the vast majority (more than 99%) that cannot be isolated or are difficult to culture with the available media [24]. Microorganisms naturally reside in communities where they exchange nutrients, metabolites, and signalling molecules. Although the classical pure-culture paradigm is still crucial for a thorough characterisation of a species, its historically exclusive use restricts research into the microbiome. Traditional clonal culture microbiology requires the addition of culture-independent microbiology that can directly characterise microbes in natural environments and answer key biological questions about those environments, such as the diversity of microbes in various environments [25], microbial interactions (including microbe–host interactions), and environmental and evolutionary processes [26]. Applications for metagenomic projects range widely, from ecology and environmental sciences [27] to the chemical industry [28] and human health (such as the metagenomics of the human gut microbiome) [29,30].

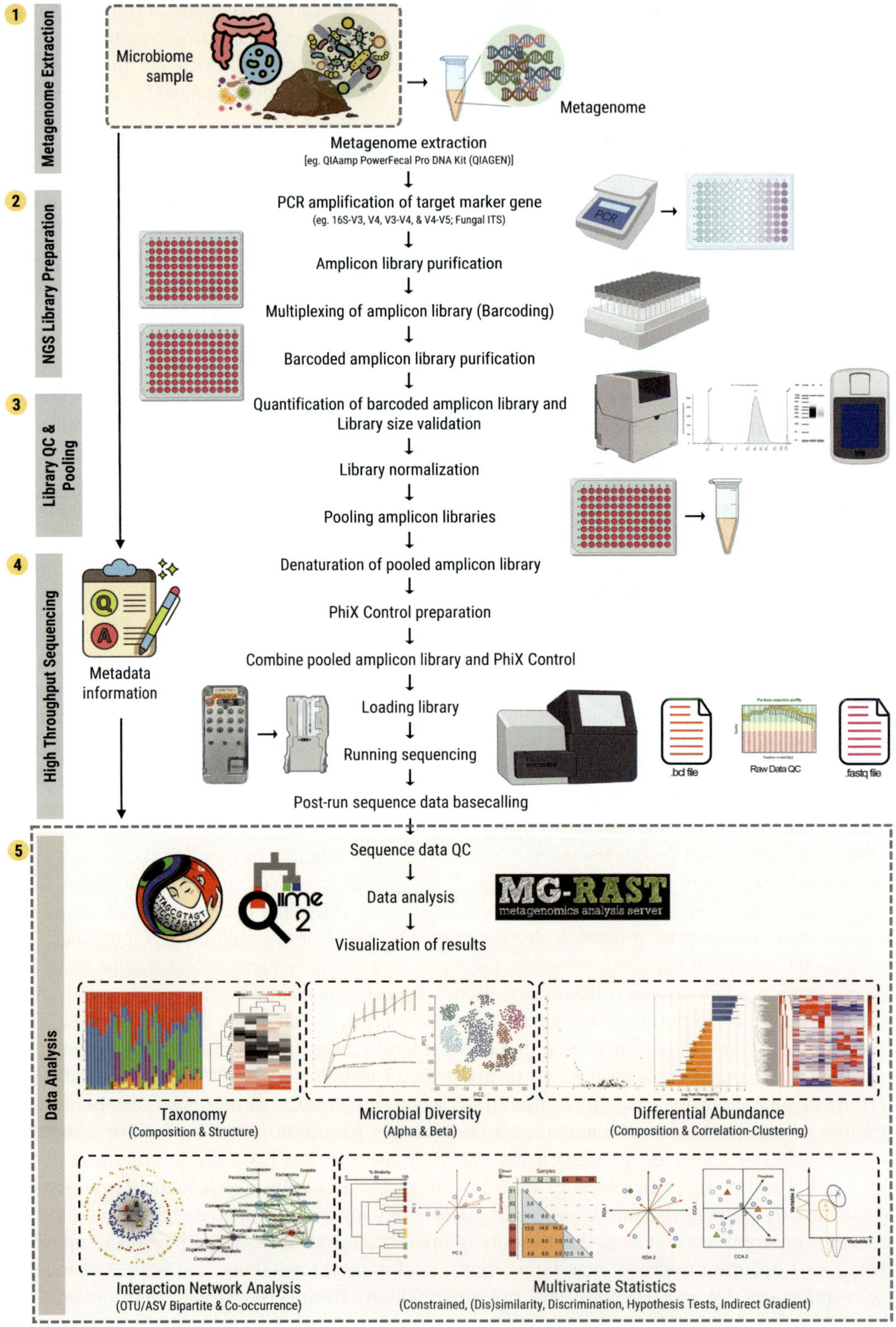
1
Metagenome Extraction
Microbiome sample
Metagenome
Metagenome extraction
[eg. QIAamp PowerFecal Pro DNA Kit (QIAGEN)]
2
NGS Library Preparation
PCR amplification of target marker gene
(eg. 16S-V3, V4, V3-V4, & V4-V5; Fungal ITS)
PCR
Amplicon library purification
Multiplexing of amplicon library (Barcoding)
Barcoded amplicon library purification
3
Library QC & Pooling
Quantification of barcoded amplicon library and Library size validation
Library normalization
Pooling amplicon libraries
4
High Throughput Sequencing
Denaturation of pooled amplicon library
PhiX Control preparation
Combine pooled amplicon library and PhiX Control
Metadata information
Loading library
Running sequencing
Post-run sequence data basecalling
.bcl file
Raw Data QC
.fastq file
5
Data Analysis
Sequence data QC
Data analysis
Visualization of results
MG-RAST
metagenomics analysis server
Taxonomy
(Composition & Structure)
Microbial Diversity
(Alpha & Beta)
Differential Abundance
(Composition & Correlation-Clustering)
Interaction Network Analysis
(OTU/ASV Bipartite & Co-occurrence)
Multivariate Statistics
(Constrained, (Dis)similarity, Discrimination, Hypothesis Tests, Indirect Gradient)

FIGURE 9.1 Schematic summary of the study design and experimental protocols involved in a typical amplicon metagenomics workflow for microbiome studies. The workflow basically involves five steps: (1) metagenome extraction; (2) amplicon library preparation; (3) library QC, normalisation and pooling; (4) sequencing and generation of raw data; and (5) sequence data analysis. This workflow is based on the Illumina MiSeq platform. The differences while using Nanopore platforms will be in the library preparation, sequencing, and base calling of raw data into the file '`fastq.gz`'.

(Images of a sample, microcentrifuge tubes, thermal cycler, 96-well plates, barcoding plates, and equipment (TapeStation, Qubit, MiSeq flow cell, and MiSeq platform) were created with BioRender.com. Images of icons were obtained from Flaticon.com. The images describing the multivariate statistics were reproduced from GUSTA ME web-based resource website https://sites.google.com/site/mb3gustame/home/visualisations?authuser=0 [last accessed on 19 November 2024] with permission from the creator and owner Buttigieg PL [31].)

BIOINFORMATICS TOOLS FOR ANALYSIS OF METAGENOMICS DATA

Currently, a variety of techniques are used to infer varying amounts of microbiome information. These techniques include whole-genome shotgun (WGS; metagenome), whole-transcriptome shotgun (metatranscriptome), and 16S ribosomal RNA (rRNA) analysis. The conservation of the 16S rRNA gene allows for the identification of microorganisms through analysis of 16S rRNA. In order to evaluate microbial identities down to the level of species or strain, the WGS analysis requires data from all genes. The whole-transcriptome shotgun sequencing (WTSS) analysis enables the investigation of microbial community functionality and gene expression patterns. An extensive list of the chemicals present in the study environment is provided by the whole-metabolite analysis, which also enables the correlation between the abundance of bacteria and the downstream chemicals. In several population-based microbiome studies, with an emphasis on the nasal, oral, cutaneous, gastrointestinal, and urogenital regions, microbial communities that inhabit the human body of healthy individuals are studied. One such study is the Human Microbiome Project [32]. The goal of the Interactive Human Microbiome Project is to better understand how a microbiome interacts with its human host through long-term research that collects numerous omics datasets from both microbiome and human [33]. Furthermore, Metagenomics of the Human Intestinal Tract (MetaHIT) concentrates on comprehending the connection between human health/disease and intestinal microbiota [34]. MetaHIT also investigates inflammatory bowel disease (IBD) and obesity. The Earth Microbiome Project (EMP) aims to characterise the variety, distribution, and structure of microbial ecosystems on Earth and has already collected more than 30,000 samples from various ecosystems, including people, animals, and plants from terrestrial, marine, and built environments. In one of the first microbiome investigations, EMP established certain guidelines for other studies. Metagenomics and metatranscriptomics have been studied more frequently in recent research due to the rising limits in comprehending an individual microbe's mechanisms on a global scale and the challenges involved with cultivating individual microbial species [19,35–37].

The rapid advances in the ease of metagenome data acquisition using massively parallel high-throughput NGS technology have led to a dramatic increase in research groups trying to analyse big sequence data. However, the vast amount of data generated by NGS technologies, coupled with the particularly flexible nature of the metagenomics approach (in terms of choices spanning sample type, metagenome extraction principles, targeted vs. shotgun sequencing approach), presents significant challenges for data analysis, interpretation, and reporting of the results. This is due to the complexity of the data. Therefore, metagenomics analysis pipelines have been developed to facilitate the processing and interpretation of the data. Significant improvement in bioinformatics tools and computational pipelines, their availability in the public free-to-use domain, and the flexibility in combining individual data with what is already known from other studies, have offered considerable potential to analyse such big data at spatial, temporal, and global scales.

Targeted amplicon metagenomic sequencing (AMS) of rRNA operon regions is one of the most popular methods, and a relatively inexpensive one (low-cost per-sample), for microbiome investigations. Examples include (i) the hypervariable regions (e.g., V3, V4, V3–V4, V4–V5, etc.) of the 16S rRNA gene in bacteria and (ii) the internal transcribed spacer (ITS) region for fungal and yeast communities. AMS data analysis involves several steps, including quality control of the sequence reads, extraction of taxonomic features, taxonomic classification, functional annotation, statistical analysis, and data visualisation. Here we describe two bioinformatics software applications, viz., QIIME 2 and mothur, for metagenomic data analysis. QIIME 2 is a powerful, adaptable, and decentralised microbiome analysis tool with an emphasis on data and analysis openness. Researchers can now begin an investigation with raw DNA sequence data and end it with publication-quality graphics and images and statistical findings. Some of the important features include an integrated and autonomous data provenance tracking system, a system of semantics, a plug-in system to increase the functionality of microbiome analysis, and support for multiple-user interfaces [15]. mothur is developed by Patrick Schloss at the University of Michigan for the study of microbial ecology data [18]. It was initially launched in 2009 and has since gained widespread acceptance in the field of metagenomics. mothur intends to provide a comprehensive set of tools for processing and analysing microbial community sequencing data. It focuses on raw sequence processing, creating taxonomic and diversity profiles, and enabling sophisticated statistical analyses of microbial communities [38].

USING QUANTITATIVE INSIGHTS INTO MICROBIAL ECOLOGY 2 (QIIME 2) FOR ANALYSING 16S-AMS DATA

QIIME 2 is a freely available, open-sourced, powerful, adaptable, and decentralised microbiome bioinformatics analysis tool developed with an emphasis on data openness and reproducibility that expedites comprehensive analysis of microbiome data. Researchers can now begin an investigation with raw metagenome sequence data (`fastq.gz` format) and end it with publication-quality graphics and images and statistical findings. Some of the important features include an integrated and autonomous data provenance tracking system, a system of semantics, a plug-in system to increase the functionality of microbiome analysis, and support for multiple-user interfaces [15]. There are various potential approaches for conducting QIIME 2 analyses, influenced by factors such as experimental objectives, data analysis goals, and data collection methodologies. At present, QIIME 2 exists in three usable interfaces: the command-line interface (q2cli), the graphical user interface (q2galaxy), and the Python 3 application programming interface (Artifact API). Throughout this tutorial, we will engage with the QIIME 2 command-line interface (q2cli) to specifically process and assess a subset of samples. QIIME 2 q2cli can be installed and used on a personal computer (server, desktop computer, or laptop), a cluster computer (an institute's high-performance computing or a research computing office), or cloud computing resources where one uses it on an hourly rental basis while the task of maintaining the hardware is done by the resource provider (e.g., Amazon Web Services [AWS] Elastic Compute Cloud [EC2]). While we begin with raw sequence files (`fastq.gz`) and adopt a singular analysis pipeline for lucidity, we acknowledge instances where alternative techniques are viable and elucidate the rationale for their potential utilisation.

This protocol provides step-by-step guidelines on how to install the QIIME 2 q2cli interface on a personal computer and use it for importing, processing, quality control, and clustering of 16S rRNA gene raw sequence reads (`fastq.gz`) and analysing the processed sequence reads for taxonomic classification, microbial community diversity, community composition, microbiome differences, and data visualisation for producing publication-ready figures and tables.

BEFORE STARTING: UNDERSTANDING THE CORE CONCEPTS AND FEATURES OF QIIME 2

QIIME 2 is developed upon and driven by certain specific concepts. These so-called 'core concepts' must be understood first by the users for a clear understanding of the analysis workflow and

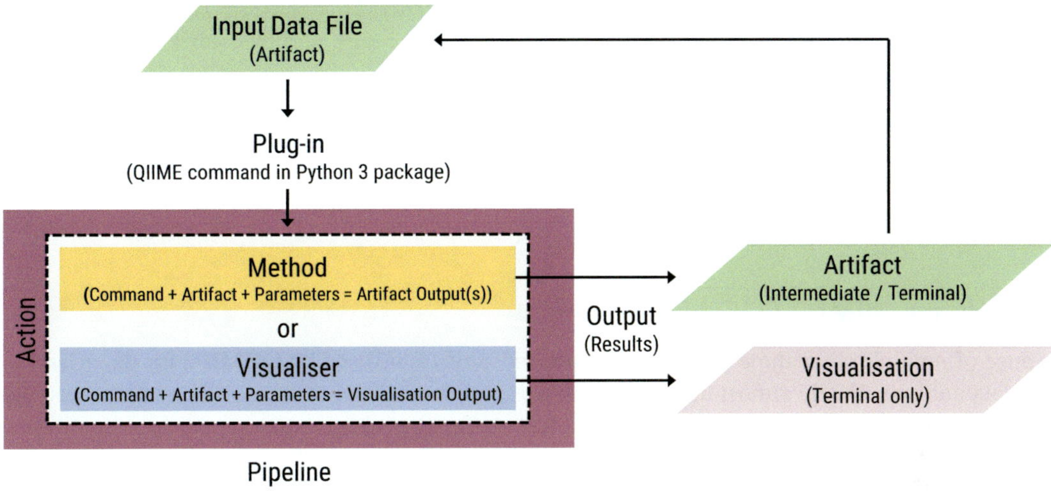

FIGURE 9.2 Schematic representation of the overview of core concepts of QIIME 2.

protocol (Figure 9.2). The core concepts include four components, viz., (i) QIIME 2-readable 'data' for analysis (referred to as either 'artifact' or 'visualisation'), (ii) identifier of type of data (referred to as 'Semantic Type'), (iii) QIIME command (referred to as 'plug-in', which is a Python 3 package that provides microbiome data analysis functionality in the form of 'action') for performing a specific bioinformatics task, and (iv) 'action' that combines one or more artifacts and multiple parameters through plug-ins and takes the artifact–parameter combination as input to generate outputs of the analysis as artifact(s) or visualisation. The action is of three types: method, visualiser, and pipeline. The 'method' takes one or more artifacts as input(s) and generates one or more artifacts as outputs, while the 'visualiser' inputs some combination of one or more artifacts and produces only one visualisation output. A 'pipeline' is a combination of two or more other actions to complete a particular aspect of the microbiome data analysis. QIIME 2 data are a compressed file that contains data and metadata about the data, enabling its decentralised provenance tracking and logging through the system while ensuring reproducible bioinformatics analysis. A QIIME 2 data artifact (file extension '`.qza`', which stands for QIIME zipped artifact) is used as input to a method or visualiser and is also generated as output of a method. Methods can generate intermediate outputs, which may subsequently serve as inputs to other methods or terminal outputs. In order to use data in QIIME 2, that data must be imported into the system as an artifact, and this can be done during any steps of the analysis, though QIIME 2 analysis typically starts by importing the raw data as an artifact. The data from an artifact can also be exported for further downstream analyses in different software/tools. QIIME 2 data visualisation (file extension '`.qzv`', which stands for QIIME zipped visualisation) is only produced as a terminal output of a visualiser or pipeline in an analysis and cannot be used as input for another method or visualiser in QIIME 2 or other software. The artifact and visualisation files are opened and viewed using the online QIIME 2 View (q2view) (https://view.qiime2.org/, last accessed on 19 November 2024) or through the q2cli interface using '`$qiime tools view`' command. We suggest readers and users of QIIME 2 access its website (https://qiime2.org/, last accessed on 19 November 2024) to find updates to the latest version of the microbiome bioinformatics platform (latest version was 2024.10 at the time of the writing of this chapter) (https://docs.qiime2.org/2024.10/, last accessed on 19 November 2024), including new plug-ins and pipelines and obsolete and improved functionalities in plug-ins (https://library.qiime2.org/, last accessed on 19 November 2024), and to learn details about QIIME 2 core concepts (https://docs.qiime2.org/2024.10/concepts/, last accessed on 19 November 2024). While we provide a generalised standard operating protocol for QIIME 2, we recommend users visit the 'QIIME 2 Forum' (https://forum.qiime2.org, last accessed on 19 November 2024). This platform serves as a hub for both users and developers, offering support for: (i) technical issues and troubleshooting, (ii) user queries and discussions, (iii) community plug-in development, (iv) sharing of tutorials and other resources, and (v) general ideas and suggestions for platform improvement.

Necessary Resources

Refer to the Supplementary Materials for this section.

Installing the QIIME 2 Q2CLI Environment

A range of multiple standalone packages and dependencies is utilised by QIIME 2 for the microbiome data analysis. While almost the same packages and dependencies are used in various versions of QIIME 2, certain changes or upgrades in these external packages may create incompatibility issues in any point of the analysis among different QIIME 2 versions. Therefore, as recommended by QIIME developers, we encourage users to install and consistently opt for the latest QIIME 2 version for an analysis. It is recommended to install QIIME 2 q2cli natively through Conda, as a Conda environment (https://docs.qiime2.org/2024.10/install/native/, last accessed on 19 November 2024). However, using a Docker container (https://www.docker.com/, last accessed on 19 November 2024) or the Windows Subsystem for Linux (WSL) (https://learn.microsoft.com/en-us/windows/wsl/install, last accessed on 19 November 2024), are good alternatives.

(Re)activating and Deactivating the QIIME 2 Environment

Refer to the Supplementary Materials for this section.

Tutorial Dataset

We validated this protocol by successfully reproducing analyses of a small 16S amplicon metagenomic sequencing dataset from our laboratory. The dataset consists of ten soil microbiome samples, consisting of five rhizospheric soil samples and five bulk soil samples. The raw sequence files (`fastq.gz`) are obtained by sequencing the V4 region of the 16S rRNA gene amplified using primer pair 515F-806R of the Earth Microbiome Project (https://earthmicrobiome.org/protocols-and-standards/16s/, last accessed on 19 November 2024) and sequencing with the Illumina MiSeq platform and paired-end 2 × 250 reads using our lab-developed barcoded-fusion-primer approach.

Refer to the Supplementary Materials for the files (raw sequence data, sample metadata, and manifest files) and folders used in the protocols.

Data Analysis Steps in QIIME 2

The core analysis workflow of QIIME 2 is shown in Figure 9.3. Based on this workflow, the protocol undertakes the following data analysis steps:

- Step 1: Make and validate sample metadata file.
- Step 2: Importing raw sequence (FASTQ) data into QIIME-readable format (i.e., as artifact, '`.qza`' file).
- Step 3: Demultiplexing sequence data (i.e., mapping the sequence to the sample it came from) (optional).
- Step 4: Quality filtering, denoising, joining reads, chimaera removal, dereplicating similar sequences into sequence variants called features (known as amplicon sequence variants (ASVs).

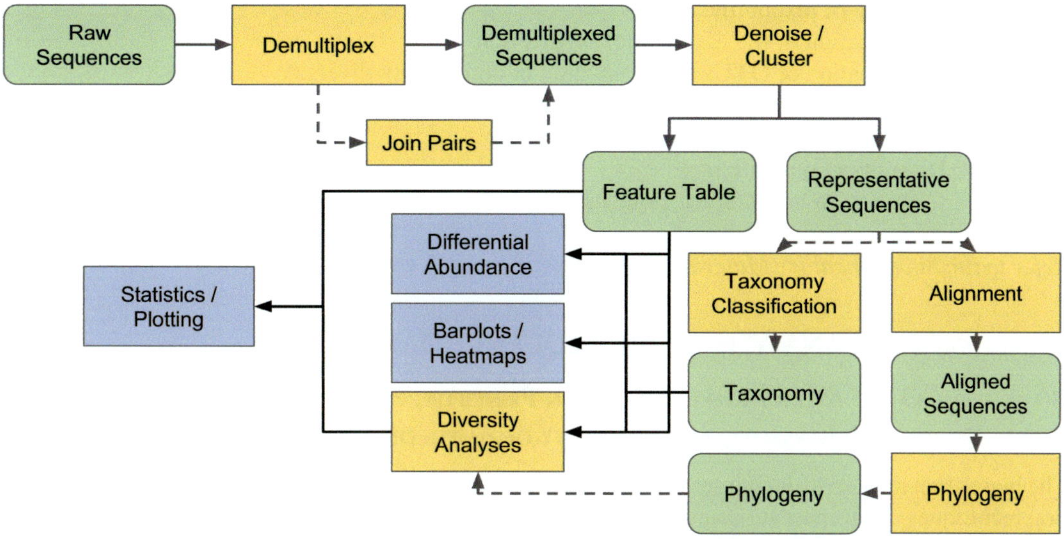

FIGURE 9.3 Schematic representation of the conceptual overview of analysis workflow of QIIME 2. The colour legends correspond to the colours of the core concepts of QIIME given in Figure 9.2.

(This figure is reproduced from https://docs.qiime2.org/2024.10/tutorials/overview/ [last accessed on 19 November 2024] with permission from the QIIME 2 developers.)

Step 5: Operational taxonomic units (OTU) clustering: clustering of features/sequence variants (ASVs) to OTUs at a user-specified percent identity threshold of their sequences.
Step 6: Taxonomic classification and assigning taxonomy.
Step 7: Filtering feature table and representative sequences.
Step 8: Merging all features with the same taxonomic assignment at the specified taxonomic level into a single feature.
Step 9: Visualising taxonomic composition of the samples with interactive plots (barplot, heatmap).
Step 10: Analysis of microbiome composition (differential abundance analysis).
Step 11: Sequence alignment and phylogenetic tree building.
Step 12: Alpha and beta diversity analyses.
Step 13: Statistical test of microbiome differences between samples.

Refer to the Supplementary Material for the details on each step.

USING mothur FOR ANALYSING 16S-AMS DATA

[Necessary Prerequisites
Refer to the Supplementary Materials for this section.
Data Analysis Steps in mothur]

Step 1: Generation of contigs from input files.
Step 2: User-defined sequence filtering.
Step 3: Creating a FASTA file with unique sequences and creating a table with the corresponding counts for each of these representative sequences.
Step 4: Generation of sequence alignment.
Step 5: Reduce sequence redundancy.
Step 6: Detection and removal of chimera.
Step 7: Sequence classification.
Step 8: Remove sequence based on taxon.

Step 9: Summary of taxonomy.
Step 10: Visualisation of the taxonomic composition.
Step 11: Identification of OTUs and evaluation of community diversity.
Step 12: Generate a file for input into visualisation applications.
Step 13: Estimation of OTU richness.
Step 14: Diversity computation.
Step 15: Comparison between the samples.

Refer to the Supplementary Material for the details on each step.

CHALLENGES IN ANALYSIS OF BIG SEQUENCE DATA

Biases from Varied Metagenomic Extraction Principles, PCR, Varied Target Regions, Universality, and Specificity of Amplicon Primers

The biases that may arise during sequencing due to the selection of metagenomic (DNA/RNA) isolation techniques can impact subsequent analysis. It is crucial to note that the extraction process must proficiently encompass all varieties of microbes. For instance, extracting DNA from Gram-positive bacteria proves challenging due to their dense peptidoglycan cell walls [39]. Two primary extraction techniques exist: (i) mechanical lysis / bead beating and (ii) chemical lysis [40]. When executed optimally, bead-beating approaches are recognised for yielding superior results. Therefore, when dealing with intricate bacterial samples, it might be beneficial to incorporate a bead-beating stage before the regular nucleic acid extraction process. Nonetheless, caution should be exercised against excessive bead beating, as it has the potential to fragment nucleic acids and impact subsequent library preparation stages [41]. The variability in polymerase chain reaction (PCR) amplification efficiency between different sequences depends on factors like sequence composition and secondary structure. Uneven amplification of sequences in PCR can occur due to a significant presence of G or C, leading to a reduction in amplification efficiency [42–44].

Short-Read vs Long-Read Platforms

While short-read NGS methods are widely used, they have limitations in achieving complete genome assemblies. Third-generation sequencing platforms, such as Pacific Biosciences RS II/ Sequel and Oxford Nanopore MinION, offer significant advantages through their extended read lengths, enabling complete microbial genome assemblies and providing benefits for various genomic applications [45,46]. Their ability to generate longer reads, combined with the absence of PCR amplification biases, improves assembly quality and resolving power.

Data Volume and Storage

Researchers are grappling with the considerable volume of data being produced through various NGS platforms. For instance, a single 30X human whole-genome sample yields a BAM file (a semi-compressed alignment file) of approximately 90 GB. A modest project involving 100 samples could therefore amass 9 TB of BAM files. Considering the capacity of a single Illumina HiSeq X instrument to generate over 130 TB of data annually, storage concerns arise rapidly. The Broad Institute exemplifies this, producing sequencing data at a rate of one 30X genome every 12 minutes—equivalent to almost 4,000 TB of BAM files each year. Although BAM files can be transformed into variant call format (VCF) files, which document only the differing bases from the reference sequence, retaining the raw sequence files remains imperative if future data reprocessing is intended. With the widespread decrease in sequencing costs, resequencing readily available samples can be a practical alternative. In terms of data analysis, researchers have an abundance of options.

Knowledge of Linux Systems and Various Interfaces

The MetAMOS (modular and open source metagenomic assembly and analysis) pipeline can integrate numerous tools for a thorough examination of metagenomic datasets, encompassing raw sequencing reads, contigs, and scaffold data. This is similar to how QIIME 2 commands can be integrated into a complete pipeline, while processes can potentially be automated. However, MetAMOS's lack of a user-friendly interface and reliance on Linux command-line operations adds complexity to utilising its wide array of tools. Installation is handled through a Python script, '`INSTALL.py`', which streamlines the process by fetching and executing the latest version [47].

Requirement of Compute-Intensive Systems

The analysis of metagenomic data requires substantial computational resources due to its high data volume and computational demands. The majority of existing metagenomic data analysis software was originally intended for deployment on individual computers or small clusters, however these setups have become inadequate for the demands of an ever-growing number of metagenomics projects. To mitigate this, it is important to devise sophisticated computational techniques and workflows for efficient analyses [48].

BEST PRACTICES FOR METAGENOMIC ANALYSIS PIPELINES AND REPORTING MICROBIOME

Study Design and Guidelines for Sample Collection

When designing a metagenomic study, the context, supporting data, or theory that guided the design must be explained in detail by the researcher. The hypothesis must be supported by preliminary data, results from related studies or topics, or a postulated biologically plausible mechanism. When doing an exploratory study, certain goals must be given. The study should define the population of interest as well as the environment, dietary habits, lifestyle choices, biological interventions, demographics, and geography. This is due to the possibility that the aforementioned parameters correspond to variations in the microbiome. The predetermined qualities used to choose study participants are called inclusion criteria and exclusion criteria. The exclusion criteria should take into account any information on recent antibiotic use. The beginning and ending dates for data collection, follow-up, and recruitment should be specified. Details of how follow-ups were completed should be stated if individuals were lost to follow-up or unable to complete all assessments in the longitudinal study. Time-point-specific sample sizes should also be reported [49]. The aforementioned study protocol requires ethics committee approval, with specific requirements varying by jurisdiction. In most countries, all participants (patients and healthy volunteers) must provide written informed consent in accordance with local regulations and institutional protocols. This consent usually includes specific authorisation for the use of samples and personal data for research purposes. To gather participant information systematically, a standardised data collection tool based on the study objectives must be created. This document, known as a case report form (CRF), is the internationally recognised format for collecting participant data in clinical research. A special code created by the researcher de-identifies the CRF and the labels. Two methods of de-identification are available: full anonymisation, in which the sample and data from the participant cannot be identified because the key to the code has been irrevocably destroyed; and pseudonymisation, in which the participant is concealed by a code but is identifiable to anyone with access to the code's key [50]. The details of the standard procedure for gathering and processing faecal and environmental samples for metagenomic analysis are provided on the websites of the International Human Microbiome Standard (IHMS) [51] and Earth Microbiome Project (https://earthmicrobiome.org/protocols-and-standards, last accessed on 19 November 2024).

Choice of Metagenome Extraction Principles

Metagenomic studies have multiplied dramatically in recent years. The challenge in obtaining accurate microbiome community profiles is affected by a wide range of factors, including community complexity and unpredictability. It is now extremely difficult to compare the findings of various studies due to the numerous solutions that have been presented to address these issues. A variation in the data gained reflects the major variations in the methods utilised in metagenomic research. This is particularly prominent in the case of metagenome extraction principles. Commonly used extraction methods based on enzymatic, mechanical, and chemical lysis principles generate differences in microbial DNA yield and differential recovery of OTUs from the same sample type [20]. Main factors that dictate the choice of extraction principles are sample type, composition of the sample matrices, ecological niche from where the sample originates, and target taxa of the microbiota. This highlights the necessity of standardising the process to remove confounding variables resulting from DNA isolation, sequencing, and bioinformatics analysis and to confirm that the variations in microbiome composition are indeed due to biological origins. With these confounding factors, one can follow sequential metagenome extraction by combining different extraction principles [20]. The IHMS project has generated multiple publications describing best practices for metagenomics studies, however a standardised process for producing and analysing metagenomic data is still a long way off. As an example, a study by Szóstak et al. demonstrated that the homogenisation duration is the primary variable influencing sample variety and suggested a shorter homogenisation period (ten minutes). The Gram-positive/Gram-negative ratio of bacteria can be more accurately reflected after ten minutes of homogenisation, and the results are the least heterogeneous in terms of the beta diversity of the samples' microbial makeup [52].

Importance of Mock Community Control

It is necessary to employ a mock community with known bacterial species and their corresponding abundances as a baseline in order to determine the accuracy of the metagenome extraction, sequencing library preparation, and sequencing procedures [53]. While this gives us a baseline to work against, cultivating, mixing, and accurately estimating the abundance of such a community are challenging tasks. The expected abundance patterns have hitherto been difficult to reconstruct using metagenomic or 16S rRNA gene amplicon sequencing [54–56]. Gram-positive and Gram-negative bacteria recovery can be the focus of mock communities, which can serve as a significant source of diversity between extraction techniques [57–59]. The imitation community must be made up of bacteria that are often missing from the gut of a healthy host. Before adding the actual sample, the number of bacteria must first be precisely measured using optical density, cell counting, and fluorescence-activated cell sorting. It is possible to assess extraction biases in the context of interindividual microbiome variation using the mock spike-in as a baseline.

Controlling Contamination and Bias

At every stage of sample collection, nucleic acid extraction, or library preparation, the environment, reagents, handlers, or equipment can introduce contamination [60–65]. This may give results that differ greatly between laboratories, reagent kits, or extraction batches [61,66,67]; cause assessments to be falsely positive or negative [68–71]; or give erroneous information about microbiological habitats [72–74]. In order to avoid inaccurate results, particularly from low biomass samples, contamination must be addressed following basic microbiology and molecular biology lab routines throughout the microbiome study workflow, starting from sample collection through sequencing. While actions can be taken to minimise contamination, current best practices cannot totally eliminate it or control for it [58,59,63,66,68,69,74–78].

Selection of Common Marker Gene and Target

Certain unique marker/target genes are widely used for amplicon sequencing for bacteria, archaea, fungus, and mycobacteria. Because the majority of marker genes retain their functional properties across phylogenetic boundaries, they can also be used as a molecular clock to track evolutionary transitions and changes. The gold standard in microbial typing and the most often utilised target gene for bacterial identification is the 16S rRNA gene (or 16S rDNA) [79,80]. In the majority of bacteria and archaea, the 16S rRNA gene encodes the prokaryotic small 30S subunit of the 70S ribosomal complex. Interestingly, the tiny eukaryotic ribosomal subunit-encoding 18S rRNA gene is separate from the bacterial 16S rRNA gene (40S). The highly conserved 16S rRNA gene suggests that it is essential for cellular survival and function, and as such, it serves as the foundation for determining the precise genomic categorisation of both known and undiscovered microbial taxa. Additionally, the 16S rRNA gene's comparatively small size (1542 bp) makes it simpler to sequence, even for extremely large sample volumes. The gene sequence consists of nine variable regions (V1–V9) and highly conserved primer binding sites. The majority of 16S rRNA-based genotyping procedures [81,82] identify and categorise microbial profiles using V5–V6, V3–V4, or V4 hypervariable regions. Alternatively, the V3 region is a superior option for PCR-denaturing gradient gel electrophoresis community profiling of archaea. Archaeal species in complex microbial communities have been genotyped using other variable areas, such as V1–V2 and V3–V4 [83]. Unlike bacteria, it is still difficult to identify the gene targets in pathologically significant yeast and fungus. Coding and noncoding spacer regions make up the fungal rDNA [84,85]. Along with many noncoding sections made up primarily of internal transcribed spacers (ITSs) and intergenic sequences, the coding region is made up of 18S, 5.8S, and 28S units. For fungi, ITS variable sections have been the most popular gene targets. Although ITS regions are commonly targeted for selective amplification and sequencing, their unequal lengths can introduce errors and biases, often leading to inaccurate abundance estimates [84].

Selecting Appropriate Sequencing Platforms and Protocols

When choosing the best sequencing platform for a metagenomics project for a microbiome study, there are many considerations [86]:

- Utilisation techniques for NGS: This covers RNA sequencing, whole genome sequencing, targeted sequencing, and whole exome sequencing.
- Depth of coverage for sequencing: This provides details on the typical number of reads in a sequenced sample that is needed to cover each base. The confidence in the sequenced bases increases as coverage increases. The Lander/Waterman equation, $C = LN/G$, is frequently used to compute coverage. It takes into account C (coverage), L (read length), and G (haploid genome length).
- Sequencing read length: There are two kinds of read lengths, those used for short-read sequencing and those for long-read sequencing. The most common technology for high-throughput sequencing is short-read sequencing. It is inexpensive and offers data that are highly accurate. Thousands of base pairs between 10 and 100 kbp can be produced using long-read sequencing, and the nucleic acids remain in their natural condition without the need for PCR. In contrast with short-read sequencing, long-read sequencing can detect base modifications, including methylation, deletion, duplication, insertions, inversions, translocation, and the detection of particular RNA transcript isoforms.
- Sequencing with single or paired ends: The choice of using either single or paired ends mostly depends on the size of the sequencing library and the sequencing kit used. However, paired-end read sequencing is often preferred, as it offers read alignments with a high degree of confidence, helps with relative read position detection, and identifies gene insertions, deletions, repetitive sequences, etc.

Various companies offer sequencing platforms; for instance, Illumina plays a significant role in the market for short-length read sequences with its MiSeq, NextSeq and NovaSeq products, among others. Thermo Fisher Scientific's Ion Torrent and BGI's DNBSEQ (formerly MGISEQ) platform are two other significant players. Single-molecule real-time (SMRT) sequencing from Pacific Biosciences (PacBio) and Nanopore sequencing from Oxford Nanopore Technologies (ONT) are leading technologies for lengthy read sequences. Your research programme should take all relevant criteria into account when choosing the appropriate NGS method and technology. Integrating NGS into preclinical research provides opportunities to optimise early-stage drug development, improving the likelihood of advancing investigational medications successfully into clinical trials.

Validating Results

Metagenome assemblies are often incomplete and error-prone due to the inherent complexity of assembling the data, regardless of the assembly strategy or sequencing technique used. In order to inform subsequent studies of the assembled data and to enable researchers to compare various assembly technologies, methods for assessing the quality and completeness of assemblies are essential. The two main types of assembly validation techniques are *de novo* and reference based. An assembly is validated using reference-based approaches that compare it to a database of previously assembled genes or genomes [87,88]. Any discrepancies found between the compiled data and the reference collection are evaluated as errors. Reference-based approaches have limited impact on real datasets, but they are particularly useful in benchmarking studies that aim to reproduce communities with known composition. For instance, a reference-based technique cannot be used to verify metagenomic segments coming from a genome for which there is no reference sequence available. Furthermore, it might be challenging to distinguish between assembled contig changes that are caused by mistakes and actual variations between the reference sequence and its relative in the metagenomic mixture.

FUTURE PERSPECTIVES OF MICROBIOME SCIENCE AND DATA ANALYSIS

The field of microbiome science has witnessed remarkable advancements in recent years, propelled by the confluence of cutting-edge technologies and innovative analytical approaches. The convergence of these developments will shape the landscape of metagenomic analysis pipelines and enhance the accuracy, scope, and depth of microbiome studies, enabling a deeper understanding of microbial communities and their impact on health and the environment. The following points outline a way forward in harnessing these advancements for comprehensive microbiome studies.

Integration of Artificial Intelligence and Machine Learning

As the wealth of microbiome data continues to expand, it becomes imperative to harness the power of advanced big data analytics to uncover meaningful insights. Artificial intelligence (AI) and machine learning (ML) have the potential to revolutionise how we analyse and interpret microbiome data. The abundance of publicly available metagenomic datasets presents a unique opportunity for metanalysis, where multiple studies are combined to derive more robust conclusions. The deployment of AI/ML techniques for standardisation of data processing, metadata annotation, and statistical methods can facilitate integration of microbiome data from multiple sources, thus enabling identification of patterns, relationships, and trends within large-scale and complex microbial metadatasets. Tools such as Large Language Models (LLMs) could serve as a bridge between researchers and advanced data analysis, enabling more accessible and insightful exploration and discovery of previously unnoticed associations within the microbiome data.

Harnessing the Power of New Sequencing Technology

The emergence of new sequencing technologies, such as long-read sequencing and single-cell sequencing, offers unparalleled resolution in characterising microbial genomes and their functions. These technologies can enable a more comprehensive understanding of complex microbial communities, including the identification of rare species and intrapopulation structure, the elucidation of functional pathways, and dissecting genome and functional heterogeneity within the same species. Integrating these technologies into metagenomic analysis pipelines will enhance accuracy and provide a finer-grained perspective on microbiome composition and dynamics.

Integration of Pan-Genome Analysis in Metagenome (Meta-pangenome)

The concept of the pan-genome (encompassing all genes present across a species) extends to the microbial world, offering insights into the genetic diversity and functional potential of microbial communities. Integrating pan-genome analysis into metagenomic pipelines can allow the identification of shared and unique genes within microbial populations. Meta-pangenome analysis involves the comparison of the totality of genes of a species and their distribution and abundance across multiple metagenomic samples in the environment. These approaches shed light on the adaptive strategies, metabolic pathways, and potential interactions that shape microbiome dynamics across various environments. Characterising the core metagenome (the set of genes consistently present across samples) provides insight into the fundamental functions that underpin the stability and functionality of a microbial ecosystem. It helps identify key players responsible for core metabolic processes, allowing researchers to target interventions that maintain ecosystem health and resilience. However, the intricate challenge of characterising genes within a community persists due to the constrained coverage of metagenomes and genomes in environmental communities, as well as the inherent complexity of assembly algorithms.

Scaling Up Bioinformatics Capacity

The growth of microbiome data presents both opportunities and challenges. Computationally, working with metagenomic data is a resource intensive task. While the volume of data generated is vast and more complex, the scalability of bioinformatics pipelines and tools must be prioritised. This entails developing efficient and versatile analysis pipelines capable of handling diverse datasets, extracting meaningful information, and accommodating the nuances of different experimental designs and platforms. Cloud-based platforms and high-performance computing infrastructure can empower researchers to process, analyse, and visualise massive datasets efficiently.

Collaborative Initiatives and Standardisation

As the integration of these advanced analysis methods becomes central to metagenomics research, collaborative initiatives are essential. Establishing standardised protocols, benchmark datasets, and quality control measures ensures reproducibility and comparability across studies. Collaborative efforts among microbiologists, data scientists, bioinformaticians, computer scientists, and domain experts are crucial for developing user-friendly pipelines. The development of user-friendly software tools for metanalysis, pan-genome analysis, meta-pangenome analysis, and core metagenome characterisation will empower researchers coming from diverse backgrounds to leverage these techniques effectively to unlock deeper insights into microbial community dynamics and functions.

The future of metagenomics analysis pipelines for microbiome studies will be exciting and transformative. Leveraging recent advances in big data analytics, AI/ML capabilities, new sequencing methods, multi-omics integration, and bioinformatics scalability will undoubtedly drive groundbreaking discoveries in microbiome science. Embracing collaborations between experts in microbiome science, bioinformatics, and AI/ML will be required to fully realise the potential of these innovative approaches. As the understanding of microbiomes becomes increasingly nuanced, these pipelines will play a pivotal role in translating knowledge from metagenomics studies into actionable insights that benefit human health, ecosystems, and industries.

CONCLUSIONS

This chapter describes two popular microbial community analysis applications, QIIME 2 and mothur, which offer comprehensive suites of tools for analysing metagenomics data. The chapter provides the step-by-step protocol to use the software suites for quality control, taxonomic classification, functional annotation, microbial community statistical analysis, and data visualisation for microbiome data analysis. In addition to the standard operating procedures, the chapter also discusses the challenges involved in analysing big sequence data, provides best practices to optimise the performance of metagenomic analysis pipelines, including guidelines for sample collection, optimising metagenome extraction and library preparation, the importance of mock community controls, selecting appropriate sequencing platforms and protocols, controlling for contamination and bias, and validating results. The protocols and best practices described in this chapter are relevant for researchers and analysts working with metagenomic data and will promote research consistency and contribute to the standardisation and reproducibility of metagenomic studies. Though significant progress has been made in NGS data analysis, challenges remain. Advances in AI and ML are likely to play a critical role in overcoming these challenges and further advancing our understanding of microbial communities.

REFERENCES

1. Shi H, Grodner B, De Vlaminck I. Recent advances in tools to map the microbiome. Curr Opin Biomed Eng. 2021;19:100289. doi: 10.1016/j.cobme.2021.100289
2. Segre J. Microbiome. Translational and functional genomics branch. National Human Genome Research Institute of NIH, Bathesta; 2023. Accessed 19 November 2024. https://www.genome.gov/genetics-glossary/Microbiome
3. Hair M, Sharpe J. Fast facts about the human microbiome. The Center for Ecogenetics and Environmental Health, University of Washington; 2014. Accessed 19 November 2024. https://depts.washington.edu/ceeh/downloads/FF_Microbiome.pdf
4. Ottman N, Smidt H, de Vos WM, Belzer C. The function of our microbiota: Who is out there and what do they do? Front Cell Inf Microbio. 2012;2:104. doi: 10.3389/fcimb.2012.00104
5. Huffnagle GB, Noverr MC. The emerging world of the fungal microbiome. Trends Microbiol. 2013;21(7): 334–41. doi: 10.1016/j.tim.2013.04.002
6. Porras-Alfaro A, Bayman P. Hidden fungi, emergent properties: Endophytes and microbiomes. Annu Rev Phytopathol. 2011;49:291–315. doi: 10.1146/annurev-phyto-080508-081831
7. Rodriguez RJ, White JF Jr, Arnold AE, Redman RS. Fungal endophytes: Diversity and functional roles. New Phytol. 2009;182(2):314–330. doi: 10.1111/j.1469-8137.2009.02773.x
8. Song E-J, Lee E-S, Nam Y-D. Progress of analytical tools and techniques for human gut microbiome research. J Microbiol. 2018 Sep;56:693–705. https://doi.org/10.1007/s12275-018-8238-5
9. Galloway-Peña J, Hanson B. Tools for analysis of the microbiome. Dig Dis Sci. 2020 Mar; 65:674–685. https://www.ncbi.nlm.nih.gov/pmc/articles/PMC7598837/
10. Tringe SG, Rubin EM. Metagenomics: DNA sequencing of environmental samples. Nat Rev Genet. 2005;6:805–814. https://www.nature.com/articles/nrg1709/
11. Thomas T, Gilbert J, Meyer F. Metagenomics – A guide from sampling to data analysis. Microb Inf Exp. 2012;2:3.https://doi.org/10.1186/2042-5783-2-3
12. Burke C, Kjelleberg S, Thomas T. Selective extraction of bacterial DNA from the surfaces of macroalgae. Appl Environ Microbiol. 2009;75(1):252–256. doi: 10.1128/AEM.01630-08
13. Delmont TO, Robe P, Clark I, Simonet P, Vogel TM. Metagenomic comparison of direct and indirect soil DNA extraction approaches. J Microbiol Methods. 2011;86(3):397–400. doi: 10.1016/j.mimet.2011.06.013
14. Venter JC, Remington K, Heidelberg JF, Halpern AL, Rusch D, Eisen JA, et al. Environmental genome shotgun sequencing of the sargasso sea. Science. 2004;304(5667):66–74. doi: 10.1126/science.1093857

15. Bolyen E, Rideout JR, Dillon MR, Bokulich NA, Abnet CC, Al-Ghalith GA, et al. Reproducible, interactive, scalable and extensible microbiome data science using QIIME 2. Nat Biotechnol. 2019;37(8):852–857. doi: 10.1038/s41587-019-0209-9
16. Keegan KP, Glass EM, Meyer F. MG-RAST, a metagenomics service for analysis of microbial community structure and function. Methods Mol Biol. 2016;1399:207–33. doi: 10.1007/978-1-4939-3369-3_13
17. Meyer F, Paarmann D, D'Souza M, Olson R, Glass EM, Kubal M, et al. The metagenomics RAST server – a public resource for the automatic phylogenetic and functional analysis of metagenomes. BMC Bioinform. 2008;9:386. doi: 10.1186/1471-2105-9-386
18. Schloss PD, Westcott SL, Ryabin T, Hall JR, Hartmann M, Hollister EB, et al. Introducing mothur: Open-source, platform-independent, community-supported software for describing and comparing microbial communities. Appl Environ Microbiol. 2009;75(23):7537–7541. doi: 10.1128/AEM.01541-09
19. Handelsman J, Rondon MR, Brady SF, Clardy J, Goodman RM. Molecular biological access to the chemistry of unknown soil microbes: A new frontier for natural products. Chem Biol. 1998;5(10). doi: 10.1016/s1074-5521(98)90108-9
20. Keisam S, Romi W, Ahmed G, Jeyram K. Quantifying the biases in metagenome mining for realistic assessment of microbial ecology of naturally fermented foods. Sci Rep. 2016;6:34155. doi: 10.1038/srep34155
21. Romi W, Ahmed G, Jeyaram K. Three-phase succession of autochthonous lactic acid bacteria to reach a stable ecosystem within 7 days of natural bamboo shoot fermentation as revealed by different molecular approaches. Mol Ecol. 2015;24(13):3372–3389. doi: 10.1111/mec.13237
22. Sarkar P, Kandimalla R, Bhattacharya A, Wahengbam R, Dehingia M, Kalita MC, et al. Multi-omics analysis demonstrates the critical role of non-ethanolic components of alcoholic beverages in the host microbiome and metabolome: A human- and animal-based study. Microorganisms. 2023;11(6):1501. doi: 10.3390/microorganisms11061501
23. Mardis ER. Anticipating the 1,000 dollar genome. Genome Biol. 2006;7(7):112. doi: 10.1186/gb-2006-7-7-112
24. Wooley JC, Ye Y. Metagenomics: Facts and artifacts, and computational challenges. J Comput Sci Technol. 2009;25(1):71–81. doi: 10.1007/s11390-010-9306-4
25. Tringe SG, von Mering C, Kobayashi A, Salamov AA, Chen K, Chang HW, et al. Comparative metagenomics of microbial communities. Science. 2005;308(5721):554–557. doi: 10.1126/science.1107851
26. Hooper SD, Raes J, Foerstner KU, Harrington ED, Dalevi D, Bork P. A molecular study of microbe transfer between distant environments. PLoS One. 2008;3(7). doi: 10.1371/journal.pone.0002607
27. Dinsdale EA, Pantos O, Smriga S, Edwards RA, Angly F, Wegley L, et al. Microbial ecology of four coral atolls in the Northern line islands. PLoS One. 2008;3(2). doi: 10.1371/journal.pone.0001584
28. Lorenz P, Eck J. Metagenomics and industrial applications. Nat Rev Microbiol. 2005;3(6):510–516. doi: 10.1038/nrmicro1161.
29. Turnbaugh PJ, Hamady M, Yatsunenko T, Cantarel BL, Duncan A, Ley RE, et al. A core gut microbiome in obese and lean twins. Nature. 2009;457(7228):480–484. doi: 10.1038/nature07540
30. Turnbaugh PJ, Ley RE, Hamady M, Fraser-Liggett CM, Knight R, Gordon JI. The human microbiome project. Nature. 2007;449(7164):804–810. doi: 10.1038/nature06244
31. Buttigieg PL, Ramette A. A guide to statistical analysis in microbial ecology: a community-focused, living review of multivariate data analyses. FEMS Microbiol Ecol. 2014;90(3):543–550. doi: 10.1111/1574-6941.12437.
32. Aagaard K, Petrosino J, Keitel W, Watson M, Katancik J, Garcia N, et al. The human microbiome project strategy for comprehensive sampling of the human microbiome and why it matters. FASEB J. 2013;27(3):1012–1022. doi: 10.1096/fj.12-220806
33. Integrative HMP (iHMP) Research Network Consortium. The Integrative Human Microbiome Project: dynamic analysis of microbiome-host omics profiles during periods of human health and disease. Cell Host Microbe. 2014;16(3):276–289. doi: 10.1016/j.chom.2014.08.014
34. Qin J, Li R, Raes J, Arumugam M, Burgdorf KS, Manichanh C, et al. A human gut microbial gene catalogue established by metagenomic sequencing. Nature. 2010;464(7285):59–65. doi: 10.1038/nature08821
35. Handelsman J. Metagenomics: Application of genomics to uncultured microorganisms. Microbiol Mol Biol Rev. 2004;68(4):669–685. doi: 10.1128/mmbr.68.4.669-685.2004
36. Riesenfeld CS, Schloss PD, Handelsman J. Metagenomics: Genomic analysis of microbial communities. Annu Rev Genet. 2004;38(1):525–552. doi: 10.1146/annurev.genet.38.072902.091216
37. Streit WR, Schmitz RA. Metagenomics – The key to the uncultured microbes. Curr Opin Microbiol. 2004;7(5):492–498. doi: 10.1016/j.mib.2004.08.002
38. Chappidi S, Villa EC, Cantarel BL. Using mothur to determine bacterial community composition and structure in 16S ribosomal RNA datasets. Curr Protoc Bioinform. 2019;67(1). doi: 10.1002/cpbi.83

39. Lu Y, Hugenholtz P, Batstone DJ. Evaluating DNA extraction methods for community profiling of pig hindgut microbial community. PLoS One. 2015;10(11). doi: 10.1371/journal.pone.0142720
40. Psifidi A, Dovas CI, Bramis G, Lazou T, Russel CL, Arsenos G, et al. Comparison of eleven methods for genomic DNA extraction suitable for large-scale whole-genome genotyping and long-term DNA banking using blood samples. PLoS One. 2015;10(1). doi: 10.1371/journal.pone.0115960
41. Bharti R, Grimm DG. Current challenges and best-practice protocols for microbiome analysis. Brief Bioinform. 2019;22(1):178–193. doi: 10.1093/bib/bbz155
42. Aird D, Ross MG, Chen WS, Danielsson M, Fennell T, Russ C, et al. Analyzing and minimizing PCR amplification bias in Illumina sequencing libraries. Genome Biol. 2011;12(2):R18. doi: 10.1186/gb-2011-12-2-r18
43. Dohm JC, Lottaz C, Borodina T, Himmelbauer H. Substantial biases in ultra-short read data sets from high-throughput DNA sequencing. Nucleic Acids Res. 2008;36(16):e105. doi: 10.1093/nar/gkn425
44. Ross MG, Russ C, Costello M, Hollinger A, Lennon NJ, Hegarty R, et al. Characterizing and measuring bias in sequence data. Genome Biol. 2013;14(5):R51. doi: 10.1186/gb-2013-14-5-r51
45. Koren S, Phillippy AM. One chromosome, one contig: Complete microbial genomes from long-read sequencing and assembly. Curr Opin Microbiol. 2015;23:110–120. doi: 10.1016/j.mib.2014.11.014
46. Nakano K, Shiroma A, Shimoji M, Tamotsu H, Ashimine N, Ohki S, et al. Advantages of genome sequencing by long-read sequencer using SMRT technology in medical area. Human Cell. 2017;30(3):149–161. doi: 10.1007/s13577-017-0168-8
47. Treangen TJ, Koren S, Sommer DD, Liu B, Astrovskaya L, Ondov B, et al. MetAMOS: A modular and open source metagenomic assembly and analysis pipeline. Genome Biol. 2013;14:R2. doi: 10.1186/gb-2013-14-1-r2
48. Yang C, Chowdhury D, Zhang Z, Cheung WK, Lu A, Bian Z, et al. A review of computational tools for generating metagenome-assembled genomes from metagenomic sequencing data. Comput Struct Biotechnol J. 2021;19:6301–6314. doi: 10.1016/j.csbj.2021.11.028
49. Mirzayi C, Renson A; Genomic Standards Consortium. Reporting guidelines for human microbiome research: The STORMS checklist. Nat Med. 2021;27(11):1885–1892. doi: 10.1038/s41591-021-01552-x
50. Guarner F, Manichanh C, Santiago D, Ehrlich J, Levenez SD, Pelletier F, et al.; IHMS SOP 01 V2: Standard operating procedure for fecal samples identification HUVH. International Human Microbiome Standards. 2015. Accessed 19 November 2024. https://human-microbiome.org/index.php?id=Sop&num=001
51. IHMS SOPS. International Human Microbiome Standards. Accessed 19 November 2024. https://human-microbiome.org/index.php#SOPS
52. Szóstak N, Szymanek A, Havránek J, Tomela K, Rakoczy M, Samelak-Czajka A, et al. The standardisation of the approach to metagenomic human gut analysis: From sample collection to microbiome profiling. Sci Rep. 2022;12(1). doi: 10.1038/s41598-022-12037-3
53. Costea PI, Zeller G, Sunagawa S, Pelletier E, Alberti A, Levenez F, et al. Towards standards for human fecal sample processing in metagenomic studies. Nat Biotechnol. 2017;35(11):1069–1076. doi: 10.1038/nbt.3960
54. Ariefdjohan MW, Savaiano DA, Nakatsu CH. Comparison of DNA extraction kits for PCR-DGGE analysis of human intestinal microbial communities from fecal specimens. Nutr J. 2010;9(1). doi: 10.1186/1475-2891-9-23
55. Santiago A, Panda S, Mengels G, Martinez X, Azpiroz F, Dore J, et al. Processing faecal samples: A step forward for standards in microbial community analysis. BMC Microbiol. 2014;14(1). doi: 10.1186/1471-2180-14-112
56. Yuan S, Cohen DB, Ravel J, Abdo Z, Forney LJ. Evaluation of methods for the extraction and purification of DNA from the human microbiome. PLoS One. 2012;7(3). doi: 10.1371/journal.pone.0033865
57. Henderson G, Cox F, Kittelmann S, Miri VH, Zethof M, Noel SJ, et al. Effect of DNA extraction methods and sampling techniques on the apparent structure of cow and sheep rumen microbial communities. PLoS One. 2013;8(9). doi: 10.1371/journal.pone.0074787
58. Kennedy K, Hall MW, Lynch MD, Moreno-Hagelsieb G, Neufeld JD. Evaluating bias of Illumina-based bacterial 16S rRNA gene profiles. Appl Environ Microbiol. 2014;80(18):5717–5722. doi: 10.1128/AEM.01451-14
59. Kennedy NA, Walker AW, Berry SH, Duncan SH, Farquarson FM, Louis P, et al. The impact of different DNA extraction kits and laboratories upon the assessment of human gut microbiota composition by 16S rRNA gene sequencing. PLoS One. 2014;9(2). doi: 10.1371/journal.pone.0088982
60. Adams RI, Bateman AC, Bik HM, Meadow JF. Microbiota of the indoor environment: A meta-analysis. Microbiome. 2015;3:49. doi: 10.1186/s40168-015-0108-3
61. de Goffau MC, Lager S, Salter SJ, Wagner J, Kronbichler A, Charnock-Jones DS, et al. Recognizing the reagent microbiome. Nat Microbiol. 2018;3(8):851–853. doi: 10.1038/s41564-018-0202-y

62. Kim D, Hofstaedter CE, Zhao C, Mattei L, Tanes C, Clarke E, et al. Optimizing methods and dodging pitfalls in microbiome research. Microbiome. 2017;5(1):52. doi: 10.1186/s40168-017-0267-5
63. Salter SJ, Cox MJ, Turek EM, Calus ST, Cookson WO, Moffatt MF, et al. Reagent and laboratory contamination can critically impact sequence-based microbiome analyses. BMC Biol. 2014;12:87. doi: 10.1186/s12915-014-0087-z
64. Sinha R, Abnet CC, White O, Knight R, Huttenhower C. The microbiome quality control project: Baseline study design and future directions. Genome Biol. 2015;16:276. doi: 10.1186/s13059-015-0841-8
65. Weiss S, Amir A, Hyde ER, Metcalf JL, Song SJ, Knight R. Tracking down the sources of experimental contamination in microbiome studies. Genome Biol. 2014;15(12):564. doi: 10.1186/s13059-014-0564-2
66. Glassing A, Dowd SE, Galandiuk S, Davis B, Chiodini RJ. Inherent bacterial DNA contamination of extraction and sequencing reagents may affect interpretation of microbiota in low bacterial biomass samples. Gut Pathog. 2016;8:24. doi: 10.1186/s13099-016-0103-7
67. Willerslev E, Hansen AJ, Poinar HN. Isolation of nucleic acids and cultures from fossil ice and permafrost. Trends Ecol Evol. 2004;19(3):141–147. doi: 10.1016/j.tree.2003.11.010
68. Bittinger K, Charlson ES, Loy E, Shirley DJ, Haas AR, Laughlin A, et al. Improved characterization of medically relevant fungi in the human respiratory tract using next-generation sequencing. Genome Biol. 2014;15(10):487. doi: 10.1186/s13059-014-0487-y
69. Laurence M, Hatzis C, Brash DE. Common contaminants in next-generation sequencing that hinder discovery of low-abundance microbes. PLoS One. 2014;9(5):e97876. doi: 10.1371/journal.pone.0097876
70. Lee D, Das Gupta J, Gaughan C, Steffen I, Tang N, Luk KC, et al. In-depth investigation of archival and prospectively collected samples reveals no evidence for XMRV infection in prostate cancer. PLoS One. 2012;7(9):e44954. doi: 10.1371/journal.pone.0044954
71. van der Zee A, Peeters M, de Jong C, Verbakel H, Crielaard JW, Claas EC, et al. Qiagen DNA extraction kits for sample preparation for legionella PCR are not suitable for diagnostic purposes. J Clin Microbiol. 2002;40(3):1126.
72. Herrera JJ, Cabo ML, González A, Pazos I, Pastoriza L. Adhesion and detachment kinetics of several strains of Staphylococcus aureus subsp. aureus under three different experimental conditions. Food Microbiol. 2007;24(6):585–591. doi: 10.1016/j.fm.2007.01.001
73. Naccache SN, Greninger AL, Lee D, Coffey LL, Phan T, Rein-Weston A, et al. The perils of pathogen discovery: origin of a novel parvovirus-like hybrid genome traced to nucleic acid extraction spin columns. J Virol. 2013;87(22):11966–11977. doi: 10.1128/JVI.02323-13
74. Wilson MR, O'Donovan BD, Gelfand JM, et al. Chronic meningitis investigated via metagenomic next-generation sequencing [published correction appears in JAMA Neurol. 2018 Aug 1;75(8):1028]. JAMA Neurol. 2018;75(8):947–955. doi: 10.1001/jamaneurol.2018.0463
75. Aho VTE, Pereira PAB, Haahtela T, Pawankar R, Auvinen P, Koskinen K. The microbiome of the human lower airways: A next generation sequencing perspective. World Allergy Organ J. 2015;8(1):23. eCollection 2015.
76. Eisenhofer R, Minich JJ, Marotz C, Cooper A, Knight R, Weyrich LS. Contamination in low microbial biomass microbiome studies: Issues and recommendations. Trends Microbiol. 2019;27(2):105–117. doi: 10.1016/j.tim.2018.11.003
77. Lauder AP, Roche AM, Sherrill-Mix S, Bailey A, Laughlin AL, Bittinger K, et al. Comparison of placenta samples with contamination controls does not provide evidence for a distinct placenta microbiota. Microbiome. 2016;4(1):29.
78. Minich JJ, Zhu Q, Janssen S, Hendrickson R, Amir A, Vetter R, et al. KatharoSeq enables high-throughput microbiome analysis from low-biomass samples. mSystems. 2018;3(3):1–16. doi: 10.1128/mSystems.00218-17.
79. Baker GC, Smith JJ, Cowan DA. Review and re-analysis of domain-specific 16S primers. J Microbiol Methods. 2003;55(3):541–555. doi: 10.1016/j.mimet.2003.08.009
80. Pel J, Leung A, Choi WWY, Despotovic M, Ung WL, Shibahara G, et al. Rapid and highly-specific generation of targeted DNA sequencing libraries enabled by linking capture probes with universal primers. PLoS One. 2018;13(12):e0208283. doi: 10.1371/journal.pone.0208283
81. Janda JM, Abbott SL. 16S rRNA gene sequencing for bacterial identification in the diagnostic laboratory: Pluses, perils, and pitfalls. J Clin Microbiol. 2007;45(9):2761–2764. doi: 10.1128/jcm.01228-07
82. Woo PCY, Lau SKP, Teng JLL, Tse H, Yuen KY. Then and now: Use of 16S rDNA gene sequencing for bacterial identification and discovery of novel bacteria in clinical microbiology laboratories. Clin Microbiol Infect. 2008;14(10):908–934. doi: 10.1111/j.1469-0691.2008.02070.x
83. Yu Z, García-González R, Schanbacher FL, Morrison M. Evaluations of different hypervariable regions of archaeal 16S rRNA genes in profiling of methanogens by archaea-specific PCR and denaturing gradient gel electrophoresis. Appl Environ Microbiol. 2008;74(3):889–893. doi: 10.1128/aem.00684-07

84. De Filippis F, Laiola M, Blaiotta G, Ercolini D. Different amplicon targets for sequencing-based studies of fungal diversity. Appl Environ Microbiol. 2017;83(17). doi: 10.1128/aem.00905-17
85. Raja HA, Miller AN, Pearce CJ, Oberlies NH. Fungal identification using molecular tools: A primer for the natural products research community. J Nat Prod. 2017;80(3):756–770. doi: 10.1021/acs.jnatprod.6b01085
86. Maves YK. Factors to consider when selecting a next generation sequencing (NGS) technology. Accessed 19 November 2024. https://blog.crownbio.com/selecting-a-next-generation-sequencing-technology
87. Mikheenko A, Saveliev V, Gurevich A. MetaQUAST: Evaluation of metagenome assemblies. Bioinformatics. 2015;32(7):1088–1090. doi: 10.1093/bioinformatics/btv69
88. Parks DH, Imelfort M, Skennerton CT, Hugenholtz P, Tyson GW. CheckM: Assessing the quality of microbial genomes recovered from isolates, single cells, and metagenomes. Genome Res. 2015;25(7): 1043–1055. doi: 10.1101/gr.186072.114

BIBLIOGRAPHY

Anderson MJ. A new method for non-parametric multivariate analysis of variance. Austral Ecology. 2001; 26(1):32–46. doi: 10.1111/j.1442-9993.2001.01070.pp.x

Bokulich NA, Kaehler BD, Rideout JR, Dillon M, Bolyen E, Knight R, et al. Optimizing taxonomic classification of marker-gene amplicon sequences with QIIME 2's q2-feature-classifier plugin. Microbiome. 2018;6(1):90. doi: 10.1186/s40168-018-0470-z

Callahan BJ, McMurdie PJ, Rosen MJ, Han AW, Johnson AJA, Holmes SP. DADA2: High-resolution sample inference from Illumina amplicon data. Nature Methods. 2016;13(7):581. doi: 10.1038/nmeth.3869

Caporaso JG, Kuczynski J, Stombaugh J, Bittinger K, Bushman FD, Costello EK, et al. QIIME allows analysis of high-throughput community sequencing data. Nature Methods. 2010;7(5):335–6. doi: 10.1038/nmeth.f.303

Coordinators NR. Database resources of the national center for biotechnology information. Nucleic Acids Res. 2017;45(D1):D12–7. doi: 10.1093/nar/gkw1071

Hamady M, Lozupone C, Knight R. Fast UniFrac: Facilitating high-throughput phylogenetic analyses of microbial communities including analysis of pyrosequencing and PhyloChip data. ISME J. 2010;4:17–27. doi: 10.1038/ismej.2009.97

Hunter JD. Matplotlib: A 2D graphics environment. Comput Sci Eng. 2007;9(3):90–5. doi: 10.1109/MCSE.2007.55

Johnson M, Zaretskaya I, Raytselis Y, Merezhuk Y, McGinnis S, Madden TL. NCBI BLAST: A better web interface. Nucleic Acids Res. 2008;36(suppl_2):W5–9.

Katoh K, Standley DM. MAFFT multiple sequence alignment software version 7: Improvements in performance and usability. Mol Biol Evol. 2013;30(4):772–780. doi: 10.1093/molbev/mst010

Kruskal WH, Wallis WA. Use of ranks in one-criterion variance analysis. J Am Stat Assoc. 1952;47(260):583–621. doi: 10.1080/01621459.1952.10483441

Lane DJ. 16S/23S rRNA sequencing. In: Stackebrandt E, Goodfellow M, editors. Nucleic acid techniques in bacterial systematics. John Wiley and Sons; 1991:115–175.

Lozupone C, Knight R. UniFrac: A new phylogenetic method for comparing microbial communities. Appl Environ Microbiol. 2005;71(12):8228–8235. doi: 10.1128/AEM.71.12.8228-8235.2005

Lozupone C, Lladser ME, Knights D, Stombaugh J, Knight R. UniFrac: An effective distance metric for microbial community comparison. ISME J. 2011;5:169–172. doi: 10.1038/ismej.2010.133

Lozupone CA, Hamady M, Kelley ST, Knight R. Quantitative and qualitative Î² diversity measures lead to different insights into factors that structure microbial communities. Appl Environ Microbiol. 2007;73(5): 1576–1585. doi: 10.1128/AEM.01996-06

Mandal S, Van Treuren W, White RA, EggesbÃ M, Knight R, Peddada SD. Analysis of composition of microbiomes: A novel method for studying microbial composition. Microb Ecol Health Dis. 2015;26(1):27663. doi: 10.3402/mehd.v26.27663

McDonald D, Clemente JC, Kuczynski J, Rideout JR, Stombaugh J, Wendel D, et al. The biological observation matrix (BIOM) format or: How I learned to stop worrying and love the ome-ome. GigaScience. 2012;1(1):7. doi: 10.1186/2047-217X-1-7

McDonald D, Vãzquez-Baeza Y, Koslicki D, McClelland J, Reeve N, Xu Z, et al. Striped UniFrac: Enabling microbiome analysis at unprecedented scale. Nature Methods. 2018;15:847–848. doi: 10.1038/s41592-018-0187-8

McKinney W. Data Structures for Statistical Computing in Python. Proceedings of the 9th Python in Science Conference. Accessed 19 November 2024. https://proceedings.scipy.org/articles/Majora-92bf1922-00a

Pedregosa F, Varoquaux G, Gramfort A, Michel V, Thirion B, Grisel O, et al. Scikit-learn: Machine learning in python. J Mach Learn Res. 2011;12(Oct):2825–2830.

Pielou EC. The measurement of diversity in different types of biological collections. J Theor Biol. 1966;13: 131–144. doi: 10.1016/0022-5193(66)90013-0

Price MN, Dehal PS, Arkin AP. FastTree 2–approximately maximum-likelihood trees for large alignments. PLoS One. 2010;5(3):e9490. doi: 10.1371/journal.pone.0009490

Pruesse E, Quast C, Knittel K, Fuchs BM, Ludwig W, Peplies J, et al. SILVA: A comprehensive online resource for quality checked and aligned ribosomal RNA sequence data compatible with ARB. Nucleic Acids Res. 2007;35(21):7188–7196.

Quast C, Pruesse E, Yilmaz P, Gerken J, Schweer T, Yarza P, et al. The SILVA ribosomal RNA gene database project: Improved data processing and web-based tools. Nucleic Acids Res. 2013;41(Database issue):D590–D596.

Quince C, Walker AW, Simpson JT, Loman NJ, Segata N. Shotgun metagenomics, from sampling to analysis [published correction appears in Nat Biotechnol. 2017 Dec 8;35(12):1211]. Nat Biotechnol. 2017;35(9): 833–844. doi: 10.1038/nbt.3935

Rideout JR, He Y, Navas-Molina JA, Walters WA, Ursell LK, Gibbons SM, et al. Subsampled open-reference clustering creates consistent, comprehensive OTU definitions and scales to billions of sequences. PeerJ. 2014;2:e545. doi: 10.7717/peerj.545

Robeson MS, O'Rourke DR, Kaehler BD, Ziemski M, Dillon MR, Foster JT, et al. RESCRIPt: Reproducible sequence taxonomy reference database management. 2021. doi: 10.1371/journal.pcbi.1009581

Rognes T, Flouri T, Nichols B, Quince C, Mahé F. VSEARCH: A versatile open source tool for metagenomics. PeerJ. 2016;4:e2584. doi: 10.7717/peerj.2584

Sørensen T. A method of establishing groups of equal amplitude in plant sociology based on similarity of species and its application to analyses of the vegetation on Danish commons. Biologiske Skrifter. 1948;5:1–34.

Shotgun Metagenomic Sequencing. Accessed 19 November 2024. https://sapac.illumina.com/areas-of-interest/microbiology/microbial-sequencing-methods/shotgun-metagenomic-sequencing.html

Weiss S, Xu ZZ, Peddada S, Amir A, Bittinger K, Gonzalez A, et al. Normalization and microbial differential abundance strategies depend upon data characteristics. Microbiome. 2017;5(1):27. doi: 10.1186/s40168-017-0237-y

10 Standard Operating Procedure and Applications in Single-Cell Transcriptomics

Mandar Bedse and R Shyama Prasad Rao

INTRODUCTION

Single-cell transcriptomics (SCT) has revolutionised the field of genomics by enabling the analysis of gene expression at the single-cell level. It has emerged as a powerful tool to investigate cellular heterogeneity, gene expression variability, and regulatory mechanisms in complex biological systems. However, the high variability, technical noise, and bias associated with SCT pose significant challenges in interpreting the data accurately. Thus, standardised protocols for SCT are essential to obtain reliable and reproducible results. While recent studies have highlighted the need for standardisation in SCT, a survey by Svensson et al. (2017) showed that technical differences in sample preparation, sequencing, and data analysis could significantly impact the results of single-cell RNA sequencing (scRNA-seq) experiments. Similarly, a study by Ziegenhain et al. (2018) demonstrated that batch effects could substantially affect the interpretation of SCT data. These studies highlight the importance of standardisation in SCT to enable reliable and reproducible results across different labs and experiments.

In recent years, scientists have developed several standard operating procedures (SOPs) and technologies to address the technical challenges of SCT. One of the most widely used technology platforms is the 10x Genomics Chromium System, which enables high-throughput SCT with high sensitivity and accuracy (Zheng et al., 2017). Other popular SOPs include the Smart-seq2 (Picelli et al., 2014) and Smart-seq3 (Hagemann-Jensen et al., 2020) protocols, Drop-seq (Macosko et al., 2015), and inDrops (Klein et al., 2015). These procedures have had a profound effect on the field of SCT by enabling the analysis of thousands of single cells in a single experiment. However, despite the advances in SCT, several challenges remain. The abundant cells in the sample can mask rare cell types and subpopulations, which is one of the biggest challenges in identifying them. To overcome this, cell hashing (Stoeckius et al., 2017) and sample multiplexing (Cao et al., 2017) have been introduced to enable the analysis of multiple samples in a single experiment, increasing the throughput and reducing the cost of SCT. Another challenge is the interpretation of the data, especially in the context of cell-type identification and annotation. To address these analytical challenges, various computational tools have been developed, including Cell Ranger (Zheng et al., 2017) for processing 10x Genomics data, and more general-purpose tools such as Seurat (Butler et al., 2018) and Scanpy (Wolf et al., 2018) for downstream analysis. These tools enable the clustering and visualisation of SCT data and provide a framework for cell-type annotation and identification.

The scope of this chapter is to provide an overview of the current state of SCT, its associated challenges, and the need for standardised protocols. We will discuss the most common protocols, their advantages and disadvantages, and their applications in different biological contexts. Additionally, we will explore emerging technologies and their potential to improve the accuracy and reliability of SCT data. Furthermore, we will provide a comprehensive overview of SCT and discuss its potential for advancing our understanding of cellular heterogeneity and gene expression variability.

 DOI: 10.1201/9781003354062-10

SAMPLE PREPARATION

Sample preparation is a critical step in SCT, as it directly affects the quality and quantity of the RNA obtained from single cells. The process involves tissue dissociation, cell isolation, and RNA extraction, which require careful consideration of several factors, including the choice of dissociation method, the type of tissue, the cell isolation technique, and the RNA extraction method.

Tissue dissociation is a crucial step in sample preparation, as it enables the release of single cells from the tissue. Several methods are available, including enzymatic and mechanical. While enzymatic methods involve using enzymes (e.g., collagenase, trypsin, or dispase) to digest the extracellular matrix and release the cells, mechanical methods involve the physical disruption of the tissue (e.g., through grinding or sonication). The choice of dissociation method depends on the tissue type and the downstream applications. For example, enzymatic methods are suitable for soft tissues like the liver or brain, while mechanical methods are more suitable for hard tissues like bone or cartilage. Enzymatic methods are also preferred for downstream applications, such as single-cell sequencing, as they result in higher cell viability and yield compared to mechanical methods (Reichard and Asosingh, 2019; Vieira Braga and Miragaia, 2019). Over the last few years, several SOPs for sample preparation in SCT have been used. For tissue dissociation, the most commonly used methods include the papain-based dissociation protocol, which involves the use of papain to digest the extracellular matrix and release the cells (Sekiguchi et al., 2019), and the Neural Tissue Dissociation Kit, which uses a combination of enzymes and mechanical dissociation to remove cells from the nervous system (Trujillo et al., 2021).

Cell isolation is another crucial step in SCT, as it separates single cells from a heterogeneous cell population. Several methods are available for cell isolation, including fluorescence-activated cell sorting (FACS), microfluidics, and magnetic-activated cell sorting (MACS). FACS is the most widely used technique for single-cell isolation, as it enables high-throughput and high-purity cell sorting based on fluorescence signals. However, FACS can be expensive and time consuming and may cause cell stress or damage. Microfluidic platforms, such as the Fluidigm C1 system, enable single-cell isolation and processing in a highly automated and controlled manner but are limited by their lower throughput and higher cost. MACS is a more straightforward and affordable technique that uses magnetic beads to selectively isolate cells based on surface markers but results in lower purity than FACS (Wu et al., 2014).

RNA extraction is the final step in sample preparation, and it involves the isolation of high-quality RNA from single cells. Several methods are available for RNA extraction, including column-based methods, such as the RNeasy Micro Kit, and bead-based methods, such as the Dynabeads mRNA DIRECT Kit (Attar et al., 2018; Geoghegan et al., 2020; Koiwai et al., 2021). For downstream applications, column-based RNA isolation methods are preferred, as they yield higher RNA purity and concentration. Bead-based methods are simpler and more affordable but produce lower RNA yield and purity than column-based methods. Moreover, the choice of RNA extraction method depends on the downstream applications, as some techniques may introduce biases or affect the RNA quality (Burgess, 2015).

LIBRARY CONSTRUCTION

Single-cell RNA sequencing (scRNA-seq) has revolutionised transcriptomics by enabling gene expression analysis at the individual cell level. An essential step in scRNA-seq is library construction, which involves converting the RNA molecules from single cells into a format compatible with high-throughput sequencing. Scientists have developed several approaches and protocols for scRNA-seq library construction, including full-length and 3′-end sequencing, single-cell combinatorial indexing, and droplet-based methods (Svensson et al., 2018). Full-length sequencing protocols, such as Smart-seq3 and Quartz-seq, capture the entire length of the RNA molecule, allowing for the analysis of alternative splicing and identification of novel transcripts (Picelli et al., 2014;

Sasagawa et al., 2017). However, these methods are relatively low throughput and can have a higher rate of technical noise. In contrast, 3′-end sequencing protocols, such as CEL-seq and Drop-seq, capture only the 3′ end of the transcript, resulting in higher throughput and lower technical noise (Macosko et al., 2015; Hashimshony et al., 2016). Furthermore, these methods may need to include important information, such as alternative polyadenylation sites.

Single-cell combinatorial indexing (SCI) methods enable high-throughput scRNA-seq by assigning unique barcodes to individual cells and RNA molecules within each cell. Two examples of such methods are split-pool ligation-based transcriptome sequencing (SPLiT-seq) and indexing-based single-cell 3′ RNA-seq by combinatorial barcoding and parallel sequencing (FIPRESCI) (Rosenberg et al., 2018; Li et al., 2023). These methods allow for the analysis of many cells but may suffer from a higher rate of technical noise due to barcode misassignment or index hopping.

Droplet-based methods offer another approach to high-throughput single-cell RNA sequencing. These methods, including InDrop, 10x Genomics Chromium, and Drop-seq, assign unique barcodes to individual cells and RNA molecules by encapsulating single cells and barcoded beads together in nanolitre-sized droplets (Klein et al., 2015; Macosko et al., 2015; Zilionis et al., 2017). Similar droplet-based approaches have also been adapted for other single-cell assays, such as ATAC-seq for measuring chromatin accessibility (Buenrostro et al. 2015). These methods enable high-throughput scRNA-seq and have become increasingly popular due to their scalability and ease of use. However, droplet-based methods can suffer from lower sensitivity and capture efficiency than other methods (Ziegenhain et al., 2018).

The choice of an scRNA-seq library construction method depends on the specific research question and experimental design. Each method has its advantages and disadvantages, and it is crucial to carefully evaluate the SOPs and tools used for library construction to ensure that the data and downstream analysis are of high quality. For example, the RNA isolation, cell lysis, and cDNA amplification steps can introduce technical noise and affect the accuracy and reproducibility of the scRNA-seq data (Han et al., 2018). Therefore, it is essential to use well-established and optimised SOPs and to carefully assess the quality of the resulting libraries using quality control (QC) metrics, such as read depth, gene coverage, and library complexity (Stoeckius et al., 2017; Cao et al., 2019).

In summary, scRNA-seq library construction is a critical step in SCT, and researchers have developed several methods to enable the high-throughput and high-quality analysis of gene expression at the single-cell level and even further at assaying chromatin accessibility. It is necessary to carefully analyse the SOPs and library preparation methods to guarantee reliable and reproducible data. Of course, the approach that you choose is dependent on the specific research problem and experimental design. Figure 10.1 illustrates the workflow of single-cell RNA sequencing (scRNA-seq). Cells are first dissociated and isolated individually, followed by RNA extraction and reverse transcription to generate cDNA. The cDNA is then amplified and sequenced, allowing for the analysis of gene expression in individual cells.

QC AND DATA PREPROCESSING

QC and the preprocessing of SCT data are crucial to ensuring high-quality downstream analysis. The following are the essential steps and metrics for these tasks.

The first step in QC is filtering the cells and genes based on specific quality metrics. Commonly used metrics for cell filtering include the number of detected genes and counts, mitochondrial gene expression, and doublet detection. Genes can be filtered based on expression threshold or variability across cells. Various software tools, such as Cell Ranger (Zheng et al., 2017) and Seurat (Satija et al., 2015), can be used for cell and gene filtering.

Normalisation is another crucial step in preprocessing SCT data. The process aims to correct for two main sources of variation: differences in sequencing depth between cells and technical variability introduced during library preparation and sequencing. Scientists have developed various normalisation methods for SCT data, including total counts normalisation, trimmed mean of

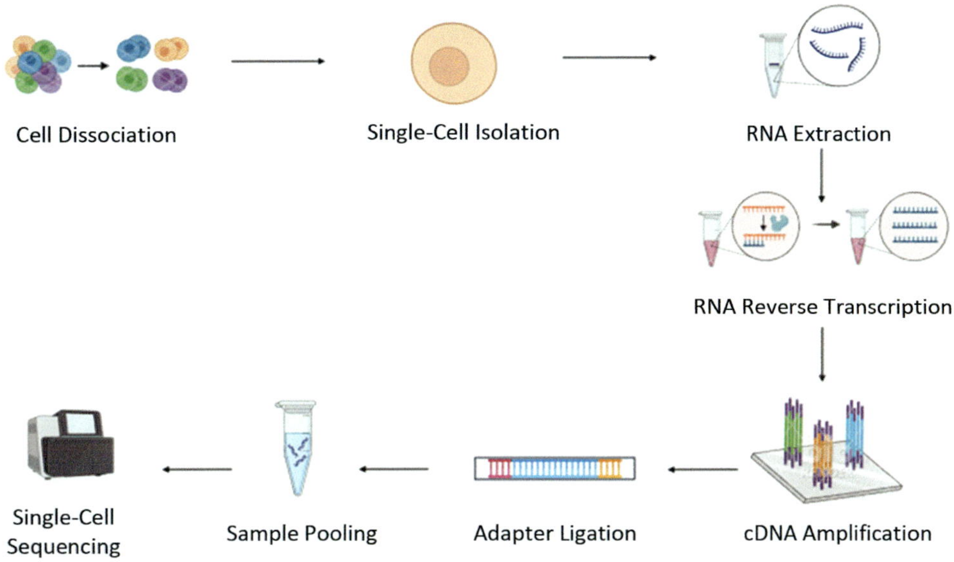

FIGURE 10.1 Library preparation steps.

M-values (TMM), and size factor normalisation. Several software tools, such as scran (Lun et al., 2016) and edgeR (Robinson et al., 2010), can also be used.

Batch effects can occur when samples are processed and sequenced at different times or using different protocols. Batch correction is necessary to remove these effects and improve the comparability of data across batches. Several batch correction methods have been developed, including ComBat (Johnson et al., 2007), MNN (Haghverdi et al., 2018), and Seurat v3 integration (Stuart et al., 2019).

Feature selection is the process of identifying the most informative genes for downstream analysis, which improves both accuracy and efficiency. Commonly used methods include variance filtering and principal component analysis (PCA). Dimensionality reduction techniques, such as t-distributed stochastic neighbour embedding (t-SNE) and uniform manifold approximation and projection (UMAP), are often used in conjunction with feature selection to visualise high-dimensional data in two or three dimensions. Software packages like Seurat and Scanpy (Wolf et al., 2018) provide tools for performing these feature selection methods and subsequent analyses, including dimensionality reduction.

In summary, QC and preprocessing are essential steps in SCT data analysis. The selection of appropriate methods and metrics for these steps can significantly impact the quality and comparability of a downstream analysis. Several software tools are available to perform these steps, and the effectiveness of these tools can vary depending on the dataset and the specific research question (Table 10.1).

ANALYSIS AND INTERPRETATION

One of the primary goals of SCT analysis is to identify cell types, states, and functions. Researchers have developed several methods for identifying and characterising cell types based on their gene expression profiles. Clustering is a common approach for identifying cell types based on their expression profiles. The Seurat pipeline includes several clustering algorithms, such as hierarchical clustering, *k*-means clustering, and graph-based clustering. Trajectory inference is another important method for characterising cell states and transitions. Developers have created several tools, such as Monocle (Trapnell et al., 2014), Slingshot (Street et al., 2018), and Wishbone (Setty et al., 2016) for inferring developmental trajectories from SCT data. Differential gene expression analysis is another critical step in analysing SCT data. This process aims to identify genes that are expressed at significantly different levels between

TABLE 10.1
Tools Available for Data Processing with Its Pros and Cons

Tool	Description	Type	Input Data	Pros	Cons	Supported Platforms	References
Cell Ranger	10x Genomics' official pipeline for processing and analysing single-cell RNA-seq data	Preprocessing and analysis	Single-cell gene expression data generated from 10x Genomics platforms	Easy-to-use interface supports batch effects removal and provides cell clustering and gene expression analysis	Limited options for customisation; only works with data generated from 10x Genomics platforms	Linux, Mac, Windows	Zheng et al., 2017
Seurat	R package for QC, normalisation, clustering, and differential gene expression analysis of single-cell RNA-seq data	Analysis	Single-cell gene expression data in various formats, including 10x Genomics, CEL-seq2, and SMART-Seq2	Flexible and customisable pipeline with a variety of statistical methods, widely used in the single-cell community	The steep learning curve for beginners requires familiarity with the R programming language	Linux, Mac, Windows	Satija et al., 2015; Stuart et al., 2019
Scanpy	Python package for preprocessing, clustering, visualisation, and differential expression analysis of single-cell RNA-seq data	Analysis	"	User-friendly interface with interactive plotting and annotation tools, extensive set of preprocessing and analysis functions	Limited options for batch effect removal. Some functions can be slow for large datasets	Linux, Mac, Windows	Wolf et al., 2018
Monocle3	R package for cell trajectory inference and differential expression analysis of single-cell RNA-seq data	Analysis	"	Powerful cell trajectory inference tools integrate well with other R packages for gene expression analysis	Limited options for preprocessing and normalisation, requires familiarity with R programming language	Linux, Mac, Windows	Cao et al., 2019
Scater	R package for QC, normalisation, and visualisation of single-cell RNA-seq data	Preprocessing	"	A flexible and customisable QC pipeline provides visualisations for various QC metrics	Limited downstream analysis functions require familiarity with the R programming language	Linux, Mac, Windows	McCarthy et al., 2017

(Continued)

TABLE 10.1 *(Continued)*
Tools Available for Data Processing with Its Pros and Cons

Tool	Description	Type	Input Data	Pros	Cons	Supported Platforms	References
RaceID3	R package for clustering and lineage inference of single-cell RNA-seq data	Analysis	"	Accurately reconstructs cell lineages and differentiation trajectories, customisable parameter settings for clustering and trajectory inference	Limited preprocessing and normalisation functions require familiarity with the R programming language	Linux, Mac, Windows	Herman et al., 2018
DESTINY	R package for single-cell trajectory analysis and visualisation	Analysis	"	Accurate and scalable trajectory inference integrates well with other R packages for gene expression analysis	Limited preprocessing and normalisation functions require familiarity with the R programming language	Linux, Mac, Windows	Angerer et al., 2016
M3Drop	Python package for filtering and normalisation of droplet-based single-cell RNA-seq data	Preprocessing	Single-cell gene expression data	1. Correct measurement errors in droplet-based scRNA-seq data 2. Improves the accuracy of downstream analysis. 3. Uses a probabilistic model to account for technical noise	1. Computationally intensive and requires high computational resources 2. May not be suitable for large datasets	Linux, Mac	Andrews et al., 2019

distinct cell types or experimental conditions. Over time, researchers have developed several methods to accurately detect these differentially expressed genes (DEGs) in single-cell data. Some popular tools for differential gene expression analysis include edgeR (Robinson et al., 2010), DESeq2 (Love et al., 2014), and MAST (Finak et al., 2015). Functional enrichment analysis often identifies biological pathways or processes enriched in DEGs. Several tools, such as GSEA (Subramanian et al., 2005), Enrichr (Kuleshov et al., 2016), and clusterProfiler (Yu et al., 2012), have been developed for this.

Again, the choice of analysis pipeline and tools depends on the research question, the experimental design, and the characteristics of the data. While several standardised pipelines and tools are available, it is crucial to carefully evaluate and optimise them to obtain meaningful biological insights from the data.

APPLICATIONS AND CASE STUDIES

SCT has already contributed significantly to various fields of biology and medicine, allowing researchers to better understand complex biological systems. Here are some examples of the application of SCT in different areas:

- *Immunology*: SCT has been instrumental in advancing our understanding of the immune system. An scRNA-seq study revealed that different T-cell subsets exist in the tumour microenvironment and have distinct transcriptional profiles (Zheng et al., 2017). Another study identified new dendritic cell subsets and their role in the immune response (Villani et al., 2017). These findings have implications for cancer immunotherapy and the development of new treatments for autoimmune diseases.
- *Cancer research*: Researchers have gained insights into the heterogeneity of cancer cells thanks to SCT, which has identified novel therapeutic targets. For example, scRNA-seq has revealed the existence of rare subpopulations of cells that are resistant to chemotherapy and contribute to relapse (Kan et al., 2022). Another study identified novel transcriptional regulators that drive cancer progression (Jia et al., 2018). These findings have the potential to improve cancer diagnosis and treatment.
- *Developmental biology*: Researchers have been allowed to analyse the dynamic changes in gene expression throughout development and differentiation because of SCT. An scRNA-seq study revealed the lineage relationships between different types of cells in the developing mouse brain (Pollen et al., 2014). Another study identified novel transcription factors that regulate cell fate decisions during embryonic development (Cao et al., 2017). These findings have implications for regenerative medicine and the development of new therapies for developmental disorders.
- *Neuroscience*: SCT has uncovered previously unknown information on the various cell types found in the brain, including their functions. For example, an scRNA-seq study identified novel cell types in the mouse visual cortex and their roles in visual processing (Shekhar et al., 2016). Another study identified the molecular signatures of different neuronal subtypes in the human brain (Darmanis et al., 2015). These findings have implications for the treatment of neurological disorders.

The preceding examples highlight the potential of SCT in advancing our understanding of complex biological systems. SOPs and analysis pipelines have been crucial in advancing these applications. They provide researchers with standardised protocols for generating, processing, and analysing data. They have facilitated data comparison across different studies, improving the reproducibility and reliability of the results. Developing new tools and algorithms for data analysis has expanded the utility of SCT in various fields. However, there is still room for improvement in data quality, scalability, and interpretation. As the area continues to evolve, new SOPs and tools are expected to address these challenges and further advance our understanding of biological systems.

Here are some of the future directions in scRNA-seq:

- *Improved sequencing technologies*: Sequencing technologies are constantly improving, leading to the development of new and more powerful scRNA-seq methods. For example, new sequencing technologies are being developed to sequence longer RNA reads, allowing researchers to study more genes in individual cells.
- *Development of new analytical tools*: New analytical tools are also being developed to allow researchers to use scRNA-seq data better. For example, new tools are being developed to identify cell types and states from scRNA-seq data and predict the function of genes from their expression patterns.
- *Application of scRNA-seq to new biological problems*: scRNA-seq is still a relatively new technique, and it is only beginning to be applied to new biological problems. For example, scRNA-seq is now being used to study the development of the brain, the immune system, and cancer.

CONCLUSIONS AND FUTURE DIRECTION

The importance of standardised protocols and SOPs in SCT has been highlighted in this chapter by discussing the advancements in sample preparation, library construction, QC, data preprocessing, and data analysis that have enabled the generation of high-quality and reliable SCT data. Case studies and examples have been presented of the applications of SCT in different fields of biology and medicine, demonstrating the potential impact of this technology in various domains.

Several challenges and limitations still need to be addressed to fully leverage the potential of SCT wherein a high degree of technical variation and noise in the data can affect the accuracy and reproducibility of downstream analyses. Therefore, further improvements in SOPs for sample preparation, library construction, and data preprocessing are needed to minimise technical variation and ensure data comparability across experiments and platforms.

Another challenge is integrating SCT data with other omics and imaging data to understand cellular processes and functions comprehensively. This requires the development of new computational methods and tools that can effectively integrate and analyse multimodal data.

Overall, the future of SCT is promising, potentially revolutionising our understanding of complex biological systems and diseases. The adoption of standardised protocols and SOPs will play a crucial role in ensuring the reproducibility and comparability of data across experiments and platforms and in accelerating the translation of this technology to clinical applications.

REFERENCES

Andrews TS, Hemberg M. M3Drop: Dropout-based feature selection for scRNASeq. Bioinformatics. 2019; 35(16):2865–7.

Angerer P, Haghverdi L, Büttner M, Theis FJ, Marr C, Buettner F. Destiny: Diffusion maps for large-scale single-cell data in R. Bioinformatics. 2016;32(8):1241–3.

Attar M, Sharma E, Li S, Bryer C, Cubitt L, Broxholme J, Lockstone H, Kinchen J, Simmons A, Piazza P, Buck D. A practical solution for preserving single cells for RNA sequencing. Scientific Reports. 2018;8(1):2151.

Buenrostro JD, Wu B, Chang HY, Greenleaf WJ. ATAC-seq: A method for assaying chromatin accessibility genome-wide. Current Protocols in Molecular Biology. 2015 Jan 5;109:21.29.1–9. doi: 10.1002/0471142727.mb2129s109. PMID: 25559105; PMCID: PMC4374986.

Burgess DJ. Bead capture for single-cell transcriptomics. Nature Reviews Genetics. 2015;16(4):195.

Butler A, Hoffman P, Smibert P, Papalexi E, Satija R. Integrating single-cell transcriptomic data across different conditions, technologies, and species. Nature Biotechnology. 2018;36(5):411–20.

Cao J, Packer JS, Ramani V, Cusanovich DA, Huynh C, Daza R, Qiu X, Lee C, Furlan SN, Steemers FJ, Adey A. Comprehensive single-cell transcriptional profiling of a multicellular organism. Science. 2017 Aug 18;357(6352):661–7.

Cao J, Spielmann M, Qiu X, Huang X, Ibrahim DM, Hill AJ, Zhang F, Mundlos S, Christiansen L, Steemers FJ, Trapnell C. The single-cell transcriptional landscape of mammalian organogenesis. Nature. 2019 Feb 28;566(7745):496–502.

Darmanis S, Sloan SA, Zhang Y, Enge M, Caneda C, Shuer LM, Hayden Gephart MG, Barres BA, Quake SR. A survey of human brain transcriptome diversity at the single cell level. Proceedings of the National Academy of Sciences. 2015;112(23):7285–90.

Finak G, McDavid A, Yajima M, Deng J, Gersuk V, Shalek AK, Slichter CK, Miller HW, McElrath MJ, Prlic M, Linsley PS. MAST: A flexible statistical framework for assessing transcriptional changes and characterizing heterogeneity in single-cell RNA sequencing data. Genome Biology. 2015;16:1–3.

Geoghegan IA, Emes RD, Archer DB, Avery SV. Method for RNA extraction and transcriptomic analysis of single fungal spores. MethodsX. 2020;7:100760.

Hagemann-Jensen M, Ziegenhain C, Chen, P, et al. Single-cell RNA counting at allele and isoform resolution using Smart-seq3. Nature Biotechnology. 2020;38:708–14 doi: 10.1038/s41587-020-0497-0

Haghverdi L, Lun AT, Morgan MD, Marioni JC. Batch effects in single-cell RNA-sequencing data are corrected by matching mutual nearest neighbors. Nature Biotechnology. 2018;36(5):421–7.

Han X, Wang R, Zhou Y, Fei L, Sun H, Lai S, Saadatpour A, Zhou Z, Chen H, Ye F, Huang D. Mapping the mouse cell atlas by microwell-seq. Cell. 2018 Feb 22;172(5):1091–107.

Hashimshony T, Senderovich N, Avital G, Klochendler A, De Leeuw Y, Anavy L, Gennert D, Li S, Livak KJ, Rozenblatt-Rosen O, Dor Y. CEL-seq2: Sensitive highly-multiplexed single-cell RNA-seq. Genome Biology. 2016;17(1):1–7.

Herman JS, Sagar N, Gruen D. FateID infers cell fate bias in multipotent progenitors from single-cell RNA-seq data. Nature Methods. 2018;15(5):379–86.

Hu P, Zhang W, Xin H, Deng G. Single cell isolation and analysis. Frontiers in Cell and Developmental Biology. 2016 Oct 25;4:116.

Jia G, Preussner J, Chen X, Guenther S, Yuan X, Yekelchyk M, Kuenne C, Looso M, Zhou Y, Teichmann S, Braun T. Single cell RNA-seq and ATAC-seq analysis of cardiac progenitor cell transition states and lineage settlement. Nature Communications. 2018 Nov 19;9(1):4877.

Johnson WE, Li C, Rabinovic A. Adjusting batch effects in microarray expression data using empirical bayes methods. Biostatistics. 2007;8(1):118–27.

Kan T, Zhang S, Zhou S, Zhang Y, Zhao Y, Gao Y, Zhang T, Gao F, Wang X, Zhao L, Yang M. Single-cell RNA-seq recognized the initiator of epithelial ovarian cancer recurrence. Oncogene. 2022;41(6):895–906.

Klein AM, Mazutis L, Akartuna I, Tallapragada N, Veres A, Li V, Peshkin L, Weitz DA, Kirschner MW. Droplet barcoding for single-cell transcriptomics applied to embryonic stem cells. Cell. 2015 May 21;161(5):1187–1201.

Koiwai K, Koyama T, Tsuda S, Toyoda A, Kikuchi K, Suzuki H, Kawano R. Single-cell RNA-seq analysis reveals penaeid shrimp hemocyte subpopulations and cell differentiation process. Elife. 2021 Jun 16;10:e66954.

Kuleshov MV, Jones MR, Rouillard AD, Fernandez NF, Duan Q, Wang Z, Koplev S, Jenkins SL, Jagodnik KM, Lachmann A, McDermott MG. Enrichr: A comprehensive gene set enrichment analysis web server 2016 update. Nucleic Acids Research. 2016;44(W1):W90–7.

Love MI, Huber W, Anders S. Moderated estimation of fold change and dispersion for RNA-seq data with DESeq2. Genome Biology. 2014;15:1–21.

Lun ATL, Bach K, Marioni JC. Pooling across cells to normalize single-cell RNA sequencing data with many zero counts. Genome Biology. 2016;17(1):1–4.

Li Y, Huang Z, Zhang Z, Wang Q, Li F, Wang S, Ji X, Shu S, Fang X, Jiang L. FIPRESCI: Droplet microfluidics based combinatorial indexing for massive-scale 5′-end single-cell RNA sequencing. Genome Biology. 2023;24(1):70.

Macosko EZ, Basu A, Satija R, Nemesh J, Shekhar K, Goldman M, Tirosh I, Bialas AR, Kamitaki N, Martersteck EM, Trombetta JJ. Highly parallel genome-wide expression profiling of individual cells using nanoliter droplets. Cell. 2015 May 21;161(5):1202–14.

McCarthy DJ, Campbell KR, Lun AT, Wills QF. Scater: pre-processing, quality control, normalization and visualization of single-cell RNA-seq data in R. Bioinformatics. 2017;33(8):1179–86.

Picelli S, Faridani OR, Björklund ÅK, Winberg G, Sagasser S, Sandberg R. Full-length RNA-seq from single cells using Smart-seq2. Nature Protocols. 2014;9(1):171–81.

Pollen AA, Nowakowski TJ, Shuga J, Wang X, Leyrat AA, Lui JH, Li N, Szpankowski L, Fowler B, Chen P, Ramalingam N. Low-coverage single-cell mRNA sequencing reveals cellular heterogeneity and activated signaling pathways in developing cerebral cortex. Nature Biotechnology. 2014;32(10):1053–8.

Reichard A, Asosingh K. Best practices for preparing a single cell suspension from solid tissues for flow cytometry. Cytometry Part A. 2019;95(2):219–26.

Robinson MD, McCarthy DJ, Smyth GK. edgeR: A bioconductor package for differential expression analysis of digital gene expression data. Bioinformatics. 2010;26(1):139–40.

Rosenberg AB, Roco CM, Muscat RA, Kuchina A, Sample P, Yao Z, Graybuck LT, Peeler DJ, Mukherjee S, Chen W, Pun SH. Single-cell profiling of the developing mouse brain and spinal cord with split-pool barcoding. Science. 2018 Apr 13;360(6385):176–82.

Sasagawa Y, Nikaido I, Hayashi T, Danno H, Uno KD, Imai T, Ueda HR. Erratum to: Quartz-seq: A highly reproducible and sensitive single-cell RNA sequencing method, reveals non-genetic gene-expression heterogeneity. Genome Biology. 2017;18. https://www.ncbi.nlm.nih.gov/pmc/articles/PMC5242007/

Satija R, Farrell JA, Gennert D, Schier AF, Regev A. Spatial reconstruction of single-cell gene expression data. Nature Biotechnology. 2015;33(5):495–502.

Setty M, Tadmor MD, Reich-Zeliger S, Angel O, Salame TM, Kathail P, Choi K, Bendall S, Friedman N, Pe'er D. Wishbone identifies bifurcating developmental trajectories from single-cell data. Nature Biotechnology. 2016;34(6):637–45.

Shekhar K, Lapan SW, Whitney IE, Tran NM, Macosko EZ, Kowalczyk M, Adiconis X, Levin JZ, Nemesh J, Goldman M, McCarroll SA. Comprehensive classification of retinal bipolar neurons by single-cell transcriptomics. Cell. 2016 Aug 25;166(5):1308–23.

Stoeckius M, Hafemeister C, Stephenson W, Houck-Loomis B, Chattopadhyay PK, Swerdlow H, Satija R, Smibert P. Simultaneous epitope and transcriptome measurement in single cells. Nature methods. 2017;14(9):865–8.

Street K, Risso D, Fletcher RB, Das D, Ngai J, Yosef N, Purdom E, Dudoit S. Slingshot: Cell lineage and pseudotime inference for single-cell transcriptomics. BMC Genomics. 2018;19:1–6.

Stuart T, Butler A, Hoffman P, Hafemeister C, Papalexi E, Mauck WM, Hao Y, Stoeckius M, Smibert P, Satija R. Comprehensive integration of single-cell data. Cell. 2019 Jun 13;177(7):1888–902.

Subramanian A, Tamayo P, Mootha VK, Mukherjee S, Ebert BL, Gillette MA, Paulovich A, Pomeroy SL, Golub TR, Lander ES, Mesirov JP. Gene set enrichment analysis: A knowledge-based approach for interpreting genome-wide expression profiles. Proceedings of the National Academy of Sciences. 2005;102(43): 15545–50.

Svensson V, Natarajan KN, Ly LH, Miragaia RJ, Labalette C, Macaulay IC, Cvejic A, Teichmann SA. Power analysis of single-cell RNA-sequencing experiments. Nature Methods. 2017;14(4):381–7.

Svensson V, Vento-Tormo R, Teichmann SA. Exponential scaling of single-cell RNA-seq in the past decade. Nature Protocols. 2018;13(4):599–604.

Trapnell C, Cacchiarelli D, Grimsby J, Pokharel P, Li S, Morse M, Lennon NJ, Livak KJ, Mikkelsen TS, Rinn JL. The dynamics and regulators of cell fate decisions are revealed by pseudotemporal ordering of single cells. Nature biotechnology. 2014 Apr;32(4):381–6.

Trujillo M, McElroy T, Brown T, Simmons P, Ntagwabira F, Allen AR. Combined mechanical and enzymatic dissociation of mouse brain hippocampal tissue. JoVE (Journal of Visualized Experiments). 2021; 21(176):e63007.

Vieira Braga FA, Miragaia RJ. Tissue handling and dissociation for single-cell RNA-seq. In Single Cell Methods: Sequencing and Proteomics. 2019:9–21. Springer Nature. https://link.springer.com/protocol/10.1007/978-1-4939-9240-9_2

Villani AC, Satija R, Reynolds G, Sarkizova S, Shekhar K, Fletcher J, Griesbeck M, Butler A, Zheng S, Lazo S, Jardine L. Single-cell RNA-seq reveals new types of human blood dendritic cells, monocytes, and progenitors. Science. 2017;356(6335):eaah4573.

Wolf FA, Angerer P, Theis FJ. SCANPY: Large-scale single-cell gene expression data analysis. Genome Biology. 2018;19:1–5.

Wu AR, Neff NF, Kalisky T, Dalerba P, Treutlein B, Rothenberg ME, Mburu FM, Mantalas GL, Sim S, Clarke MF, Quake SR. Quantitative assessment of single-cell RNA-sequencing methods. Nature Methods. 2014;11(1):41–6.

Yu G, Wang LG, Han Y, He QY. clusterProfiler: an R package for comparing biological themes among gene clusters. Omics: A Journal of Integrative Biology. 2012;16(5):284–7.

Zheng GX, Terry JM, Belgrader P, Ryvkin P, Bent ZW, Wilson R, Ziraldo SB, Wheeler TD, McDermott GP, Zhu J, Gregory MT. Massively parallel digital transcriptional profiling of single cells. Nature Communications. 2017;8(1):14049.

Ziegenhain C, Vieth B, Parekh S, Hellmann I, Enard W. Quantitative single-cell transcriptomics. Briefings in Functional Genomics. 2018;17(4):220–32.

Zilionis R, Nainys J, Veres A, Savova V, Zemmour D, Klein AM, Mazutis L. Single-cell barcoding and sequencing using droplet microfluidics. Nature Protocols. 2017;12(1):44–73.

11 Benchmarking and Evaluation of *de novo* Assembly Tools for Prokaryotic Long Reads from Oxford Nanopore Technologies

Megha S Kumar and Satheesh Babu T G

INTRODUCTION

Background

Long-read sequencing (LRS) technologies, particularly Oxford Nanopore Technologies (ONT), have transformed genomics by generating real-time reads spanning thousands of nucleotides, the maximum reported being >2 Mb in length. This enhances complex genome assembly with improved accuracy and contiguity (1). Even though the bottleneck of sequencing the genome has been eliminated, the analysis of long reads presents unique challenges, including higher error rates and the need for specialised assembly algorithms.

Although smaller and less complex than their eukaryotic counterparts, prokaryotic genomes still span several million base pairs and incorporate extrachromosomal components known as plasmids. This simpler genetic makeup allows for more straightforward investigations of fundamental cellular mechanisms. Generating an accurate and complete genome sequence is essential in genomic applications like analysis of the evolutionary history of genes, means of genetic diversification, impact of horizontal gene transfer, and many more (2). However, accurate and efficient assembly of prokaryotic genomes from ONT long reads requires robust assembly tools, and benchmarking and evaluating these tools is essential to determine their performance and identify the most suitable approaches for different applications.

This chapter reviews the characteristics of ONT long reads and discusses the challenges in assembly while outlining various tools available for prokaryotic long reads, namely *de novo* approach. The chapter further explores the benchmarking process, evaluation metrics, and datasets used for assessing the performance of assembly tools. Additionally, it elaborates on the strengths, limitations, and future directions in prokaryotic long-read assembly.

Overview of LRS

In contrast to short-read sequencing (SRS) technologies, LRS technologies produce reads that are substantially longer, often spanning thousands of bases. These long reads offer several advantages, including the ability to resolve repetitive regions, structural variations, and complex genomic architectures more accurately. In microorganisms, the repeat regions and the structural variations play an important role in developing and propagating antimicrobial resistance and virulence factors.

 DOI: 10.1201/9781003354062-11

Oxford Nanopore Technologies

ONT is a leading provider of LRS platforms and was launched in an independent beta-testing programme, the MinION Access Programme (MAP), in 2014 and commercialised in 2015 (3). The nanopore sequencing technique is a third-generation sequencing technology based on single-molecule detection, in which a single strand of DNA/RNA is ratcheted through the nanopore molecule implanted in a synthetic bilayer under an applied voltage. As the nucleotides pass through the pore, a sensor detects changes in ionic current for the nucleotides occupying the pore, and the resulting electrical signal is used to determine the DNA sequence. This sequencing technology provides rapid and real-time LRS of nucleic acids. ONT platforms, such as MinION, GridION, and PromethION, have become popular due to their portability, real-time sequencing capabilities, high throughput, and the ability to generate long reads. The downstream bioinformatics annotation is employed further to check for candidate genes and mutations (1).

Characteristics of ONT Long Reads

ONT long reads exhibit certain characteristics, including:

- Variable read lengths ranging from kilobases to megabases, which can be generated without the need for polymerase chain reaction (PCR) amplification.
- Higher error rates compared to short-read technologies, typically in the range of 10–15%, which has been improving with flow cell chemistry and base-calling algorithms.
- Direct detection of epigenetic modifications on native DNA.
- Base-calling errors associated with homopolymers and repetitive regions.

Challenges in Assembly of ONT Long Reads

The assembly of ONT long reads poses specific challenges, including:

- Higher error rates that require specialised error correction and consensus algorithms.
- The presence of context-specific errors, which necessitates tailored error models.
- Complex and repetitive regions that may lead to fragmented assemblies or misassemblies.
- Assembly algorithms that need to be scalable to handle large volumes of long reads efficiently.

Factors Influencing Assembly Quality

Read quality and length: The quality and length of ONT reads are fundamental factors influencing assembly quality. Higher quality and longer reads contribute to more accurate and contiguous assemblies. Longer reads are particularly advantageous in resolving repetitive regions and complex genomic structures.

k-mer size: Choosing *k*-mer size in De Bruijn Graph (DBG)-based assemblers can impact assembly quality. A smaller *k*-mer size improves sensitivity in resolving repeats but may increase memory usage and other computational demands. On the other hand, a larger *k*-mer size can lead to challenges in resolving repetitive regions but might require fewer resources.

Preprocessing and correction of raw reads: Preprocessing and error correction of raw ONT reads are critical for improving assembly quality. Techniques addressing base-calling errors, context-specific errors, and homopolymer-associated errors in ONT data enhance the accuracy of base calling, resulting in higher-quality assemblies.

Coverage depth: The depth of coverage, or the number of reads covering specific genomic positions, impacts the accuracy and completeness of the assembly. Higher coverage can improve consensus accuracy and resolve complex genomic regions but may also lead to potential misassembly and the overcorrection of errors.

Genome complexity: The inherent complexity of prokaryotic genomes, which include repetitive regions, structural variations, and mobile genetic elements, affects assembly quality. Assemblers need to manage complex genomic architectures and accurately resolve repetitive elements.

Assembly algorithm and strategy: Different assembly algorithms and strategies can impact assembly quality. Some assemblers are better suited for specific genome characteristics, and careful selection based on dataset characteristics can improve assembly outcomes.

Error correction strategies: The error correction strategies employed, such as using additional short-read data or polishing algorithms, can help refine the assembly and improve accuracy.

Contig merging and scaffolding: Post-assembly steps, such as contig merging and scaffolding using additional long-range information, can improve the contiguity of the assembly.

Consensus generation: Generating the consensus sequence from the raw reads can influence assembly accuracy. Employing robust consensus algorithms helps in producing more accurate genome sequences.

IMPORTANCE OF BENCHMARKING IN BIOINFORMATICS

The availability of computational methods for data analysis has been growing daily, and nowadays researchers face the challenge of selecting the most suitable tools. Benchmarking plays a crucial role in this by addressing the need for a standardised way to evaluate the performance of tools and algorithms (4). Accurate, efficient, and reliable approaches to a specific study can be identified by comparing the performance of tools, using shared datasets and evaluation metrics. This standardised approach promotes reproducibility and ensures that results obtained by different researchers, using different tools, can be compared and validated. Carefully designed and well-implemented benchmarking studies can provide accurate and unbiased results, pointing out the strengths and weaknesses of the tools or algorithms under investigation (4). This also increases the scope of innovations and improvements that can be made to bioinformatics tools and algorithms.

KEY COMPONENTS OF A GENOME ASSEMBLY BENCHMARKING PIPELINE

The quality of the input data is critical for successful genome assembly. When analysing LRS data, several key metrics must be considered to ensure reliable results. These metrics include read length, read accuracy, and coverage. Careful consideration of these metrics helps to minimise biases and avoid artefacts in the final genome assembly (5). Choosing datasets with a range of read lengths and error rates is crucial to evaluating how the assembler performs with different data types. The quality control (QC) step is critical to ensuring the integrity and reliability of the data. Tools like NanoPlot, Nanopack, and FastQC (https://www.bioinformatics.babraham.ac.uk/projects/fastqc/ last accessed 23 January 2025) are used for QC analysis (Figure 11.1). Parameters are assessed, including the total number of reads, the quality score distribution, and the guanine-cytosine (GC) content of the raw reads.

Data preprocessing is another crucial step in LRS analysis. It prepares raw reads for subsequent tasks, primarily genome assembly, by improving data quality and reliability. This enhancement ensures more accurate genome reconstruction and supports further genomic analyses. Preprocessing includes quality trimming, adapter removal, filtering out low-quality reads, and handling technical artefacts. It will help reduce false-positive and false-negative results and decrease the cost and time for the computational analysis, producing reliable data. Nanofilt, Filtlong, Porechop, Cutadapt (6), Trimmomatic (7), and Fastp (8) are some of the tools available for data preprocessing.

The preprocessing step is followed by an assembly process, with the assemblers chosen for benchmarking. The benchmarking can be done using the default parameters on a standard dataset. Alternatively, parameters like the *k*-mer length, coverage cutoff, resource requirement, and minimum overlap can be fine-tuned to achieve the best assembly quality. The assembly quality can then

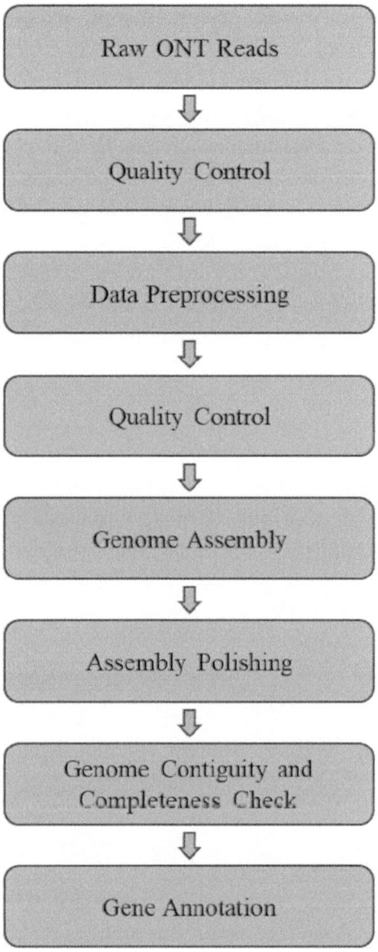

FIGURE 11.1 An outline of key steps involved in processing ONT long reads.

be assessed with the help of specific metrics like the N50, L50, and misassembly rate. The assemblies are also 'polished' to reduce remaining errors with tools like Medaka, Nanopolish, and Racon (9).

The assemblies can be evaluated by comparing various features with those of the reference genome. This comparison can include gene annotations, assessing gene coverage and completeness. Other important aspects to evaluate are the resolution of repetitive regions, accuracy of functional annotations and ontologies, and detection of structural variations and rearrangements.

ASSEMBLY METHODS

Assembly methods for prokaryotic genomes typically involve two main approaches: *de novo* and reference-guided assembly. *De novo* assembly reconstructs genomes without using any reference, making it ideal for novel or highly variable genomes. This method relies on algorithms like Overlap-Layout-Consensus (OLC), which identifies overlaps between sequencing reads to construct contigs; De Bruijn Graph (DBG), which uses *k*-mer-based nodes to represent sequences; and hybrid approaches that combine the strengths of both OLC and DBG. Reference-guided assembly, on the other hand, leverages an existing reference genome to align reads, enabling efficient reconstruction in cases where a closely related genome is available. These methods offer complementary strategies for genome reconstruction depending on the research goals and data characteristics.

This study selected *de novo* assembly as the preferred approach due to its capability to work directly with raw reads from ONT data without requiring prior reference genomes. The benchmarking study specifically assessed tools employing diverse algorithms: Flye, Canu, Miniasm, Shasta, Redbean, NextDenovo, HINGE, NECAT, Raven, and SMARTdenovo. Tools like Flye, Canu, and Miniasm use OLC strategies; Redbean employs a DBG-inspired fuzzy Bruijn graph approach; and hybrid assemblers like NextDenovo integrate multiple methodologies to enhance accuracy and contiguity. Shasta employs marker-based representation to prioritise computational efficiency, while NECAT uses progressive error correction. Raven, HINGE, and SMARTdenovo bring unique approaches to resolving repetitive regions and improving contiguity. Details on each tool and its unique methodology are briefly explained in the following section. The tools and algorithms were evaluated to determine their suitability for prokaryotic genome assembly, emphasising their performance in handling ONT data.

BENCHMARKING METRICS

Assembly Evaluation Metrics

The benchmarking of assembly tools requires defining and calculating evaluation metrics that capture different aspects of assembly quality. Commonly used metrics include:

- *Contiguity metrics*: N50, L50, NG50, and genome coverage, which assess the contiguity and fragmentation of the assembly.
- *Accuracy metrics*: Base-level accuracy, error rate, and misassembly rate evaluate the accuracy of the assembly.
- *Scalability metrics*: Memory usage, run time, and parallelisation efficiency measure the computational performance and resource requirements of assembly tools.

Datasets and Simulated Reads

Benchmarking studies utilise reference genomes and simulated long reads to evaluate assembly tools. The selection of diverse prokaryotic reference genomes and carefully designed simulated reads ensures the evaluation of assembly tools under various conditions and scenarios.

Prokaryotic Reference Genomes

A collection of reference genomes from different prokaryotic species, including bacteria and archaea, is used as a benchmarking resource. These genomes cover a range of sizes, complexities, and genomic features.

Simulated Long Reads

Simulated long reads are generated using realistic error models that mimic the characteristics of ONT long reads. These simulated reads allow for controlled benchmarking experiments and the assessment of assembly tools under specific error profiles.

Performance Metrics

Contiguity Metrics

Contiguity metrics evaluate the assembly quality in terms of contig lengths and the overall representation of the reference genome. Common contiguity metrics include:

N50: The contig length at which equal or longer length contigs cover half of the genome.
L50: The number of contigs required to reach the N50 value.
NG50: Similar to N50 but considers the size of the reference genome.

Accuracy Metrics

Accuracy metrics assess the correctness of the assembled sequences. These metrics include:

Base-level accuracy: The percentage of correctly assembled bases.
Error rate: The rate of base-calling errors in the assembly.
Misassembly rate: The proportion of misassembled regions or structural variations in the assembly.

Scalability Metrics

Scalability metrics focus on the computational efficiency and resource requirements of assembly tools, including:

Memory usage: The amount of memory required for the assembly process.
Run time: The time taken to complete the assembly.
Parallelisation efficiency: The effectiveness of parallelisation strategies in improving assembly speed and resource utilisation.

EVALUATION AND COMPARATIVE ANALYSIS

This section presents the evaluation and comparative analysis of popular *de novo* assembly tools for prokaryotic long reads. A discussion on the strengths and limitations of each tool is provided in the following list.

Flye

Flye is a *de novo* assembler designed to handle long, noisy reads from PacBio and ONT (10). It utilises the OLC approach with a variation of the DBG called the A-Bruijn graph. This allows Flye to efficiently identify overlaps between reads without the need for error correction beforehand. The assembly process of Flye involves combining reads into error-prone disjointing, constructing a repeat graph to represent repetitive regions accurately, and resolving the repeats to generate the contigs, followed by polishing. Flye also incorporates specific features that benefit prokaryote assemblies, such as the recovery of small plasmids and handling uneven depth of coverage. This enhances the accuracy and quality of the assembly results, particularly for prokaryotic genomes. Flye can be accessed through a command-line interface and accepts input data in the FASTA format.

Canu

Canu is a long-read assembler based on the OLC approach derived from the Celera Assembler (CA) (11). Developed in 2015, Canu has a long history and is specifically designed for handling high-noise, long-read data. The Canu pipeline consists of three stages: overlap detection, correction, and assembly, which can be performed together or separately, allowing flexibility in the analysis process. Canu incorporates various features to enhance usability and efficiency with single-molecule sequence data, including resource discovery, adaptive *k*-mer weighting, error rate estimation, sparse graph construction, and graphical fragment assembly outputs. It automates the reconstruction of high-quality reference genomes and can be accessed through a command-line interface, supporting various grid options for distributed computing. Canu accepts input data in either FASTQ or FASTA formats.

HINGE

HINGE is a *de novo* assembler tool designed explicitly for long-read data, utilising the OLC approach with repeat-resolution capabilities of DBG assemblers (12). It introduces a concept called 'hinging' to handle repetitive regions in the genome. The assembly process in HINGE involves several steps: pairwise overlaps are detected using DAligner, chimeric reads are filtered, hinges are placed, repeat regions are annotated, the overlap graph is constructed using hinging, a hinge-aided greedy assembly is performed, and alignment and consensus generation follow. Notably, HINGE also generates an assembly graph that can be visually interpreted, aiding in understanding the structure of the assembled genome. HINGE accepts input in FASTA format and does not have an error correction step.

Shasta

Shasta is an assembler specifically designed to prioritise computational efficiency (13). Shasta operates on a reduced representation of marker *k*-mers (typically around $k \approx 10$) instead of directly working with the entire long-read sequences. These marker *k*-mers are utilised to identify overlaps between reads and construct an assembly graph. From this graph, a consensus sequence is derived. One notable feature of Shasta is its utilisation of a run-length representation of the long-read sequences. This representation helps mitigate errors in the homopolymer repeat counts commonly observed in long reads generated by technologies like ONT.

wtdbg2/Redbean

Redbean, previously known as wtdbg2, is a long-read assembler that employs a unique approach called a fuzzy Bruijn graph (FBG) (14). This approach is inspired by the concept of the DBG commonly used for short-read assembly but adapted to handle the inexact sequence matches found in noisy long reads. Redbean utilises a novel sequence alignment algorithm called *k*-mer–BIN–mapping (KBM) for read mapping and constructs an FBG assembly graph for efficient assembly of large genomes.

In the Redbean assembly process, long reads are divided into 1,024 bp segments, which are then merged into vertices based on similarity. These vertices are connected in the FBG based on the adjacency of the segments along the reads, creating a graph representation. This accounts for mismatches and gaps in the noisy long reads, preserving the read paths while collapsing *k*-mers. Using the FBG in Redbean effectively assembles long-read data, accommodating the inherent errors and variations in the sequences. By leveraging the KBM read mapping algorithm and the FBG assembly graph, Redbean produces improved contig and uniting consensus sequences, contributing to the accurate reconstruction of large genomes.

NextDenovo

NextDenovo is a high-performance assembler that prioritises efficiency in the assembly process (15). It follows a similar approach to other assemblers like Canu and NECAT (see below) by incorporating read correction as an initial step in its pipeline. NextDenovo performs the first two steps of the OLC approach, which involves identifying overlaps between reads and constructing a layout or assembly graph. However, NextDenovo differs in that it separates the final consensus step from the assembly process itself, even as it relies on a separate tool called NextPolish to generate the consensus sequence from the assembly graph (16). NextPolish is specifically designed to polish the assembled sequences, improving the accuracy and quality of the final consensus.

SMARTdenovo

SMARTdenovo is a *de novo* assembler specifically designed for long-read data. It follows a straightforward approach, performing all-versus-all raw read alignment without incorporating an error correction step (17). The assembler focuses on generating accurate consensus sequences by utilising dependent consensus polish tools. The SMARTdenovo assembly process consists of several steps. Initially, it performs read overlapping to identify overlaps between the raw reads. It then proceeds to rescue any missing overlaps that may have been initially missed.

Additionally, SMARTdenovo identifies low-quality regions and chimeric reads in the dataset. This step helps to improve the overall quality of the assembly. By following these steps, SMARTdenovo aims to produce better uniting consensus sequences, which are contiguous sequences representing a consensus of overlapping reads. While it does not include an error correction step, the assembler relies on subsequent dependent consensus polish tools to refine and enhance the accuracy of the consensus sequences.

NECAT

NECAT is an assembler specifically designed for nanopore sequencing reads. It addresses the high error rate of nanopore reads through various methods, including progressive error correction, adaptive supporting reads, read overlap selection, and a two-step assembly approach (18). The pipeline consists of four modules: preprocessing, correction, trimming, and assembly. The preprocessing module filters out short and ill-formed reads, ensuring the quality of the input data. The correction module employs a progressive strategy, correcting nanopore reads in two steps to improve their accuracy. The trimming module removes low-quality subsequences from the corrected reads, enhancing the data quality. Finally, the assembly module constructs a string graph to assemble the genome in two steps, leveraging the corrected and trimmed reads.

Miniasm

Miniasm is also based on the OLC approach, which follows a series of steps, including crude read selection, fine read selection, generation of a string graph, and merging unambiguous overlaps to produce Unitigs, formed by kmers, represent the sequences without branch in assembly graph. Several connected unitigs form a contig and one unitig can be present in multiple contigs (19). It is known for its speed, making it one of the fastest OLC assemblers available. However, Miniasm does not perform read overlapping, which needs to be done separately using tools like Minimap. Additionally, Miniasm lacks a consensus step, resulting in assembly error rates comparable to the raw read error rates. This means that the consensus sequence generated by Miniasm can have a relatively high error rate, especially for regions with repetitive or duplicated sequences. A separate polishing tool, such as Racon, can be used in conjunction with Miniasm to achieve higher sequence accuracy.

Raven

Raven, previously known as Ra, is another tool that adopts the OLC approach (20). It shares overlapping algorithms with Minimap and utilises the consensus step based on Racon. However, Raven distinguishes itself through its layout step, which incorporates innovative techniques to eliminate spurious overlaps from the graph. This optimisation enhances the assembly contiguity, resulting in improved genomic reconstruction.

The performance of each tool based on key metrics is shown in Table 11.1.

RESULTS

TABLE 11.1
The Performance of each Tool based on Key Metrics

Tool	Computational Resource Requirements	Run Time (hours)	Contiguity (N50, bp)	Gene Completeness (BUSCO Score)	Strengths	Weaknesses
Canu	High	Longest	High	Excellent	Best for high-error rate datasets	Computationally demanding
Minisam	Low	Shortest	Moderate	Moderate	Fast processing, quick preliminary analyses	Limited accuracy with low coverage data
Shasta	Moderate	Fast	Moderate	Moderate	Good balance between speed and accuracy	Moderate computational requirements
Redbean	Moderate	Fast	Moderate	Moderate	Efficient, user-friendly	Limited contiguity
NextDenovo	Moderate	Fast	High	High	Efficient error correction, good contiguity	Moderate runtime
Miniasm	Low	Fast	Moderate	Moderate	Quick assembly for less complex datasets	Lower contiguity for high-error reads
HINGE	Moderate	Moderate	Moderate	Moderate	Handles complex genomes	Average speed and contiguity
NECAT	Moderate	Moderate	High	High	Good gene completeness, effective for low error datasets	Moderate computational demands
Raven	Moderate	Moderate	Moderate	Moderate	Effective in handling repetitive regions	Moderate contiguity
Flye	Moderate	Moderate	Moderate	Moderate	Suitable for circular genomes and plasmids	Lower contiguity for low coverage reads
SMARTdenovo	Moderate	Slow	High	High	Efficient for high-contiguity assemblies	Slow processing speed

DISCUSSION

Comparative Analysis of Tools

Our analysis of the *de novo* assembly tools listed above compares their strengths and weaknesses. While we evaluated the tools based on factors such as computational resource requirements, accuracy, and contiguity, all of the assemblers chosen were open source, easily installable, and relatively user-friendly. Computational resource requirements are essential for performing assemblies because a more extensive run time and high computational demand can limit a tool's usability. Under limiting resource assignment, Canu had the longest run time, whereas Minisam had the shortest. Shasta, Redbean, NextDenovo, and Miniasm were in the fastest category, while HINGE, NECAT, Raven, and Flye were middling, and Canu and SMARTdenovo were in the slowest.

Contiguity analysis of the genome was based on the metric N50, although it does not provide an accurate assessment of the performance of the assembler. The N50 of the data was more or less the same for different assemblers. The assembly with the largest N50 may still have misconnections, resulting in low-quality assemblies. Therefore, another metric, 'gene completeness' using BUSCO, was used to evaluate the quality of the assemblies. NECAT, Canu, NextDenovo, and SMARTdenovo showed excellent results in this regard.

It is important to note that the preprocessing and postprocessing of the assemblies significantly impacts the quality. Preprocessing would include eliminating low-quality reads for assembly and error correction based on consensus. Some tools included the error correction module, such as in Canu, NECAT, and NextDenovo. Postprocessing involves correcting the assemblies with the polishing process, as done by other tools.

Overall Performance of Assembly Tools

No single assembly tool outperformed all others in all aspects, reflecting the various challenges posed by prokaryotic genome assembly. Each assembler was created by its developers with specific problems in mind. To achieve our objectives, we had to make trade-offs while taking into account the characteristics of the data. Thus, the choice of assembly tool should be informed by the dataset characteristics and the research objectives.

CONCLUSIONS AND FUTURE DIRECTIONS

The assembly tools discussed above are important components of prokaryotic bioinformatics pipelines. With the advent of third-generation sequencing, these tools have become essential for developing reproducible and reusable approaches to keep pace with faster next-generation sequencing (NGS) technologies. Their ongoing improvement is crucial for maintaining accuracy and reliability in genomic studies. The following are future directions that we can envision:

1. There is a need for standardised approaches to infer bacterial communities using molecular tools accurately. These tools should be able to integrate data from various sequencing platforms and provide reliable insights into the composition and dynamics of microbial communities. Developing robust algorithms that can handle complex datasets and interpret mixed microbial populations will be crucial for advancing our understanding of microbial ecosystems.
2. There is an opportunity to develop toolsets for processing pipelines that can elucidate alpha diversity (within a single genome or sample) and beta diversity (between multiple genomes or samples) of genomes. These toolsets should be capable of analysing large datasets efficiently and provide accurate measures of genetic diversity, enabling researchers to explore the evolutionary dynamics of microbial species. Improving these processing pipelines will help in identifying novel genes, understanding genetic diversification, and assessing the impact of horizontal gene transfer on microbial communities.

ACKNOWLEDGEMENT

I would like to express my sincere gratitude to Dr. Prashanth N. Suravajhala at Amrita School of Biotechnology for his unwavering support, guidance, and mentorship throughout this research work.

REFERENCES

1. Wang Y, Zhao Y, Bollas A, Wang Y, Au KF. Nanopore sequencing technology, bioinformatics and applications. Nat Biotechnol. 2021 Nov;39(11):1348–65.
2. Weissenbach J. The rise of genomics. Comptes Rendus Biologies. 2016 Jul 1;339(7):231–9.

3. Chen P, Sun Z, Wang J, Liu X, Bai Y, Chen J, et al. Portable nanopore-sequencing technology: Trends in development and applications. Front Microbiol [Internet]. 2023 Feb 1 [cited 2024 Dec 18];14. Available from: https://www.frontiersin.org/journals/microbiology/articles/10.3389/fmicb.2023.1043967/full
4. Weber LM, Saelens W, Cannoodt R, Soneson C, Hapfelmeier A, Gardner PP, et al. Essential guidelines for computational method benchmarking. Genome Biology. 2019 Jun 20;20(1):125.
5. Zhang Y, Lu HW, Ruan J. GAEP: a comprehensive genome assembly evaluating pipeline. Journal of Genetics and Genomics. 2023 Oct 1;50(10):747–54.
6. Martin M. Cutadapt removes adapter sequences from high-throughput sequencing reads. EMBnet.journal. 2011 May 2;17(1):10–2.
7. Trimmomatic: a flexible trimmer for Illumina sequence data | Bioinformatics | Oxford Academic [Internet]. [cited 2024 Dec 18]. Available from: https://academic.oup.com/bioinformatics/article/30/15/2114/2390096
8. fastp: an ultra-fast all-in-one FASTQ preprocessor | Bioinformatics | Oxford Academic [Internet]. [cited 2024 Dec 18]. Available from: https://academic.oup.com/bioinformatics/article/34/17/i884/5093234
9. Comparative evaluation of Nanopore polishing tools for microbial genome assembly and polishing strategies for downstream analysis | Scientific Reports [Internet]. [cited 2024 Dec 18]. Available from: https://www.nature.com/articles/s41598-021-00178-w
10. Assembly of long, error-prone reads using repeat graphs | Nature Biotechnology [Internet]. [cited 2024 Dec 8]. Available from: https://www.nature.com/articles/s41587-019-0072-8
11. Koren S, Walenz BP, Berlin K, Miller JR, Bergman NH, Phillippy AM. Canu: scalable and accurate long-read assembly via adaptive k-mer weighting and repeat separation. Genome Res. 2017 May 1;27(5):722–36.
12. HINGE: long-read assembly achieves optimal repeat resolution [Internet]. [cited 2024 Dec 18]. Available from: https://genome.cshlp.org/content/27/5/747.long
13. Nanopore sequencing and the Shasta toolkit enable efficient de novo assembly of eleven human genomes | Nature Biotechnology [Internet]. [cited 2024 Dec 18]. Available from: https://www.nature.com/articles/s41587-020-0503-6
14. Ruan J, Li H. Fast and accurate long-read assembly with wtdbg2. Nat Methods. 2020 Feb;17(2):155–8.
15. Hu J, Wang Z, Sun Z, Hu B, Ayoola AO, Liang F, et al. NextDenovo: an efficient error correction and accurate assembly tool for noisy long reads. Genome Biology. 2024 Apr 26;25(1):107.
16. NextPolish: a fast and efficient genome polishing tool for long-read assembly | Bioinformatics | Oxford Academic [Internet]. [cited 2024 Dec 18]. Available from: https://academic.oup.com/bioinformatics/article/36/7/2253/5645175
17. Liu H, Wu S, Li A, Ruan J. SMARTdenovo: a de novo assembler using long noisy reads. GigaByte. 2021 Mar 8;2021:gigabyte15.
18. Chen Y, Nie F, Xie SQ, Zheng YF, Dai Q, Bray T, et al. Efficient assembly of nanopore reads via highly accurate and intact error correction. Nat Commun. 2021 Jan 4;12(1):60.
19. Li H. Minimap and miniasm: fast mapping and de novo assembly for noisy long sequences. Bioinformatics. 2016 Jul 15;32(14):2103–10.
20. Time- and memory-efficient genome assembly with Raven | Nature Computational Science [Internet]. [cited 2024 Dec 18]. Available from: https://www.nature.com/articles/s43588-021-00073-4

12 Best Practices for Reproducible Microbial Genomics Analyses

Venkata Satagopam and Shaman Narayanasamy

INTRODUCTION

Microbes are among the most ubiquitous and diverse life forms on earth. They are present in all types of environments, including some of the most extreme, demonstrating their versatility, adaptability, and vast diversity.[1] They are also known to play important roles in human health (e.g., human gut[2]), industry (e.g., wastewater treatment[3]), and the environment (e.g., marine[4] and soil[5]).

The microbial genome encodes all the physiological and functional properties of a given microbe. Deciphering its composition allows us to assess the potential of a given microbe (*Note*: the term 'microbes' may cover taxonomic clades of bacteria, archaea, viruses and microeukaryotes, mainly fungi and protozoans.) In this chapter, we focus specifically on bacterial and/or archaeal microbes (the former being the more common) due to their hallmark characteristics, including (i) easy culturing and (ii) small and/or simple genomes that allow for convenient genetic and physiological studies.

Sequencing is a process that 'reads' biological DNA molecules and converts that into *in silico* (i.e., machine-readable) formats. To that end, next-generation sequencing (NGS) allows for low-cost and highly accurate DNA sequence data generation, while high-performance computing capabilities combined with highly efficient genome assembly programs allow for rapid assembly of genomes. Importantly, the rapid improvements in long-read or third-generation sequencing technologies now enable complete or almost-complete draft genomes to be reconstructed[6], thus accelerating the path from bacterial culture/isolation to a finished genome sequence.

The core computational tools required for microbial genomics analysis can be broadly classified into the following categories: (i) NGS data preprocessing, (ii) genome assembly, and (iii) gene annotation. These tools will enable the reconstruction of a draft genome with associated gene function predictions to enable downstream analysis. Fortunately, bioinformatics workflow managers can be used to automate these steps and deploy them on remote computing platforms, such as high-performance computing (HPC) clusters or cloud infrastructures.[7] Development efforts of open-source and commercial software has led to a vast number of programs now being available for each of the aforementioned categories. While the users have a multitude of choices for their tools, they should also be cautious and carefully select the ones that meet their requirements. Finally, manual steps such as contamination inspection and genome finishing can be applied to improve the quality of the genome assembly.

This chapter covers the essential steps of a bioinformatics analysis procedure for research in microbial genomics. We also cover the relevant and commonly used tools for those different steps. Finally, we discuss supplementary steps that enhance the quality of analysis output, alongside techniques for automating analytical pipelines. This automation ensures that the entire process can be reliably reproduced in future studies.

A GENERAL BIOINFORMATICS WORKFLOW FOR MICROBIAL GENOMICS

Microbial genomics processing and analysis begins with the data generated from shotgun sequencing via NGS platforms. Briefly, shotgun sequencing is the process of shearing/fragmenting multiple copies of a genome and then subjecting those fragments to further processing before finally undergoing sequencing. Figure 12.1 provides a simplified depiction of the sequencing process, followed

DOI: 10.1201/9781003354062-12

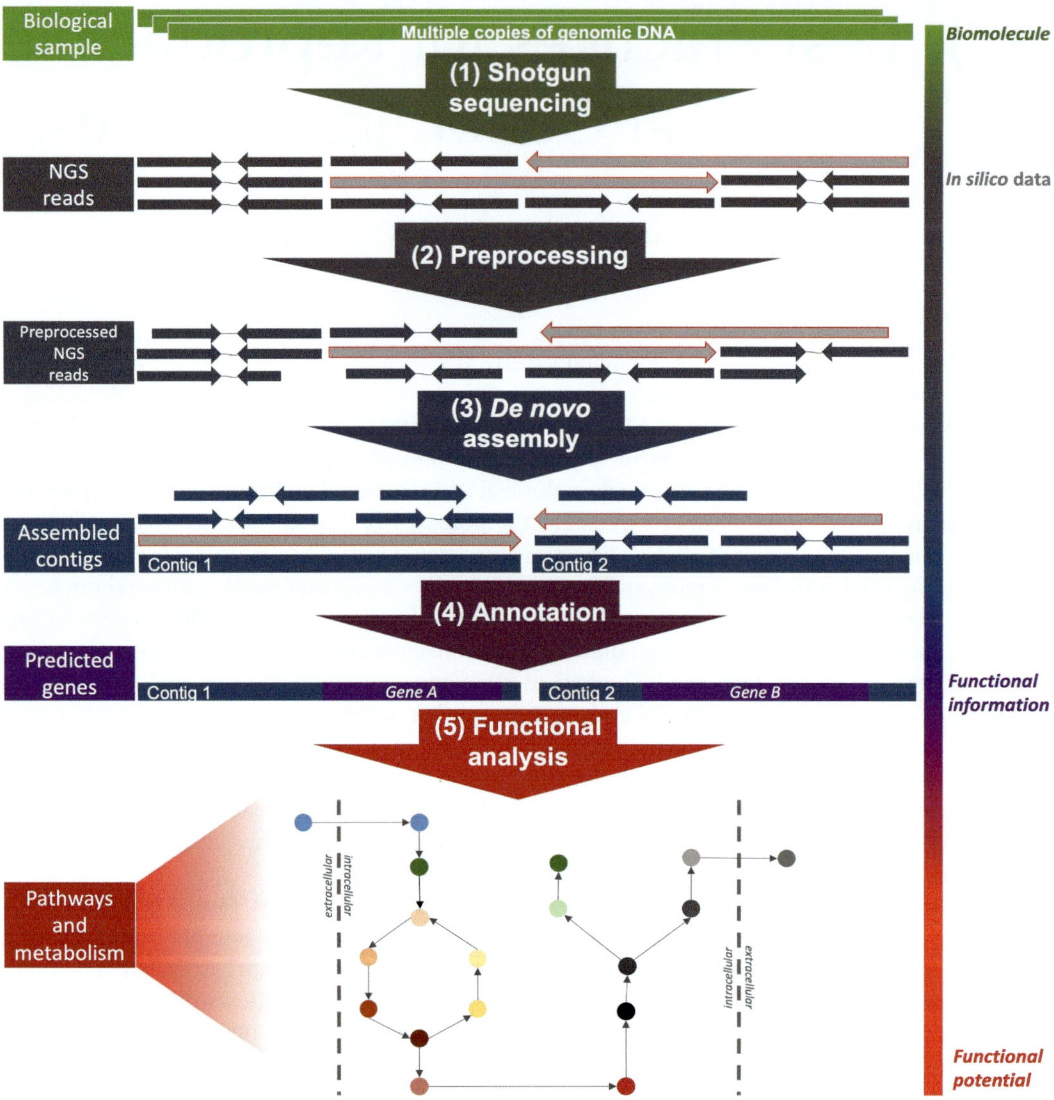

FIGURE 12.1 General workflow for microbial genomics, from sequencing to functional annotation. (1) The biological DNA sample undergoes shotgun sequencing procedure, where multiple copies of a microbial genome are processed, fragmented, amplified, and followed by NGS, yielding *in silico* representation of those fragments. NGS reads can either be in the form of paired-end NGS reads (dark grey) or long reads (light grey with red borders). (2) The NGS reads undergo preprocessing, where technical artefacts from sequencing, e.g., primer sequences or low-quality bases or reads, are either trimmed or excluded from downstream processes. (3) *De novo* assembly aligns reads against each other to form longer contiguous sequences (i.e., contigs). Contigs represent the consensus of the aligned reads. (4) Annotation is a process of predicting ORFs/genes followed by functional assignment/prediction. (5) The annotations with functional annotation are then used to assess the functional potential through the prediction of bio/geo-chemical pathways. Figure is a modified adaptation from Narayanasamy 2017.[23]

TABLE 12.1
A Non-exhaustive List of Bioinformatics Tools for NGS Genomics Data Processing and Analysis

		NGS Platform Support		
Steps	**Tools**	**Illumina**	**PacBio**	**ONT**
Quality assessment	FastQC[8] / MultiQC[9]	✓	✓	
	NanoPlot[10]		✓	✓
	LongQC[11]		✓	✓
	PycoQC[12]			✓
Preprocessing	Trimmomatic[13]	✓		
	Cutadapt[14]	✓		
	Skewer[15]	✓		
	Porechop[16]			✓
	NanoFilt[10]		✓	✓
	HiFiAdapterFilt[17]		✓	
Assembly	SPAdes[18]	✓	✓	✓
	Canu[19]		✓	✓
	miniasm[20]		✓	✓
	Unicycler[21]	✓	✓	✓
	HASLR[22]	✓	✓	✓

PacBio: Pacific Biosciences, *ONT:* Oxford Nanopore Technologies.

by standard *in silico* processing and analytical steps for microbial genomics, covering (i) NGS data preprocessing, (ii) *de novo* assembly, (iii) gene annotation, and (iv) functional annotation. Table 12.1 summarises the steps and the relevant core tools that can be used in each step of a microbial genomics workflow. We describe each of these processes in detail in the following subsections.

NGS Data Preprocessing

In general, NGS data preprocessing can be broadly separated into two parts that are carried out on the NGS read level, i.e., each step is applied on each read generated from the sequencing. These include quality control (QC) and quality-based filtering.

Quality Control of NGS Reads

Read-level QC should be conducted on the raw and preprocessed reads. This step ensures that the data are of sufficient quality before proceeding to the computationally intensive downstream steps. Table 12.1 lists some of the quality assessment tools for NGS data and the sequencing data types that they can support. Briefly, these tools produce visualisations and plots that are meant to be inspected manually by the user to assess the quality of the data.

Quality-Based Filtering of NGS Reads

The sequencing process binds synthetic DNA molecules such as primers and adapters to the fragmented genomic DNA. Specifically, adapters enable fixation of the genomic DNA fragments onto the sequencing platform, while primers allow for amplification of DNA fragments to increase signal intensity. The latter is only relevant for amplification-based sequencing such as Illumina but not for single-molecule sequencing technologies such as Oxford Nanopore Technologies (ONT) or PacBio, which do not require an amplification step.

Despite the importance of those synthetic DNA molecules in the sequencing process, they are considered technical contaminants in downstream processing and analysis. This especially applies to the genome assembly process, as these sequences may cause mis-assemblies, which may cascade into downstream analysis, producing faulty output. Therefore, these sequences should be removed so that only the portions representing true genomic DNA sequences of the sequencing reads are preserved. The preprocessing of NGS reads may differ slightly between the more widely used Illumina short-read sequencing and the more recent long-read sequencing (LRS) technologies such as ONT and PacBio. Briefly, the quality-based filtering process can be separated into two distinct parts: the first involves base-level trimming, while the second involves read-level filtering.

Base-level trimming involves the removal of low-quality bases, typically defined by the user, e.g., Phred score < 20 or unknown (i.e., N) bases. Next, sequencing artefacts such as adapter/primer sequences are removed by matching against known adapter and/or primer sequences from manufacturers. Base-level trimming is then followed up by read-level filtering, whereby bases that do not pass a certain average quality threshold or that do not reach a certain length (e.g., 50 bp) are discarded from downstream processing. Table 12.1 summarises the different quality trimming and filtering tools for different types of sequencing data, while Figure 12.1 provides a simplified depiction of the preprocessing step. Finally, it is worth referring to up-to-date analysis procedures and best practices for ONT and PacBio data processing on their respective commercial websites[24,25] due to the fast-paced development of these technologies and their relevant processing tools.

De Novo Genome Assembly

In this chapter, we assume that the sequencing depth[26] and quality is sufficient for *de novo* assembly. Briefly, the genome assembly process involves the alignment of all reads against one another (i.e., all-to-all alignment) to generate a representative consensus sequence. These consensus sequences constructed from shorter reads are referred to as contiguous sequences or contigs (Figure 12.1). There are two major categories of *de novo* assembly algorithms: the first is the overlap layout consensus (OLC), and the more recent is de Bruijn graph (DBG).[27] Table 12.1 lists popular *de novo* assembly programs used for microbial genome assemblies.

OLC *de novo* assemblers were developed for classic Sanger sequencing data,[28,29] while the more modern DBG procedures were initially developed for short-sequencing technologies but have recently been adapted to support LRS. Briefly, a consensus sequence is generated based on the agreement/similarity of overlapping reads.[27,29] An example of a classic OLC assembler is the Celera Assembler3,[30] while miniasm is an example of a modern one.[20]

DBG[31] assemblers work in three stages: (i) identification of all *k*-mers (i.e., stretches of nucleotides of length *k*) within a collection of NGS reads, (ii) construction of the DBG based on the identified *k*-mers, and (iii) inferring consensus sequence from the DBG.[32–34] Popular state-of-the-art DBG assemblers include SPAdes[18,35,36] and Canu[19]. It is worth mentioning that Canu[19] is a tool capable of performing all steps from trimming up to *de novo* assembly for long-read data within a single implementation. Finally, tools such as Unicycler[16,21] (which combines miniasm, Racon, and SPAdes) and HybridSPAdes (an extension of SPAdes) can perform hybrid long- and short-read assemblies.

Open Reading Frame (ORF) Prediction and Gene Annotation

Genome annotation is the process of identifying potential ORFs and other functional elements in a microbial genome and predicting their functions. This process allows for the extrapolation of functional elements present in a genome based entirely on the genome sequence. Tools commonly used for ORF prediction include Prodigal,[37] Glimmer,[38] GeneMark,[39] and FraGeneScan.[40] Most of these modern tools predict ORFs using statistical and machine learning procedures.[37–40] It is worth noting that ORFs are typically translated to protein sequences to facilitate annotation. Predicted genes are assigned functions by comparing the gene sequence to known genes within a database. The methods for comparison can range from sequence-alignment-based searches, such as nucleotide/protein Basic Alignment Search Tool (BLAST),[41] to more advanced hidden Markov model (HMM)-based search methods.[42]

It is important to note several caveats when performing gene annotation. Firstly, predicted ORFs may not be expressed as genes/proteins. This can only be confirmed through additional omics measurements (e.g., RNA-seq, mass spectrometry, etc.) and/or biological validation (e.g., polymerase chain reaction, or PCR). Hence, when performing gene functional annotation, we are assuming that the ORFs predicted are expressed genes. Second, functional annotation relies upon knowledge of known genes and functions within a given database, e.g., National Center for Biotechnology Information (NCBI) non-redundant protein database, Kyoto Encyclopedia of Genes and Genomes (KEGG),[43] and Uniprot/Swissprot[44] database (among others). To that end, the selected database resource used for annotation will influence the outcome. Thirdly (and finally), the quality of the annotation is dependent on the alignment algorithm. Specifically, HMM searches are known to be more sensitive compared to sequence-alignment-based methods such as BLAST.

Fortunately, there are convenient and reliable implementations, such as Bakta,[45] Prokka,[46] and Rapid Annotation using Subsystem Technology (RAST),[47] that enable end-to-end support for the annotation process. Briefly, users simply provide the genome sequence (or contigs), and these programs will perform ORF calling and annotation within a single process, using the aforementioned tools and databases. Overall, while genome annotation is a complex process with various influencing factors, these implementations allow researchers to obtain quick and reliable results for a broad functional understanding of an assembled genome.

Quality Inspection

The quality of the draft assembly should be assessed based on standard assembly quality metrics. Quality Assessment Tool (QUAST)[48] is a convenient implementation that performs assessment of bacterial genome assemblies using standard quality metrics (e.g., N50, L50 statistic) that assess the completeness of a genome. In addition, one may assess the purity of the genome assembly as a measure of quality. To that end, a simple yet effective means of quality assessment includes remapping sequencing reads to the assembly using standard sequence alignment tools such as Burrow-Wheeler Aligner (BWA)[49] and Bowtie2.[50] In fact, this is a standard step carried out especially when using *k*-mer-based (i.e., non-alignment) assemblies, as read placement within the assembly is not transparently available. It provides an overview about data usage, where a higher-quality draft genome is reflected by >90% mappable reads, while lower-quality assemblies may have > 10% reads that are unmappable, either due to low sequencing data quality and/or contamination. Furthermore, read recruitment via mapping also serves as a proxy for how efficiently the sequencing data were utilised for a given assembly. Alternatively, genome assembly purity can be further assessed using multiple strategies. For instance, CheckM uses a set of 'universal single copy orthologs', which are genes expected to be present in most prokaryotic genomes in single copies.[51] But QUAST can also be used to compare the draft assembly to reference genomes (if available) or genomes available on NCBI. A pure assembly tends to be similar to genomes that are closely related (based on taxonomy), while contaminated genomes may exhibit 'scattered' similarities with genomes from diverse taxa. It is also worth noting that contamination detection may also be performed in read-level processing using tools such as NanoLyse,[10] if required.

VIRTUAL ENVIRONMENTS AND CONTAINERS

Virtual environments and containers aid in the portability and management of bioinformatics software.[52] These platforms allow developers to create self-contained and isolated software environments to run bioinformatics tools/software, ensuring that all the dependencies of a given implementation can be installed and configured within the user's operating environment.[53] Conda and Mamba are popular Python package managers with virtual environment capabilities,[54] while

Docker and Singularity are containerisation software.[55] The practical difference between using a virtual environment and a container is that the former is isolated at the application level within the host operating system (OS),[54] while the latter is isolated at the OS level (it includes its own minimal OS layer but works within the host OS kernel).[55] It must be pointed out that the ability to use either virtual environments or containers depends on the user's computing environment. For instance, if a user is working on a local machine or server with full admin capabilities, then containerisation may be a suitable choice,[55] but if they are working in a computing cluster without support for containerised software, then they may opt for virtual environments.[53]

AUTOMATION USING WORKFLOW MANAGEMENT TOOLS

Workflow automation tools such as Nextflow,[56] Snakemake,[57] and Galaxy[58] are designed to streamline the process of running bioinformatics analyses by automating the execution of multiple commands/steps and tools.[7,59] These tools allow users to create reproducible and scalable workflows that can be easily shared, deployed, and executed across different computational platforms, ranging from local machines, computing clusters, and cloud platforms. While Nextflow and Snakemake are command-line-based workflow management systems,[56,57] Galaxy is a web-based platform for bioinformatics that provides a user-friendly drag-and-drop interface for creating and executing workflows, offering wide range of tools and functionalities, including data management, visualisation, and analysis, and is widely used by researchers and bioinformaticians.[58] The use of these automation tools can save time and reduce potential errors, increasing the reproducibility and reliability of the results. To that end, it is worth noting various prebuilt workflows such as Bacterial Genome Pipeline (BAGEP)[60] and bacass,[61] implemented using Snakemake and Nextflow, respectively, are maintained by a community of users and can be used as-is or customised. Importantly, it allows new users to implement complex microbial genomics workflows with relatively low hurdles. Similarly, users can also replicate workflows published in draft genome articles.[62] In summary, users should choose the best workflow management tool for their needs and implement their pipeline accordingly, especially if they intend to perform genome assemblies routinely.

GOING BEYOND GENOME ASSEMBLY AND ANNOTATION

Producing a draft assembly and gene annotations (addressed in the previous sections) is considered the core or minimum requirement for releasing and/or publishing draft genomes. The value of a genome assembly project can be increased by including laboratory experimentation data (e.g., from growth experiments) or additional omics data that have been generated (e.g., transcriptomics and proteomics data for functional expression). However, conducting laboratory experiments may not always be feasible due to constraints. Fortunately, researchers can add value to their genome assemblies by implementing various computational analyses, such as follows.

Phylogenetic Analysis

Phylogenetic analysis can be performed on either one or a set of selected marker genes from the newly assembled genome(s) and other related and unrelated genomes. Such analysis is relatively straightforward but provides important insight into taxonomic classification and the placement of the new genome within the tree of life.

Genome-Scale Metabolic Modelling

Genome-scale metabolic models are emerging as effective *in silico* methods to add functional context beyond simple gene annotations.[63] Tools such as gapseq,[64] METABOLIC,[65] and CarveMe[66] enable genome-wide metabolic models.

Assessment Using Environmental Meta-Omics

Users may align publicly available (or self-generated) metagenomic and/or metatranscriptomic NGS reads onto their newly assembled genomes. In the era of the microbiome, such information is considered important, as it may offer insight into the behaviours (i.e., dynamics and/or functions) of the newly sequenced microbe in real-world environments.[61]

Microbial Epigenetics

It is worth noting that single-molecule LRS technologies such as PacBio and ONT offer the opportunity of detecting methylated DNA bases,[67,68] Methylation detection requires processing raw sequencing data at the base-calling stage, even though most modern protocols typically provide users directly with pre-base-called data. Tools such as KineticTools[68] and MultiMotifMaker[69] allow for modified base detection on PacBio data, while tools such as fast5mod,[70] Nanopolish,[71] Megalodon,[72] and METEORE,[73] among others,[67] are capable of the equivalent processing on ONT data. A description of the microbial epigenome will most definitely add value to a given genome assembly.

Genome Finishing: Beyond the Draft Genome

Researchers may also choose to polish the draft genome through a genome finishing process, which involves closing gaps, eliminating mis-assemblies, and ensuring that the consensus sequence is represented by high-quality bases.[74,75] Briefly, the process of finishing requires scaffolding information to understand the orientation of the contigs. This can be achieved by using long-range mate pairs (e.g., 1,000–10,000 bp). If those longer-range mate pairs were not produced as part of the sequencing, then combinatorial PCR experiments using primers designed at the end of contigs could be employed.[75] Next, a manual *in silico* curation is required to close the gaps between the contigs. As one would expect, genome finishing is a time-consuming process due to the additional experiments and *in silico* work.[76] Fortunately, the availability of high-quality LRS these days typically results in complete, or almost complete, draft genomes requiring little to no genome finishing.[76]

GENERAL CRITERIA FOR SELECTING BIOINFORMATICS TOOLS

The rapid improvement of LRS technologies introduces a plethora of bioinformatics tools, most of which perform the same or similar tasks. Importantly, some of the tools discussed in this chapter may become either obsolete or significantly improved in several years. Thus, bioinformatics pipelines will require frequent updates to reflect the state of the art. However, such a large selection of available tools may overwhelm a person who is new to microbial genomics. Therefore, it is worth addressing the general criteria and considerations for the selection of such tools in a way that readers of this chapter can make informed decisions to construct their microbial genomics workflow. Table 12.2 addresses seven tenets for the selection of bioinformatics tools, in general.

CONCLUSIONS

This book chapter discussed the core processing steps in microbial genomics, namely NGS read preprocessing, *de novo* genome assembly, and gene annotation. We then expanded on ways that users can leverage software best practices, such as automation, environment virtualisation, and containerisation, to improve their analysis. We included examples of prebuilt community-maintained microbial genomics pipelines that users can implement with minimal hurdles. Additionally, we covered several *in silico* analysis procedures that could add value to their microbial genomics projects. Given the fast-paced development of bioinformatics software and sequencing technologies, we also advised users on how to evaluate bioinformatics software on their own, beyond those introduced

TABLE 12.2
Seven Tenets for Good Bioinformatics Software Selection

Tenets	Descriptions	Example Use Cases
Suitability	A tool suitable to your requirements, e.g., data type, goal	Using the right quality trimmer for PacBio or Illumina reads
Quality	Tools that provide the best output quality. Existing reviews and benchmarks can be very helpful for this purpose	Selecting a *de novo* assembler that produces the most contiguous assemblies
Usability	Easy to install and run	Tools implemented within virtual environments (e.g., Conda) or containers (e.g., Docker and Singularity)
Flexibility	Parameters should be customisable, if required	Using a read quality trimmer that can trim customised sequences (e.g., different Illumina adapters)
Efficiency	Lower memory and/or CPU requirements, fast and/or parallelised implementations	Using a *de novo* assembler with CPU threading for speed and low memory footprint
Support	Tool is actively maintained long term by authors/developers/maintainers	Check code repository (e.g., GitHub) issues and observe if they are regularly addressed and resolved, with active engagement by the maintainers*
Interoperability	Tools that support standard input/output formats	Avoid tools that generate exotic/custom file formats that cannot be used by other tools

* Most open-source bioinformatics tools are maintained by researchers, thus users should be considerate and understand that maintaining the software may not be their sole responsibility

in this chapter. Finally, we hope that this chapter provides a primer on the state-of-the-art tools for microbial genomics, but more importantly, has educated potential users on how they could build their own microbial genomics workflows through strong foundational knowledge.

CONFLICT OF INTEREST

Authors declare no conflict of interest.

AUTHOR CONTRIBUTIONS

SN reviewed literature. SN and VS wrote and revised the manuscript.

ACKNOWLEDGEMENTS

We would like to thank the European Eureka funding agency for the Eurostars ReDiRECt (E! 113726) grant.

REFERENCES

1. Nayfach, S. *et al.* A genomic catalog of Earth's microbiomes. *Nature Biotechnology* **39**, 499–509 (2021).
2. Greenhalgh, K., Meyer, K. M., Aagaard, K. M. & Wilmes, P. The human gut microbiome in health: Establishment and resilience of microbiota over a lifetime. *Environmental Microbiology* **18**, 2103–2116 (2016).
3. Muller, E. E. L., Sheik, A. R. & Wilmes, P. Lipid-based biofuel production from wastewater. *Current Opinion in Biotechnology* **30C**, 9–16 (2014).
4. Sunagawa, S. *et al.* Ocean plankton. Structure and function of the global ocean microbiome. *Science (New York, N.Y.)* **348**, 1261359 (2015).
5. Hultman, J. *et al.* Multi-omics of permafrost, active layer and thermokarst bog soil microbiomes. *Nature* **521**, 208–212 (2015).

6. Chin, C.-S. *et al.* Nonhybrid, finished microbial genome assemblies from long-read SMRT sequencing data. *Nature Methods* **10**, (2013).
7. Wratten, L., Wilm, A. & Göke, J. Reproducible, scalable, and shareable analysis pipelines with bioinformatics workflow managers. *Nature Methods* **18**, 1161–1168 (2021).
8. Babraham Bioinformatics – FastQC, a Quality Control Tool for High Throughput Sequence Data. https://www.bioinformatics.babraham.ac.uk/projects/fastqc/.
9. Ewels, P., Magnusson, M., Lundin, S. & Käller, M. MultiQC: Summarize analysis results for multiple tools and samples in a single report. *Bioinformatics* **32**, 3047–3048 (2016).
10. De Coster, W., D'Hert, S., Schultz, D. T., Cruts, M. & Van Broeckhoven, C. NanoPack: Visualizing and processing long-read sequencing data. *Bioinformatics* **34**, 2666–2669 (2018).
11. Fukasawa, Y., Ermini, L., Wang, H., Carty, K. & Cheung, M.-S. LongQC: A quality control tool for third generation sequencing long read data. *G3 Genes|Genomes|Genetics* **10**, 1193–1196 (2020).
12. Leger, A. & Leonardi, T. pycoQC, interactive quality control for Oxford nanopore sequencing. *JOSS* **4**, 1236 (2019).
13. Bolger, A. M., Lohse, M. & Usadel, B. Trimmomatic: A flexible trimmer for illumina sequence data. *Bioinformatics* **30**, 2114–2120 (2014).
14. Martin, M. Cutadapt removes adapter sequences from high-throughput sequencing reads. *EMBnet. journal* 17(1):10–12 (2011). doi:10.14806/ej.17.1.200.
15. Jiang, H., Lei, R., Ding, S.-W. & Zhu, S. Skewer: A fast and accurate adapter trimmer for next-generation sequencing paired-end reads. *BMC Bioinformatics* **15**, 182 (2014).
16. Wick, R. R., Judd, L. M., Gorrie, C. L. & Holt, K. E. Y. Completing bacterial genome assemblies with multiplex MinION sequencing. *Microbial Genomics* **3**, e000132 (2017).
17. Sim, S. B., Corpuz, R. L., Simmonds, T. J. & Geib, S. M. HiFiAdapterFilt, a memory efficient read processing pipeline, prevents occurrence of adapter sequence in PacBio HiFi reads and their negative impacts on genome assembly. *BMC Genomics* **23**, 157 (2022).
18. Bankevich, A. *et al.* SPAdes: A new genome assembly algorithm and its applications to single-cell sequencing. *Journal of Computational Biology: A Journal of Computational Molecular Cell Biology* **19**, 455–477 (2012).
19. Koren, S. *et al.* Canu: Scalable and accurate long-read assembly via adaptive *k*-mer weighting and repeat separation. *Genome Res.* **27**, 722–736 (2017).
20. Li, H. Minimap and miniasm: Fast mapping and de novo assembly for noisy long sequences. *Bioinformatics* **32**, 2103–2110 (2016).
21. Wick, R. R., Judd, L. M., Gorrie, C. L. & Holt, K. E. Unicycler: Resolving bacterial genome assemblies from short and long sequencing reads. *PLoS Computational Biology* **13**, e1005595 (2017).
22. Haghshenas, E., Asghari, H., Stoye, J., Chauve, C. & Hach, F. HASLR: Fast hybrid assembly of long reads. *iScience* **23**, 101389 (2020).
23. Narayanasamy, S. Development of an integrated omics in silico workflow and its application for studying bacteria-phage interactions in a model microbial community. (2017). https://orbilu.uni.lu/handle/10993/29800
24. Nanopore sequencing data analysis. *Oxford Nanopore Technologies* https://nanoporetech.com/support/nanopore-sequencing-data-analysis.
25. Open source tools – PacBio. https://www.pacb.com/products-and-services/analytical-software/open-source-tools/.
26. Sims, D., Sudbery, I., Ilott, N. E., Heger, A. & Ponting, C. P. Sequencing depth and coverage: Key considerations in genomic analyses. *Nature Reviews Genetics* **15**, 121–132 (2014).
27. Li, Z. *et al.* Comparison of the two major classes of assembly algorithms: Overlap-layout-consensus and de-Bruijn-graph. *Briefings in Functional Genomics* **11**, 25–37 (2012).
28. Sanger, F., Nicklen, S. & Coulson, A. R. DNA sequencing with chain-terminating inhibitors. *Proceedings of the National Academy of Sciences of the United States of America* **74**, 5463–5467 (1977).
29. Staden, R. A strategy of DNA sequencing employing computer programs. *Nucleic Acids Research* **6**, 2601–2610 (1979).
30. Myers, E. W. *et al.* A whole-genome assembly of drosophila. *Science* **287**, 2196–2204 (2000).
31. de Bruijn, N. G., Van der Woude, w. A combinatorial problem. *Proceedings of the Nederlandse Akademie Van Watenschappen* **49**, 758–764 (1946).
32. Idury, R. M. & Waterman, M. S. A new algorithm for DNA sequence assembly. *Journal of Computational Biology* **2**, 291–306 (1995).
33. Pevzner, P. A., Tang, H. & Waterman, M. S. An Eulerian path approach to DNA fragment assembly. *Proceedings of the National Academy of Sciences of the United States of America* **98**, 9748–53 (2001).

34. Compeau, P. E. C., Pevzner, P. A. & Tesler, G. How to apply de Bruijn graphs to genome assembly. *Nature Biotechnology* **29**, 987–991 (2011).
35. Prjibelski, A., Antipov, D., Meleshko, D., Lapidus, A. & Korobeynikov, A. Using SPAdes de novo assembler. *Current Protocols in Bioinformatics* **70**, e102 (2020).
36. Antipov, D., Korobeynikov, A., McLean, J. S. & Pevzner, P. A. hybridSPAdes: An algorithm for hybrid assembly of short and long reads. *Bioinformatics* **32**, 1009–1015 (2016).
37. Hyatt, D. *et al.* Prodigal: Prokaryotic gene recognition and translation initiation site identification. *BMC Bioinformatics* **11**, 119–119 (2010).
38. Delcher, A. L., Harmon, D., Kasif, S., White, O. & Salzberg, S. L. Improved microbial gene identification with GLIMMER. *Nucleic Acids Research* **27**, 4636–4641 (1999).
39. Lomsadze, A., Ter-Hovhannisyan, V., Chernoff, Y. O. & Borodovsky, M. Gene identification in novel eukaryotic genomes by self-training algorithm. *Nucleic Acids Res* **33**, 6494–6506 (2005).
40. Rho, M., Tang, H. & Ye, Y. FragGeneScan: Predicting genes in short and error-prone reads. *Nucleic Acids Research* **38**, e191 (2010).
41. Johnson, M. *et al.* NCBI BLAST: A better web interface. *Nucleic Acids Research* **36**, W5–9 (2008).
42. Finn, R. D., Clements, J. & Eddy, S. R. HMMER web server: Interactive sequence similarity searching. *Nucleic Acids Research* **39**, W29–37 (2011).
43. Kanehisa, M. & Goto, S. KEGG: Kyoto Encyclopedia of genes and genomes. *Nucleic Acids Research* **28**, 27–30 (2000).
44. UniProt: A hub for protein information. *Nucleic Acids Res* **43**, D204–D212 (2015).
45. Schwengers, O. *et al.* Bakta: Rapid and standardized annotation of bacterial genomes via alignment-free sequence identification. *Microbial Genomics* **7**, 000685.
46. Seemann, T. Prokka: Rapid prokaryotic genome annotation. *Bioinformatics* **30**, 2068–2069 (2014).
47. Aziz, R. K. K. *et al.* The RAST server: Rapid annotations using subsystems technology. *BMC Genomics* **9**, 75 (2008).
48. Gurevich, A., Saveliev, V., Vyahhi, N. & Tesler, G. QUAST: Quality assessment tool for genome assemblies. *Bioinformatics* **29**, 1072–1075 (2013).
49. Li, H. & Durbin, R. Fast and accurate long-read alignment with Burrows–Wheeler transform. *Bioinformatics* **26**, 589–595 (2010).
50. Langmead, B. & Salzberg, S. L. Fast gapped-read alignment with bowtie 2. *Nat Methods* **9**, 357–359 (2012).
51. Parks, D. H., Imelfort, M., Skennerton, C. T., Hugenholtz, P. & Tyson, G. W. CheckM: Assessing the quality of microbial genomes recovered from isolates, single cells, and metagenomes. *Genome Research* **25**, 1043–55 (2015).
52. Cox, R. *Bioinformatics: the machine learning approach.* (Cambridge University Press, 2018).
53. Niemeyer, E. & Perez, F. Managing software dependencies in bioinformatics: Virtual environments and containers. *Journal of Biomedical Informatics* **102**, 103358 (2020).
54. Smith, C. *et al.* Anaconda: A distribution of python for large-scale data processing and scientific computing. (2017).
55. Kapadia, A., Gao, Y. & Rho, Y. Containerization of bioinformatics workflows. *Journal of Biomedical Informatics* **78**, 1–14 (2018).
56. Di Tommaso, P. *et al.* Nextflow enables reproducible computational workflows. *Nature Biotechnology* **35**, 316–319 (2017).
57. Köster, J. & Rahmann, S. Snakemake – a scalable bioinformatics workflow engine. *Bioinformatics* **28**, 2520–2522 (2012).
58. Afgan, E. *et al.* The galaxy platform for accessible, reproducible and collaborative biomedical analyses: 2018 update. *Nucleic Acids Research* **46**, W537–W544 (2018).
59. Ahmed, A. E. *et al.* Design considerations for workflow management systems use in production genomics research and the clinic. *Scientific Reports* **11**, 21680 (2021).
60. Olawoye, I. B., Frost, S. D. W. & Happi, C. T. The bacteria genome pipeline (BAGEP): An automated, scalable workflow for bacteria genomes with snakemake. *PeerJ* **8**, e10121 (2020).
61. Peltzer, A. *et al.* nf-core/bacass: v2.0.0 nf-core/bacass: 'Navy Steel Swordfish'. (2021) doi:10.5281/zenodo.5289278.
62. Muller, E. E. E. L. L. *et al.* First draft genome sequence of a strain belonging to the *Zoogloea* genus and its gene expression in situ. *Standards in Genomic Sciences* **12**, 64, (2017).
63. Fang, X., Lloyd, C. J. & Palsson, B. O. Reconstructing organisms in silico: Genome-scale models and their emerging applications. *Nature Reviews Microbiology* **18**, 731–743 (2020).
64. Zimmermann, J., Kaleta, C. & Waschina, S. Gapseq: Informed prediction of bacterial metabolic pathways and reconstruction of accurate metabolic models. *Genome Biology* **22**, 81 (2021).

65. Zhou, Z. *et al.* METABOLIC: High-throughput profiling of microbial genomes for functional traits, metabolism, biogeochemistry, and community-scale functional networks. *Microbiome* **10**, 33 (2022).
66. Machado, D., Andrejev, S., Tramontano, M. & Patil, K. R. Fast automated reconstruction of genome-scale metabolic models for microbial species and communities. *Nucleic Acids Research* **46**, 7542–7553 (2018).
67. Liu, Y. *et al.* DNA methylation-calling tools for Oxford nanopore sequencing: A survey and human epigenome-wide evaluation. *Genome Biology* **22**, 295 (2021).
68. Clark, T. A. *et al.* Characterization of DNA methyltransferase specificities using single-molecule, real-time DNA sequencing. *Nucleic Acids Research* **40**, e29 (2012).
69. Li, T., Zhang, X., Luo, F., Wu, F.-X. & Wang, J. MultiMotifMaker: A multi-thread tool for identifying DNA methylation motifs from Pacbio reads. *IEEE/ACM Transactions on Computational Biology and Bioinformatics* **17**, 220–225 (2020).
70. Fast5Mod. (2022). https://github.com/nanoporetech/fast5mod.
71. Simpson, J. T. *et al.* Detecting DNA cytosine methylation using nanopore sequencing. *Nature Methods* **14**, 407–410 (2017).
72. Welcome to Megalodon's documentation! — Megalodon 2.3.3 documentation. https://nanoporetech.github.io/megalodon/.
73. Yuen, Z. W.-S. *et al.* Systematic benchmarking of tools for CpG methylation detection from nanopore sequencing. *Nature Communications* **12**, 3438 (2021).
74. Narayanasamy, S. De-novo assembly and finishing of the genome of neuro-toxin (anatoxin-a) producing cyanobacterium. *Anabaena Sp. Strain* **37** (2012).
75. Wang, H. *et al.* Genome-derived insights into the biology of the hepatotoxic bloom-forming cyanobacterium *Anabaena sp.* Strain 90. *BMC Genomics* **13**, 613 (2012).
76. Gordon, D., Abajian, C. & Green, P. Consed: A graphical tool for sequence finishing. *Genome Research* **8**, 195–202 (1998).

13 Single-Cell RNA-seq Data Analysis in the Era of Artificial Intelligence

Qanita Bani Baker, Sakshi Singh, and Ankit Agrawal

INTRODUCTION TO RNA SEQUENCING

Every year, the journal Nature selects a groundbreaking method from the past year that has made a significant impact on the fields of science and technology. Bestowed as the 'Nature Method of the Year' (Table 13.1) this accolade has encompassed various innovative techniques. Notable among these, the AlphaFold method (2021), which is rooted in the realm of artificial intelligence (AI), utilises a transformer model to tackle protein folding. Five of the recognised methods are related to next-generation sequencing (NGS), three have been instrumental in advancing imaging techniques, and one has focused on organoids. Each of these methodologies holds crucial importance in delving into the intricate nanoscale organisation of cells, exploring *in vitro* organoid culture to unravel the complexities of diseases, understanding tissues' three-dimensional architecture, comprehending the mechanisms behind stem cell differentiation, retracing the lineage history of cells, unravelling gene-regulatory mechanisms, and deciphering protein structure and folding.

The Method of the Year for 2013 heralded a turning point with the introduction of single-cell sequencing. This breakthrough has had a profound impact on the realms of biology and medicine. It has facilitated the exploration of cell heterogeneity, the identification of mutations and structural alterations in the genomes of cancer cells, the discovery of distinct cell types, the revelation of cell-type-specific accessible regions in chromatin, and numerous other insights. A comprehensive review of these transformative advancements can be accessed in Kashima et al. [Kashima et al., 2020]. In the following section, our focus is on single-cell sequencing methodology, which plays a pivotal role in capturing RNA present within both the cytoplasm and nuclei of individual cells.

RNA is an essential biomolecule governing the numerous biological processes that make the proteins and non-coding RNA that participate in multicellular regulatory activities. These regulatory

TABLE 13.1
Technological Development over ten Years

Nature Method of the Year	Year
Single-cell sequencing	2013
Light-sheet fluorescence microscopy	2014
Cryo-EM	2015
Epitranscriptome	2016
Organoids culture	2017
Imaging in freely behaving animals	2018
Single-cell multimodal omics	2019
Spatially resolved transcriptomics	2020
Protein structure prediction (*DeepMind AlphaFold*)	2021
Long-read sequencing	2022

 DOI: 10.1201/9781003354062-13

activities maintain tissue homeostasis and change during inflammation and disease conditions. Traditionally, the disease state is attributed to individual cell types, but new therapies utilise multicellular functions for treatment strategies. The amount and types of mRNA molecules reflect the state of a cell. Thus, at any given time a cell state is defined by a gene expression signature from thousands of transcripts and the gene regulatory network. The behaviour of functional tissue can be speculated when gene expression changes across multiple cell types are understood. Methods to investigate gene expression can be traced back to the microarray technique's discovery before the NGS method's advent. There are two main categories to capture the RNA inside the cells: the targeted and the untargeted approach. The untargeted method aims to capture and sequence short stretches of transcripts and link information back to the gene sequence to determine the gene's expression. This approach is very similar to RNA-seq technology. The biggest limitation of this technology was sequencing before NGS, as it was super slow and expensive. The targeted approach is based on hybridising gene-specific probes to the complementary sequence of a gene of interest transcript. Examples of targeted methods include RNA microarray technologies, reverse transcription polymerase chain reaction (RT-PCR), and image-based spatial transcriptomics. The targeted approach generally confirms the presence or absence of specific genes in a tissue cell. Additionally, these methods can monitor relative gene expression changes between two experimental conditions. However, a major limitation of targeted approaches is that they require prior knowledge of the gene sequences of interest. This requirement makes these technologies less suitable for studying species whose genome sequences are not yet available. Table 13.2 shows a comparison between RNA-seq and Microarray Technologies.

Bulk RNA Sequencing

Early RNA-seq methods typically used bulk samples (or tissues) to measure the abundance of RNA transcripts. In bulk RNA-seq, total RNA is extracted from a sample, and the RNA molecules are converted to complementary DNA (cDNA), which is then sequenced using high-throughput sequencing technologies. It provides an average gene expression profile of all the cells in a sample. This approach can help investigate changes in gene expression patterns in response to different experimental conditions, such as disease states, drug treatments, or developmental stages. Quantitative RT-PCR (qRT-PCR) is a method used in research to reliably find and measure RNA. It involves two main steps: first, converting RNA into cDNA; second, creating multiple copies

TABLE 13.2
Comparison between RNA-seq and Microarray Technologies

Features	RNA-seq	Microarray Technologies
Detect novel transcripts	Yes, because it does not require species or transcript-specific probes	No
Other than novel transcripts detection	Gene fusions, single nucleotide variants, indels (small insertions and deletions)	No
Gene expression	It produces discrete, digital sequencing read counts, typical range > 10^5	Measurement is limited by background at the low end and signal saturation at the high end, typical range 10^3
Specificity and sensitivity	High (it can detect a higher percentage of differentially expressed genes, especially genes with low expression)	Low
Detection of rare and low abundance transcripts	Sequencing coverage depth can easily be increased to detect rare transcripts, single transcripts per cell, or weakly expressed genes	NA

of this cDNA through PCR. The 'quantitative' aspect of the method helps researchers accurately gauge the initial amount of RNA in a sample. However, qRT-PCR has its challenges. A significant issue is its sensitivity to contaminants in the RNA sample, which can disrupt the initial steps of the process. Additionally, if the RNA is of low quality, it can lead to inaccurate measurements. To ensure precision, scientists must compare their RNA measurements to reference genes, which act as benchmarks to validate the accuracy of the process. Another concern is qRT-PCR's susceptibility to two issues: first, it can be confounded by similar genetic sequences, and second, it may amplify unintended DNA fragments. Both of these problems can affect the precision of the results. Despite these challenges, qRT-PCR has been widely used because of its speed and sensitivity. When properly controlled and optimised, it can provide reliable measurements of gene expression.

Bulk RNA-seq can also be used for various applications, including identifying differentially expressed genes, detecting novel transcript isoforms, analysing gene fusion events, and predicting functional pathways and gene networks. However, while bulk RNA-seq provides valuable information on a sample's average gene expression profile, it cannot distinguish between different cell types or spatially resolved gene expression patterns within a tissue or organ. qRT-PCR and bulk RNA-seq are both used to study gene expression, but they differ in their approach and capabilities. qRT-PCR measures the expression of specific target genes with high precision, making it suitable for a small number of genes in a limited sample. In contrast, bulk RNA-seq provides a comprehensive analysis of the entire transcriptome, offering a global view of gene expression for all genes in a sample. In other words, while qRT-PCR is highly sensitive and specific, it is more targeted. Bulk RNA-seq provides a broader overview but requires more input material and is generally less sensitive to low-abundance transcripts. These limitations of bulk techniques have led to new developments in the field. In recent years, there has been increasing interest in developing spatial transcriptomics techniques that enable the investigation of gene expression patterns in their native tissue context, including single-cell and spatially resolved transcriptomics methods.

Single-Cell RNA Sequencing (scRNA-seq)

scRNA-seq is a whole-transcriptome approach that captures the entire mRNA transcriptome of individual cells. In scRNA-seq, the mRNA from each individual cell is captured and sequenced, allowing for the measurement of the expression levels of thousands of genes in each cell. This provides a comprehensive view of gene expression in each cell, helping to identify different cell types and characterise cellular states and transitions. While scRNA-seq does not explicitly target a subset of genes, it does require careful consideration of the experimental design and data analysis to ensure high-quality data and accurate interpretation of the results. Additionally, some scRNA-seq methods, such as Smart-seq2 [Picelli et al., 2013], the C_1 Single-Cell Auto Prep System with SMARTer Ultra Low RNA library kit [Pollen et al., 2014], and multiple annealing and dC-tailing-based quantitative single-cell RNA-seq (MATQ-seq) [Sheng et al., 2017], are able to capture full-length cDNA. This allows for more precise transcript quantification compared to other methods, such as MARS-seq [Jaitin et al., 2014], Cell Expression by Linear Amplification and Sequencing (CEL-seq) [Hashimshony et al., 2012], Drop-seq [Macosko et al., 2015], InDrop [Klein et al., 2015], Chromium [Zheng et al., 2017], SEQ-well [Gierahn et al., 2017], and SPLiT-seq [Rosenberg et al., 2018]), which primarily capture only the 3′ end of the mRNA.

The methods used for single-cell isolation in these technologies can be grouped into four platform categories. (i) Plate-based platforms include Smart-seq2, MATQ-seq, MARS-seq, CEL-seq, and SPLiT-seq. (ii) Droplet-based platforms consist of Drop-seq, InDrop, and Chromium. (iii) The C_1 SMARTer method belongs to microfluidics-based platforms, while (iv) SEQ-well falls under Nanowell array-based platforms.

Additionally, in some of the techniques, a random *n*-mer oligonucleotide sequence that tags individual mRNA molecules called unique molecular identifiers (UMIs) are incorporated, allowing for the distinction between original molecules and amplification duplicates derived from the cDNA or library amplification [Kivioja et al., 2012]. The number of distinct tags is intended to significantly exceed the number of copies of each transcript species to be amplified, resulting in uniquely tagged molecules and allowing control for amplification biases. It is shown that utilising UMI information improves the quantification of mRNA molecules [Grün et al., 2014; Islam et al., 2014]. It has been implemented in several scRNA-seq protocols, such as STRT [Islam et al., 2014], CEL-seq [Grün et al., 2014], CEL-seq2 [Hashimshony et al., 2016], Drop-seq [Macosko et al., 2015], InDrop [Klein et al., 2015], and MARS-seq [Jaitin et al., 2014]. A detailed review can be seen in Haque et al., 2017 and Adil et al., 2021. Here are some of the most common methods for performing scRNA-seq, each with its advantage and limitations:

1. *CEL-seq2*: This method uses a modified version of the template-switching oligo (TSO) approach to capture full-length mRNA transcripts from single cells. It is a low-throughput method that enables high-quality gene expression profiling of small numbers of cells. CEL-seq2 is based on the Illumina sequencing platform and utilises a UMI barcoding system to minimise technical noise and enable accurate quantification of gene expression levels.
2. *Smart-seq2*: This method involves capturing the mRNA using oligo-dT primers attached to beads, followed by reverse transcription, amplification, and library preparation. Smart-seq2 allows for full-length cDNA amplification, which can provide more accurate transcript quantification, but it can also introduce biases.
3. *Drop-seq*: This method involves encapsulating single cells and beads into droplets, where the mRNA is captured and reverse-transcribed. Drop-seq enables high-throughput scRNA-seq with low cost and high sensitivity but may have lower accuracy than other methods.
4. *InDrops*: This method involves capturing cells and beads into droplets, followed by reverse transcription, amplification, and library preparation. InDrops allows for high throughput with low cost and high sensitivity but requires specialised microfluidic devices.
5. *Chromium*: This method involves isolating single cells and barcoding the resulting cDNA libraries with UMIs during the reverse transcription step. The barcoded cDNA libraries are then pooled and sequenced using Illumina sequencers. Developed by 10x Genomics, this approach offers high throughput but may have lower sensitivity and coverage compared to some other methods.
6. *Seq-Well*: This method captures cells in nanolitre-sized wells coated with oligo-dT primers, followed by reverse transcription, amplification, and library preparation. Seq-Well enables high-throughput scRNA-seq with low-input material but may have higher technical variation than other methods.

According to a comparison of different techniques by Ziegenhain et al. [Ziegenhain et al., 2017], Smart-seq2 detected the most genes per cell and across cells, and CEL-seq2, Drop-seq, MARS-seq, and SCRB-seq quantified mRNA levels with less amplification noise due to the use of UMIs. The comparison of different sequencing depths showed that Drop-seq is the most cost effective for transcript quantification for a large number of cells. At the same time, MARS-seq, SCRB-seq, and Smart-seq2 efficiently analyse fewer cells. Each of the aforementioned methods has its own advantages and limitations, and the choice of method depends on the specific research question and experimental design.

Steps Involved in the RNA-seq Protocol

Drop-seq was a pioneering scRNA-seq technology developed by researchers at Harvard Medical School. It is a microfluidics-based method that uses droplet technology to capture individual cells in oil droplets together with unique barcoded beads and reagents. Drop-seq enables the analysis of thousands of single cells in parallel and has been widely used for profiling the transcriptomes of complex tissues and organisms. Drop-seq was an early method that inspired the development of droplet-based technologies like the Chromium System by 10x Genomics, which has further advanced the field of single-cell genomics with its high-throughput capabilities. The Chromium System allows researchers to encapsulate individual cells in droplets together with unique barcoded beads and reagents, enabling high-throughput parallel processing of single cells. The Chromium System offers various applications, including scRNA-seq, single-cell immune profiling, and spatial transcriptomics. Early studies in single-cell research targeted 100–200 cells, but advancements in sorting strategies now allow the capture of up to 1 million cells. The main steps of the experimental protocol in scRNA-seq technologies typically include the following:

1. *Cell isolation*: The cells of interest are isolated from the tissue or sample of interest. The isolation method may vary depending on the type of cells and the sample source.
2. *Cell lysis and mRNA capture*: The cells are lysed to release the mRNA, which is then captured using oligo-dT primers attached to beads or a microfluidic device.
3. *Reverse transcription and cDNA amplification*: The captured mRNA is reverse-transcribed into cDNA and amplified using PCR or other amplification methods to generate enough material for sequencing.
4. *Library preparation*: The amplified cDNA is fragmented, and adapters are added to the ends of the fragments. The resulting cDNA library is then sequenced using high-throughput sequencing platforms.
5. *Postprocess*: The resulting sequencing data are then processed and analysed using specialised software and algorithms to identify and quantify gene expression levels in each individual cell.

Depending on the specific scRNA-seq protocol and experimental design, additional steps may be included, such as quality control checks and normalisation steps to account for technical variation.

Postprocess Steps Involved after scRNA-seq Experiments

The experimental protocol of scRNA-seq technology provides the transcriptome of individual cells. It involves single-cell isolation, cell lysis, transcript capture, cDNA conversion and amplification, library preparation, and sequencing. The sequencing machine generates the FASTQ file format containing the sequencing reads and their corresponding quality scores. Several steps are involved in generating a qualitative cell-by-gene count matrix from the FASTQ file:

1. *Quality control (QC)*: QC is performed to identify potential issues, such as adapter contamination and low-quality bases (contains Ns). The FastQC tool can assess the read quality, sequence content, and generate a QC report summarising quality score, base content, and statistics to spot potential problems in library preparation or sequencing protocol steps.
2. *Alignment and mapping*: Mapping millions of short cDNA fragment reads back to a reference genome or (*de novo* assembled) transcriptome is one of the critical steps in converting the raw data into quantitative data. Popular tools for aligning scRNA-seq reads include Cell Ranger, zUMIs, Kallisto-bustools, and STARsolo. The main criteria for selecting a tool are its ability to accurately quantify gene abundance, alignment accuracy, computational speed, and memory usage.

   In scRNA-seq, two main quantification schemes are typically used: gene counts and transcript counts. The gene counts scheme, similar to that used in bulk RNA-seq, can be problematic in scRNA-seq due to amplification bias. This bias refers to the uneven amplification of transcripts, which may occur during PCR amplification, cDNA synthesis, or as a result of RNA degradation. To mitigate this bias, individual transcripts are barcoded with UMIs before amplification. This approach provides a more accurate quantification of transcript abundance, as it allows for the identification and counting of unique RNA molecules, rather than PCR duplicates.

The sequenced fragments, derived from cDNA, represent either fully spliced or partially spliced transcripts. Such reads often span the boundaries of splice junctions. This results in contiguous read sequences with constituent subsequences that may be separated by large nucleotide gaps. There are two approaches to perform this task: spliced alignment against a reference genome, and unspliced alignment against a reference transcriptome. The first approach is used for gene-level quantification, which assigns fragments to genes. The second approach is used for transcript-level quantification, which assigns fragments to specific transcripts.

Depending on the sequencing protocol, the resulting raw sequence file will contain a combination of the following: cell-level information, commonly known as 'cell barcodes'; UMIs; and the raw cDNA sequence (read sequence) corresponding to the original RNA transcript.

3. *Data preprocessing*: In this step, low-quality cells need to be filtered out. This can be performed using tools such as Scanpy, Seurat, or RaceID. In an ideal scRNA-seq experiment, each droplet should contain a single cell and a single barcoded bead. Droplets meeting these criteria and yielding high-quality data are considered to contain high-quality cells. But droplets may be flagged as containing low-quality data for several reasons: poor-quality sequencing reads, detection of too few unique genes, or insufficient sequencing depth (low count of total reads per cell). Low-quality droplets can be empty, contain only ambient RNA, or have doublets or dead cells. Dead cells often show high mitochondrial gene counts due to the relative stability of mitochondrial membranes during cell death. These low-quality data points can distort downstream analysis and should be removed. There are three criteria for removing them:
   - The number of genes per barcode.
   - The fraction of counts from mitochondrial genes per barcode.
   - The number of counts per barcode.

   Some analysts recommend filtering out ribosomal genes (prefixed with RPS or RPL), haemoglobin (containing HB), and genes starting with Gm-, in addition to mitochondrial genes (prefixed with MT-), to reduce potential biases. In Scanpy, methods such as `'sc.pp.calculate_qc_metrics()'`, `'sc.pp.filter_genes()'`, and `'sc.pp.filter_cells()'` are typically used for these purposes. After the filtering, the gene-by-cell count matrix is normalised to account for differences in library size and sequencing depth. The following are two popular normalisation methods:
   - *Shifted log transformation*: $log(Y/s + 1)$, where Y is the count of a gene in a cell, and s is a size factor. This method should not be confused with log-transformed counts per million (CPM), which is sometimes incorrectly used in tutorials.
   - *SCTransform*: This method uses an analytic approximation of Pearson residuals. It fits a regularised negative binomial regression model, using sequencing depth as a covariate, to obtain transformed count matrices. This approach aims to remove technical effects while preserving biological variation.

   Shifted transformation works better for stabilising variance for subsequent dimensionality reduction, while Pearson residuals are better suited for selection of biologically variable genes and identification of rare cell identities.
4. *Data exploration and visualisation*: In Scanpy workflows, violin and scatter plots are often used to visualise key quality measures. These typically include the number of genes expressed per cell, total RNA counts per cell, and the percentage of counts from mitochondrial genes. These measures are usually stored in the `.obs` attribute of the Python object `AnnData` with user-defined names (e.g., `'n_genes_by_counts'`, `'total_counts'`, `'pct_counts_mt'`). For further exploration, dimensionality reduction techniques such as Principal Component Analysis (PCA) and neighbourhood graphs are used to visualise the cells in lower-dimensional space.
5. *Clustering*: Community detection methods are typically applied to neighbourhood graphs to identify distinct cell populations in the data. These clusters are then assigned cell type labels based on the expression of marker genes characteristic of each cluster. Leiden and

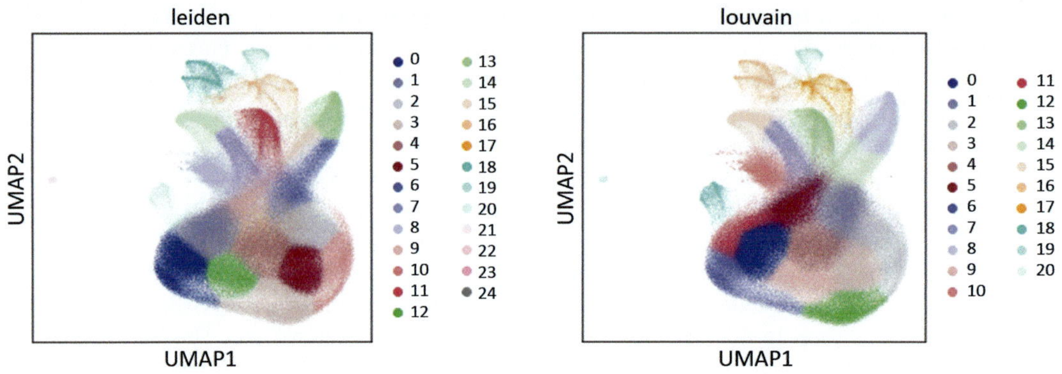

FIGURE 13.1 A typical example of Leiden and Louvain clustering for scRNA-seq data.

Louvain algorithms are commonly used clustering methods in scRNA-seq data analysis. An example is shown in Figure 13.1.

6. *Differential gene expression (DGE) analysis*: In Scanpy, DGE analysis can be performed using the Python method '`sc.tl.rank_genes_groups`'. This method offers three approaches: t-test, Wilcoxon rank-sum test, and logistic regression, to identify genes upregulated in one cluster compared to others. These differentially expressed (DE) genes often serve as marker genes to annotate clusters. The results can be visualised using Python methods like '`sc.pl.dotplot`'.
7. *Functional enrichment analysis and cell-cell communication*: Gene set enrichment analysis can be performed to group the DE genes into functionally related sets, identifying enriched biological processes or signalling pathways. Additionally, ligand-receptor analysis can be used to infer potential cell-cell communication patterns.
8. *Cell compositional analysis*: Sometimes, rather than gene expression changes, the proportions of cell types shift significantly across conditions (e.g., development, disease, perturbation). Tools like scCODA and MILO are designed to analyse these significant changes in cell-type proportions.
9. *Cell trajectory analysis*: Cell identities often exist on a continuum rather than as discrete states. This analysis is useful for studying dynamic processes such as stem cell differentiation, development, lineage progression, cellular state changes, or tumour evolution. It aims to map cell fates in this continuous space. Popular tools include Pseudotime, Monocle, Slingshot, RaceID, RNA Velocity (implemented in Velocyto and scVelo), and CellRank, among others.
10. *Multimodal analysis*: Finally, while scRNA-seq provides valuable transcriptomic information, it captures only one aspect of cellular biology. Cells in tissues interact with their neighbours to perform coordinated functions. These functions are influenced by multiple factors, including the transcriptome, epigenetics (e.g., chromatin accessibility from ATAC-seq, DNA methylation), protein abundance, and cellular context (e.g., spatial relationships from spatially resolved transcriptomics).

    The integration of these different data types (or modalities) is challenging because they often come from different samples or even different individuals of the same species, rather than from the same exact cells. Common computational methods aim to find corresponding cells across different data types. Deep learning (DL) approaches can be particularly effective at this, creating a joint lower-dimensional representation of the combined datasets, often referred to as a 'latent space' in machine learning (ML). This latent space represents inferred, hidden features that capture the essential characteristics of the integrated data across all modalities.

Building on this integrated data representation, researchers can then address fundamental questions about tissue biology and disease. For instance, how does multicellular regulation occur within tissues, particularly during processes like regeneration? In disease conditions, such as various types of cancer, how can we identify potential therapeutic targets (genes or proteins) that are effective across multiple cell types or cancer subtypes?

POPULAR ML AND AI METHODS USED IN NGS

The term 'artificial intelligence' implies that machines are programmed to think, learn, and perform tasks typically requiring human intelligence. This term encompasses a wide range of techniques, algorithms, and approaches that enable machines to process and analyse data, recognise patterns, make decisions, and solve problems. In modern biological research, AI has become a valuable tool for scientists in various ways, reshaping the field and accelerating discoveries. Here are some ways in which AI helps with new discoveries in biology:

1. *Genomics and sequencing*: AI is extensively used in genomics and DNA sequencing to analyse large volumes of genomic data quickly and efficiently. It helps in identifying genetic variations, gene mutations, and disease-causing genes.
2. *Drug discovery and development*: It is used to screen and analyse vast databases of compounds to identify potential drug candidates. It helps in predicting drug–protein interactions, simulating drug interactions, and designing novel drugs.
3. *Disease diagnosis and prognosis*: AI-powered diagnostic tools can analyse medical images, pathology slides, and patient data to assist in the early detection and accurate diagnosis of diseases, such as cancer.
4. *Personalised medicine*: AI algorithms can analyse individual patient data, including genetics and medical history, to tailor personalised treatment plans and predict treatment outcomes.
5. *Single-cell biology*: AI techniques like ML are used in analysing scRNA-seq data to identify distinct cell types, understand cell heterogeneity, and study cellular interactions.
6. *Bioinformatics*: It plays a crucial role in bioinformatics in analysing biological data, such as DNA sequences, protein structures, and molecular interactions.
7. *Drug repurposing*: It can identify existing drugs that could be repurposed to treat different diseases, potentially accelerating the development of new therapies.
8. *Systems biology*: AI models help in understanding complex biological systems and signalling pathways, revealing interactions between genes, proteins, and other biomolecules.
9. *Biomarker discovery*: It can analyse large datasets to identify potential biomarkers for various diseases, aiding in early detection and treatment monitoring.
10. *Biomedical imaging*: AI-powered image analysis tools can process medical images, such as MRI and CT scans, and histopathological images, assisting in accurate diagnoses and reducing interpretation time.

Overall, AI is transforming the field of biology by providing researchers with powerful tools to analyse and understand complex biological data, accelerating drug discovery, improving disease diagnosis and treatment, and facilitating personalised medicine. Its ability to handle vast amounts of data and identify patterns and relationships in biological systems, has the potential to unlock new insights and fundamentally alter our approach to healthcare and biotechnology. There are various tasks and methods in AI used to perform intelligent functions and mimic human-like behaviour. They will be discussed in the following subsection.

MAIN TASKS IN AI

ML is a primary task in AI. It enables machines to learn from data and improve their performance over time without being explicitly programmed. ML encompasses several subtasks, discussed in the following subsections.

Supervised Learning

Supervised learning involves the utilisation of labelled datasets to train algorithms capable of either categorising data or accurately predicting outcomes. Within the realm of supervised learning, the primary tasks encompass *classification* and *regression*. During the model's training, input data are supplied, and the data's weights are adjusted until they achieve an appropriate fit, as part of the cross-validation process. This form of learning aids organisations in tackling diverse real-world challenges at scale, such as distinguishing spam from legitimate emails or adjusting production levels of frozen products in a company, based on temperature fluctuations across different timeframes (months, weeks, or days).

In *classification* tasks, users provide training data (*X*) as feature vectors, along with labelled outputs (*Y*). ML models require numeric inputs and outputs, so ordinal or categorical *Y* labels must be encoded. Ordinal encoding assigns integers to categories (e.g., `bobcat:1`, `wolf:2`, `dog:3`), while one-hot encoding represents categories as binary vectors (e.g., `bobcat:[1,0,0]`, `wolf:[0,1,0]`, `dog:[0,0,1]`). The term 'one-hot' refers to the single '1' value in each encoded vector. One-hot encoding can increase feature count, potentially triggering the 'curse of dimensionality'—where high-dimensional data becomes sparse and harder to analyse effectively. The choice between encoding methods depends on data characteristics, algorithm requirements, and analysis goals. To ensure equal feature contribution and prevent dominance of larger-valued features, users must perform feature scaling, bringing *X* to a similar scale. Figure 13.2 illustrates a learned decision boundary from a classification algorithm.

Classification encompasses four primary tasks: (i) binary classification, which involves distinguishing between a normal state (e.g., a negative test result for a condition) and an abnormal state (e.g., a positive test result); (ii) multiclass classification, which aims to classify data into more than two classes, such as different cell types within a tissue; (iii) multilabel classification, which predicts multiple outputs (two or more class labels), treating each output as a Bernoulli probability distribution; and (iv) imbalanced classification, which pertains to situations where classes are unevenly distributed, e.g., detecting a rare pathological condition (1%) versus a normal healthy condition (99%).

Regression involves predicting a continuous outcome variable based on input features – for example, estimating a person's weight (in kilogrammes or pounds) using characteristics like height, gender, and age. Common regression techniques include linear regression, lasso regression (which uses L1 regularisation), and ridge regression (which uses L2 regularisation). Overfitting, a frequent challenge in regression, can be understood through the bias-variance trade-off, illustrated in Figure 13.3. Bias represents the difference between the model's average prediction and the true value it is trying to predict.

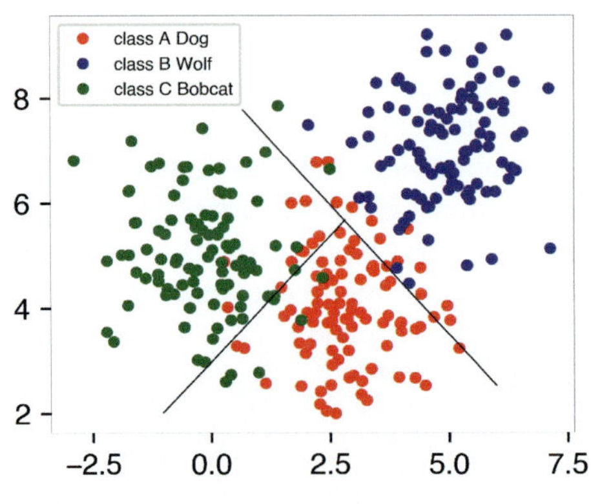

FIGURE 13.2 A typical example of a classification task where black decision boundaries separate three classes, such as dog, wolf, and bobcat.

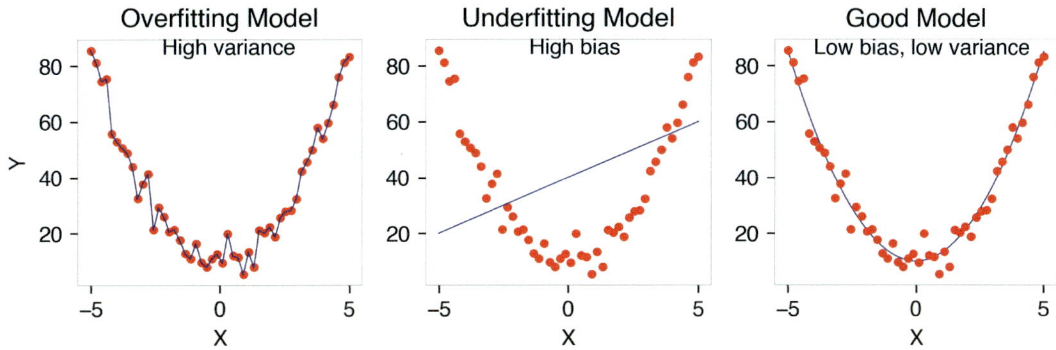

FIGURE 13.3 An example typical of overfitting, underfitting, and good model practice in ML tasks.

The bias-variance trade-off is a fundamental concept in ML. High bias leads to underfitting, where the model fails to capture the underlying patterns in the data. Conversely, high variance results in overfitting, where the model becomes too sensitive to the specific training data. Striking the right balance between these extremes is crucial for developing robust models. Generally, models with few parameters tend to have high bias and low variance, while those with many parameters exhibit high variance and low bias. To address this trade-off, techniques like lasso and ridge regression employ regularisation. Lasso and ridge regression add regularisation terms to balance model complexity: lasso for selecting important features, ridge for adjusting all features' impact, both aiming to improve performance on unseen data.

Unsupervised Learning

Unsupervised learning, a fundamental concept in ML, empowers researchers to delve into data and uncover hidden structures and relationships. Think of it as deciphering intricate patterns seen in a complex jigsaw puzzle without having to know the whole picture. When using unsupervised learning algorithms, we let the algorithms explore datasets autonomously while they unveil inherent relationships and assemble data-point clusters based on shared features. These clusters might correspond to distinct classes or categories within the data, even when we have not explicitly specified what they are to the algorithm. Later, using expert knowledge and insights, we can assign labels based on context. Unsupervised learning finds applications in various real-world scenarios. For example, in genetics and NGS data, scientists may group genes that behave similarly, even without specifying functions. This is akin to using unsupervised learning for market analysis, where it can help identify groups of customers with similar buying patterns, leading to more effective marketing strategies. Unsupervised learning models encompass three primary ML tasks: clustering, association, and dimensionality reduction, which are explained next.

Clustering is a technique used to categorise unlabelled data based on inherent similarities or dissimilarities. Various algorithms exist for this purpose, each with distinct approaches. The *k*-means and *k*-medoids algorithms, for example, are partitioning methods that assign data points to a predetermined number (k) of clusters based on their proximity to cluster centroids or medoids. These methods typically produce convex clusters, which may not always suit complex biological data but are often appropriate for applications like market analysis, document clustering, image segmentation, and compression. In contrast, hierarchical clustering builds a nested hierarchy of clusters based on data point similarities. This approach employs four main linkage types to measure similarity: Ward's, Average, Complete, and Single linkage, each offering different perspectives on data relationships. Hierarchical clustering can be implemented using two distinct approaches: agglomerative (bottom-up) and divisive (top-down). The agglomerative method begins with each data point as a separate cluster, progressively merging similar clusters until a single, comprehensive cluster remains. Conversely, the divisive approach starts with all data points in one large cluster, iteratively dividing it based on dissimilarities until reaching individual data points or a predetermined number of clusters. These processes are commonly visualised using a dendrogram, a tree-like structure illustrating the merging or splitting of data points at each stage. The final number of clusters can be determined by selecting a cutoff at different levels of the hierarchy. In contrast, probabilistic clustering techniques, such as the Gaussian mixture model, assign data points to groups based on their likelihood of belonging to specific probability distributions. In this model, clusters are defined by normal distributions with varying means and variances.

Association rules focus on uncovering relationships between variables in the data. For instance, a recommender system might predict that if a person likes an item, they will like something similar. Algorithms like *k*-nearest neighbours (KNN), Bayesian inference, *Apriori* list, Non-negative Matrix Factorisation (NMF), Latent Semantic Analysis (LSA), and Latent Dirichlet Allocation (LDA) are employed to identify these associations.

Dimensionality reduction is crucial when working with high-dimensional data. As the number of dimensions (or features) increases, data can become sparse, and models may become overly complex, potentially capturing noise instead of meaningful patterns. This phenomenon, known as the 'curse of dimensionality', can lead to overfitting in ML algorithms, especially when attempting to analyse or explain high-dimensional data. To mitigate this issue, techniques like PCA and NMF are employed. These methods transform the data into lower-dimensional spaces while retaining essential information, thereby reducing the risk of overfitting and improving model performance. PCA identifies orthogonal linear combinations of original features that capture maximum variance in the data, creating new variables called principal components. Figure 13.4 illustrates this concept using a 3D ellipsoid. The first principal component captures the largest variance, with subsequent components capturing progressively less variance. By selecting the top *k* principal components, one can reduce data dimensionality while preserving most of the variability. In contrast, NMF decomposes a non-negative data matrix into two smaller, non-negative matrices: the basis matrix (W) and

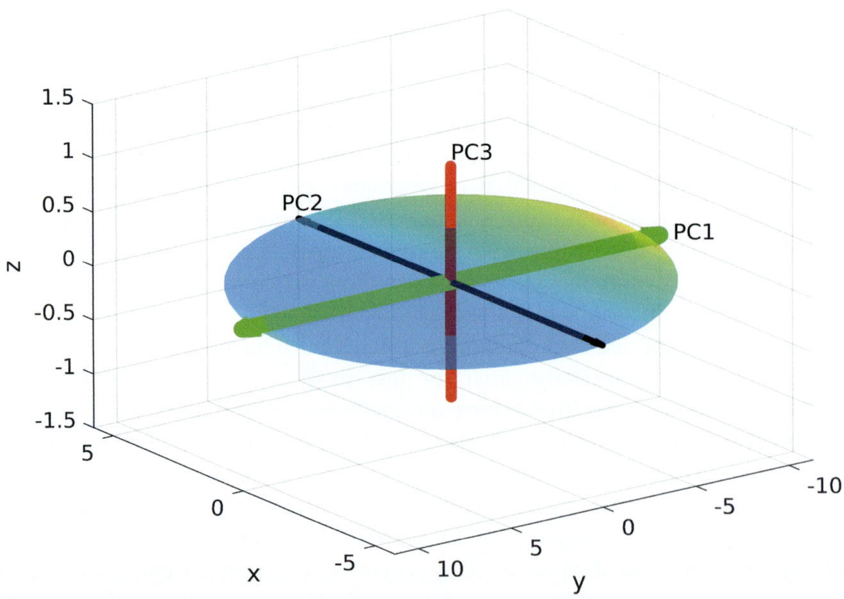

FIGURE 13.4 An example of PCA for the ellipsoid shape. If object shape is provided as point cloud data points, then the PCA principal component is equivalent to the principal axes of the object. In this case, the ellipsoid major axis PC1 is shown in green, the ellipsoid minor axis PC3 is shown in red, and the ellipsoid middle axis PC2 is shown in black.

the coefficient matrix (H). This approach approximates the original high-dimensional data in a lower-dimensional space, maintaining non-negativity constraints. It is particularly effective for data with inherent additive structures, such as images, signals, or document word counts. Unlike PCA, which optimises for axes of maximum variability, NMF focuses on learning key components or features. This results in a parts-based representation of the data, often yielding more interpretable results than PCA in certain contexts. Another dimensionality reduction technique, t-distributed Stochastic Neighbour Embedding (t-SNE), specialises in visualising high-dimensional data clusters by reducing them to two or three dimensions, thereby revealing underlying patterns and relationships in the data.

Reinforcement Learning

Reinforcement learning (RL) is inspired by how humans and animals learn from trial-and-error experiences, adjusting their actions based on environmental feedback in the form of rewards or punishments. RL focuses on teaching agents to make sequential decisions in an environment to maximise cumulative rewards. An agent interacts with its environment by performing action and receives feedback as rewards, indicating the action's effectiveness in achieving goals. The agent's objective is to learn a policy that maps environment states to actions, maximising the expected total reward over time. RL is applicable in scenarios where optimal strategies can be learned through repeated environmental interactions, such as training robots, programming self-driving cars, optimising business processes, finding hidden states between healthy and disease states, and controlling complex systems.

RL is generally slower than supervised or unsupervised learning methods due to several factors. RL involves 'sequential decision-making', where an agent interacts with the environment over multiple time steps, making decisions that affects subsequent state and rewards. This sequential nature inherently requires more time than processing single instances, as in supervised learning. RL also necessitates 'exploration' of different actions to learn an optimal policy, balancing this with exploitation of known rewards. Exploration in RL involves actions that may not yield immediate rewards but provide crucial environmental information. Balancing this exploration with exploitation of known rewards requires additional time, slowing the learning process. Often, rewards in RL are delayed, necessitating the agent to understand how its actions influence future outcomes, which extends learning times. RL's trial-and-error approach, where the agent learns from mistakes and iteratively improves its strategy, can be slower than supervised learning's direct learning from labelled data. Despite these challenges, RL excels at solving complex problems involving long-term decisions and dynamic environments, making it valuable for various applications. Researchers are continually developing more efficient RL algorithms to address these speed-related issues.

RL algorithms can be categorised as either model-based or model-free. Model-based algorithms learn environmental dynamics to predict future states and rewards, while model-free algorithms estimate optimal strategies directly through trial and error. Q-learning, a model-free approach, enables agents to learn optimal actions iteratively. It employs a Q-function to estimate expected future rewards for actions in given states. The Q-values are updated using the Bellman equation as the agent interacts with the environment, refining its decision strategy over time. Deep Q Networks (DQNs) enhance Q-learning by using deep neural networks (DNNs) to approximate Q-values, allowing the algorithm to handle complex, high-dimensional state spaces typical in real-world applications. DQNs have demonstrated remarkable success across various domains, from video game playing to robot control, by autonomously learning sophisticated reward-maximising strategies in diverse environments.

Semi-Supervised Learning

Labelling data can be challenging, particularly when distinguishing between varied images like different dog and cat breeds. In complex scenarios, such as determining cell normality or tumorigenicity in tissues, the task becomes expensive and requires experienced domain experts. Semi-supervised learning (SSL) addresses these challenges by utilising training datasets that include both labelled and unlabelled instances. This approach leverages the explicit information provided by labelled data while also benefiting from the underlying structures and relationships revealed by unlabelled data features.

SSL comprises several key steps. Initially, a subset of training data with known labels is selected, helping the model make accurate predictions for similar instances. Next, the model trains on the unlabelled data segment, capturing the underlying data distribution and enhancing its ability to generalise. Finally, SSL integrates insights from both labelled and unlabelled data, improving the model's understanding of the data's internal structure and relationships. Throughout this process, the model aims to develop a data representation that effectively distinguishes between different classes or clusters.

The inclusion of unlabelled data primarily serves to regularise the model and reduce overfitting risk. This approach helps the model better capture true underlying patterns, potentially improving performance on both labelled and unlabelled data. In essence, SSL harnesses the strengths of both data types to build more robust and accurate ML models. However, carefully designing the learning process and selecting suitable algorithms remain crucial for fully leveraging the available information.

Transfer Learning

Transfer learning applies knowledge gained from one task to enhance performance on a related task. Rather than creating models from scratch for each new problem, transfer learning utilises pretrained models that have learned representations from large datasets in an initial task. These learned representations, often referred to as features or weights, capture general patterns and high-level characteristics that can be applied across different tasks. The core principle of transfer learning is to exploit commonalities between source and target tasks. The underlying assumption is that knowledge acquired in solving one task can benefit another, even if the tasks have different data distributions or slightly varying objectives. This approach is particularly valuable when the target task has limited labelled data, as the pretrained model already contains rich information from the source task.

Transfer learning can be implemented in various ways. A common method is fine-tuning, where the pretrained model's parameters are adjusted to align with the target task's data. This enables the model to adapt to the new task's specific characteristics while maintaining the general knowledge acquired from the initial task. Another approach is feature extraction, which uses the pretrained model's layers as feature extractors for the target task. In this method, the features learned for the source task serve as input for a new classifier designed for the target task.

Transfer learning has applications in a variety of fields, including image processing, natural language processing, and even specialised fields such as medical imaging, genomics, and transcriptomics. It saves time and computational resources, accelerates model convergence, and often leads to better performance, especially when there is limited labelled data for the target task. The ability for models to leverage prior knowledge has made transfer learning an indispensable tool for solving complex ML problems efficiently and effectively.

Classical AI Methods

The field of ML algorithms has evolved significantly in recent decades, producing a diverse array of algorithms that enhance our ability to extract insights from data. Within this domain, a wide range of classical AI methods has emerged, each offering unique strengths and applications:

- *Neural networks (NNs)*: These algorithms, inspired by the human brain's structure and function, consist of interconnected nodes, or 'neurons', organised in layers. A typical NN includes an input layer receiving data, one or more hidden layers processing it, and an output layer delivering the final result. Connections between neurons have weights representing their strength. Depending on network architecture, complexity, and problem-solving capabilities, NNs can be classified into several types, including:

1. *Perceptron model*: This is one of the simplest and oldest models of an NN, proposed by Minsky et al. (1998). It accepts weighted inputs and applies an activation function to produce the final output.
2. *Feedforward*: These NNs are the simplest and consist of input, hidden, and output layers. Data information flows unidirectionally from the input layer to the output layer. They are mainly used for pattern recognition and classification tasks.
3. *Backpropagation*: In the training process, this NN feeds back error rates (i.e., loss) of the previous epoch through the network to fine-tune the weights. This process ensures lower error rates and increases the reliability and generalisation of the model.
4. *Multilayer perceptron*: This complex NN features multiple layers of interconnected neurons. Every node in one layer is connected to all nodes in the next, making it a fully connected NN. It employs both forward and backward propagation, adjusting weights to minimise loss. Nonlinear activation functions are used followed by the 'softmax' function as an output layer activation function.
5. *Convolutional (CNNs)*: These networks are designed for image and video analysis tasks. They utilise convolutional layers to automatically learn and extract hierarchical features from images, making them highly effective for classification, object detection, and image generation.
6. *Recurrent (RNNs)*: These are designed to process sequential data, making them suitable for tasks involving time-series data or sequential information, such as speech recognition, etc.. They have feedback connections that allow information to be passed from one step of the network to the next, enabling them to capture temporal dependencies.
7. *Long short-term memory networks (LSTMs)*: LSTMs are a type of RNN that addresses the vanishing gradient problem, making them more effective at capturing long-range dependencies in sequences. They are widely used in language modelling, machine translation, and speech recognition applications.
8. *DL*: DL extends the NN concept by introducing 'deep' networks with numerous hidden layers. DL models can have tens, hundreds, or even thousands of hidden layers, enabling them to learn hierarchical features and representations from raw data automatically. The depth of these networks allows them to capture intricate patterns and relationships in data, making them highly suitable for complex tasks such as image and speech recognition, etc.

   DL algorithms have shown remarkable performance improvements in various domains because they can learn and extract features directly from data without relying on explicit feature engineering. This capability is particularly advantageous in handling vast amounts of raw, unstructured data, especially relevant in today's era of Big Data.
9. *Autoencoders*: Autoencoders are designed for unsupervised learning and feature extraction. They consist of an encoder (input layer to the hidden layer) that compresses input data into a lower-dimensional representation and a decoder (hidden layer to the output layer) that reconstructs the original data from the compressed representation. Autoencoders are used for dimensionality reduction, data denoising, and anomaly detection.

- *Decision trees (DTs)*: DTs are tree-like structures used to make decisions based on a series of if-then-else rules. They are often used in classification and regression tasks.
- *Random forest (RF)*: The RF algorithm, an extension of DTs, constructs multiple DTs to form a robust ensemble model that improves prediction accuracy and deals with overfitting.
- *Support vector machines (SVMs)*: An SVM is a supervised learning algorithm used for classification and regression tasks. It works by finding the optimal hyperplane that separates the data into different classes.

- *Genetic algorithms (GAs)*: GAs are optimisation algorithms based on the principles of natural selection and genetic inheritance. They are used to solve complex optimisation problems where traditional methods may struggle. GAs are particularly useful when dealing with vast solution spaces or problems comprising multiple objectives. The foundational idea behind a GA is to mimic the process of natural evolution, leveraging concepts like selection, crossover, and mutation. The algorithm starts with an initial population of potential solutions, represented as individuals or chromosomes. Each individual representative is evaluated based on a fitness function; an evaluative metric gauging solution efficacy plays a pivotal role in this evaluation.
- *Evolutionary algorithms (EA)*: EAs are a broader realm of optimisation algorithms that encompass GAs as a subset. While GAs focus on the evolution of populations, EAs extend this concept to include other techniques such as evolution strategies and genetic programming. A distinctive feature demarcating GAs from their evolutionary counterparts is the capacity of EAs to navigate both discrete and continuous optimisation predicaments, whereas GAs are optimally tailored for discrete challenges. This inherent adaptability renders EAs amenable to a wider spectrum of optimisation scenarios.

The field of AI is constantly evolving, with researchers exploring novel techniques and applications to enhance the capabilities of these systems. Recent advancements in generative AI techniques have significantly accelerated progress in AI discovery and application. Consequently, we cannot conclude this section without an introduction to generative AI approaches.

Advanced AI Methods

Generative AI, also known as generative adversarial networks (GANs), is a subfield of AI that focuses on generating new data that resemble a given dataset. GANs can generate new content such as text, images, or audio, similar to human creativity. GANs consist of two NNs, the generator and the discriminator, which are trained together in a competitive manner. The generator produces synthetic data patterns, while the discriminator evaluates whether the patterns are real (from the original dataset) or fake (generated by the generator). Through this adversarial process, the generator improves its ability to produce realistic data, while the discriminator becomes better at distinguishing real from fake data. The key techniques of generative AI are as follows:

- *Variational autoencoders (VAEs)*: These are generative models that learn the underlying representation of the data and can generate new samples from that representation. VAEs are often used for data imputation and generation tasks.
- *Conditional GANs*: These incorporate additional information, such as data labels, into the generative process, allowing for more targeted generation of synthetic data. These models are useful for generating label-specific synthetic samples.
- *Wasserstein GANs (WGANs)*: These address some of the training challenges of traditional GANs by using the Wasserstein distance as a metric for training stability. They are less prone to mode collapse and can produce higher-quality synthetic data.
- *DL for dimensionality reduction*: Generative models can benefit from dimensionality reduction techniques such as t-SNE or uniform manifold approximation and projection (UMAP) to visualise and interpret the generated synthetic data in lower-dimensional spaces. A typical example of UMAP is shown in Figure 13.1.
- *Transformer architecture*: The transformer model was introduced by Vaswani et al. (2017) and ushered in a new era in language understanding and generation tasks. Unlike traditional sequence-to-sequence models, transformers use self-attention mechanisms to capture contextual relationships between words, enabling them to effectively process sequences of varying lengths. This has become the foundation for advanced language models such as

generative pretrained transformer (GPT) and bidirectional encoder representations from transformers (BERT). Transformers have strengthened generative AI by enabling more accurate and coherent text generation, translation, and even tasks beyond language by learning intricate patterns and dependencies within data.

- *ChatGPT*: The well-known tool ChatGPT is built upon GPT-3.5 and GPT-4 and is part of the family of large-scale language models developed by OpenAI to understand and generate human-like text. It is trained on a vast corpus of text data and learns the patterns, context, and grammar of the language to generate coherent and contextually relevant responses. Overall, ChatGPT is a generative AI archetype that demonstrates how advanced language models can understand and generate human-like text, enabling more sophisticated and natural interactions between humans and machines.

The integration of generative AI techniques into NGS data holds significant promise. These advanced methods can enhance data quality by imputing missing values and reducing noise in signals, leading to more accurate downstream analyses such as variant calling and gene expression quantification. Transformer models can predict alternative splicing events, enabling a deeper understanding of transcript diversity [Choi and Lee, 2023]. In addition, they can enhance genomic element annotation and gene network construction, facilitating pathway analysis and uncovering multicellular regulatory mechanisms. This advanced approach can expedite the identification of disease-associated variants, deepen insights into personalised medicine, and accelerate the discovery of novel therapeutic targets. Consequently, these techniques are driving significant advancements in genomics research and its biomedical applications.

AI TECHNIQUES FOR SCRNA-SEQ

AI techniques have emerged as valuable tools for analysing scRNA-seq data. Because of the increasing volume and complexity of these data, traditional analysis methods have become inadequate for fully exploring information and extracting knowledge from scRNA-seq datasets. AI techniques encompass a range of approaches, including ML [Petegrosso et al., 2020], DL [Flores et al., 2022], and data mining [Chen et al., 2023]. These techniques have been used to process and interpret the vast amount of data generated by scRNA-seq experiments. There are many AI techniques used in these analyses, such as clustering and dimensionality reduction [Sun et al., 2019], cell-type identification and annotation [Pasquini et al., 2021], trajectory analysis [Tran and Bader, 2020], differential expression analysis [Wang et al., 2019b], and integration of multimodal data [Lee and Li, 2023]. In the following subsections we explain the relevant AI techniques and their applications in scRNA-seq.

The Role of ML in scRNA-seq

Recently, ML approaches have played a crucial role in scRNA-seq data analysis [Petegrosso et al., 2020; Liu et al., 2021], including enabling extraction of valuable insights from high-dimensional data. In this section, we focus on supervised, unsupervised, and semi-supervised ML techniques for scRNA-seq data analysis.

The Role of Supervised Learning Techniques in scRNA-seq

Supervised learning techniques are a subset of ML algorithms that rely on labelled data, where input samples are paired with corresponding output labels, to learn patterns and relationships between features and the target variables [Nasteski, 2017]. Supervised learning methods have demonstrated remarkable success in several domains, including NGS data analysis [Esposito et al., 2019].

In Table 13.3, we list the tasks related to supervised learning in scRNA-seq analysis, and for each task we show the supervised learning techniques along with corresponding studies.

TABLE 13.3
The Main Tasks Related to Supervised Learning in scRNA-seq Analysis

Main Tasks	Supervised Learning Techniques	Studies (References)
Cell-type classification	• SVMs with a linear kernel • KNN • Gradient boosting machine (GBM), which is an ensemble of DTs • SVM	[Wagner and Yanai, 2018]; [Bishara et al., 2022] [Le et al., 2022] [Alquicira-Hernandez et al., 2019]
Cell-state prediction	• Partial least square (PLS) supervised ML algorithm	[Hao et al., 2023]
Drug response prediction	• LightGBM (tree-based learning algorithm) • RF algorithm	[Xia et al., 2022] [Clayton et al., 2020]
Identification of gene regulatory networks	• Several classifiers, including logical regression (LR), DT, SVM, RF, naive Bayesian (NB), gradient boosting decision trees (GBDT), and KNN • ML models (supervised learning vs. unsupervised learning) • SVM • ML algorithms, including xgboost, SVM, and deepboost	[Yang et al., 2022] [Yang et al., 2022] [Ni et al., 2016] [Turki and Taguchi, 2020]
Gene expression imputation	• LR and SVM models • KNN • RF • KNN, least absolute shrinkage and selection operator (LASSO), with other DL models such as DNN and GAN	[Yang et al., 2018] [De Silva and Perera, 2017]; [Albrecht et al., 2022]; [Mancuso et al., 2020]

Cell-type classification is a task performed in scRNA-seq analysis that uses supervised ML techniques to classify cells into different cell types, mainly based on their gene expression profiles. As shown in Table 13.3, there are many research studies that used supervised traditional techniques for this task, such as Wagner and Yanai [2018], Bishara et al. [2022], and Le et al. [2022]. The second task we show in the table is cell-state prediction, which refers to inferring the specific functional state, or the condition, of individual cells based on their scRNA-seq data.

The Role of Unsupervised Learning Techniques in scRNA-seq

Unsupervised learning techniques play a crucial role in scRNA-seq analysis by enabling the exploration of heterogeneity within the data without relying on predefined annotations or labels. In Table 13.4, we list the tasks related to unsupervised learning in scRNA-seq analysis, and for each task, we show the unsupervised learning techniques, along with corresponding studies.

The Role of DL in scRNA-seq

DL has emerged as a powerful set of techniques that can be used in omics data analysis [Martorell-Maruga´n et al., 2019] and in scRNA-seq analysis [Flores et al., 2022]. DL provides novel and efficient approaches to handle the complexity of the high dimensionality of scRNA-seq data. Several studies reviewed and surveyed the DL techniques utilised for scRNA-seq data analysis [Brendel et al., 2022; Flores et al., 2022]. DL techniques have been instrumental in advancing various aspects of scRNA-seq analysis, offering several key tasks, as shown in Table 13.5.

Limitations of DL in scRNA-seq Data Analysis

While the DL techniques have shown remarkable potential in scRNA-seq data analysis, there are many limitations related to utilising it for that purpose [Kharchenko, 2021; Ma and Xu, 2022]. One main limitation is the data labelling requirements, as the DL models typically require large amounts

TABLE 13.4
The Main Tasks Related to Unsupervised Learning in scRNA-seq Analysis

Main Tasks	Unsupervised Learning Techniques	Studies (References)
Cell clustering and cell-type discovery	• PCA + graph-based • PCA + *k*-means • Data-driven reduction + *k*-means	[Kiselev et al., 2019] [Traag et al., 2019]
Dimensionality reduction and visualisation	• Linear (PCA, ICA) • Non-linear (t-SNE, UMAP) • NN (DCA, scvis, VAE), Ensembl method (SIMLR)	[Jolliffe, 2002; Liebermeister, 2002] [Van der Maaten and Hinton, 2008; McInnes et al., 2018] [Ding et al., 2018; Hu and Greene, 2018; Eraslan et al., 2019] [Wang et al., 2017]
Trajectory inference (cell fate determination by pseudotime)	• Slingshot, PAGE, p-Creode (multibranching graphs), review on best practice	[Street et al., 2018] [Wolf et al., 2019] [Herring et al., 2018] [Saelens et al., 2019]
Cell fate inference by RNA velocity	• Splicing kinetics using unspliced and spliced reads (velocyto, scVelo) • Phase portraits on RNA velocity (Cell-Rank) • LineageOT optimal transport based • CoSpar static barcode lineage tracing	[La Manno et al., 2018; Bergen et al., 2020] [Lange et al., 2022] [Forrow and Schiebinger, 2021] [Wang et al., 2022]
Differential gene expression analysis	• Pseudobulk methods with sum aggregation (edgeR, DESeq2 and limma) • Mixed model with random effect (MAST) comparison across methods such as Wilcoxon rank-sum test and others	[Robinson et al., 2010; Love et al., 2014; Ritchie et al., 2015] [Finak et al., 2015] [Junttila et al., 2022]
Batch effect correction	• Based on linear-embedding models (canonical correlation analysis, Harmony, mnnCorrect, Scanorama)	[Butler et al., 2018; Haghverdi et al., 2018; Korsunsky et al., 2019; Hie et al., 2019]
Gene set enrichment analysis	• Popular databases (MSigDB, GO, KEGG, Reactome) • PROGENy (signalling pathways), DoRothEA (transcription factors), Pagoda2 and benchmarking with other methods	[Ashburner et al., 2000; Liberzon et al., 2011; Kanehisa et al., 2017; Gillespie et al., 2022] [Schubert et al., 2018; Garcia-Alonso et al., 2019] [Zhang et al., 2020b; Barkas et al., 2021]
Cell compositional analysis	• scDC, scCODA, tascCODA • DA-seq and MILO	[Cao et al., 2019; Buettner et al., 2021; Ostner et al., 2021] [Zhao et al., 2021; Dann et al., 2022]
Latent space learning models	• scGen and CellBox	[Lotfollahi et al., 2019; Yuan et al., 2021]

of labelled data, and having enough labelled scRNA-seq data is a challenging task [Torroja and Sanchez-Cabo, 2019]. Another major limitation is computational resource availability, as DL models need intensive computation, especially with large scRNA-seq datasets containing thousands of cells and genes [Lin et al., 2022]. In many cases, the interpretability of the DL models is a challenging task that could limit understanding the biological significance of model outputs, which could in turn limit the DL models' use in some scRNA-seq data analysis applications. More general limitations include identifying the optimal model architecture and hyperparameters, which could require expertise and be time consuming [Thompson et al., 2020].

Prospects and Future Directions of DL in scRNA-seq Data Analysis

Future research in DL for scRNA-seq analysis requires comprehensive benchmarking studies to establish standardised evaluation platforms using large datasets [Brendel et al., 2022]. Enhancing

TABLE 13.5
The Main Tasks Related to DL in scRNA-seq Analysis

Main Tasks	DL Techniques	Studies (References)
Cell-type classification	• DNN models • Multilevel convolutional NN (MulCNN) CNNs	[Torroja and Sanchez-Cabo, 2019] [Jiao et al., 2023] [Jia et al., 2023]
Cell-type identification	• scCapsNet model (DL) • DL based on transformer	[Wang et al., 2020] [Song et al., 2022]
Cell-state prediction	• Backpropagation NN	[Zhu et al., 2023]
Cell clustering	• Autoencoder • DL clustering model	[Li et al., 2020] [Yu et al., 2021]
Trajectory inference	• Autoencoders • Multitask-oriented graph autoencoder	[Du et al., 2020] [Luo et al., 2021]
Gene expression imputation	• GAN • Autoencoder • DNN • Graph-based generative adversarial network RNN	[Xu et al., 2020] [Bishara et al., 2022] [Arisdakessian et al., 2019] [Lakkis et al., 2022] [Huang et al., 2023]
Batch effect correction	• Transfer learning • Adversarial paired transfer networks	[Wang et al., 2021] [Wang et al., 2019a]
Dimensionality reduction and visualisation	• Autoencoder • Deep variational autoencoder	[Wang and Gu, 2018] [Jiang et al., 2023]
Gene regulatory network inferences	• DNN • Multiple-layer perception DL techniques	[Moerman et al., 2019] [Chen et al., 2021] [Luo et al., 2022]

DL performance through standardised approaches for scRNA-seq data integration is also crucial [Brendel et al., 2022]. Developing interpretable DL models is essential for wider adoption, particularly in applications like drug target identification and biomarker discovery [Xu et al., 2022]. Additionally, a clear need exists for user-friendly DL tools that will enable researchers with varying computational expertise to effectively utilise these models for scRNA-seq data analysis and to interpret the results.

The Role of Generative AI in scRNA-seq

Generative AI has found diverse applications in single-cell biology, serving as a powerful tool for data augmentation, imputation, and synthetic sample generation. Several methods exemplify this approach: scGAN [Marouf et al., 2020], scPhere [Ding and Regev, 2021], and scGPT [Cui et al., 2023]. These generative AI techniques enable the integration of cell data from various sources, facilitating a deeper understanding of the common principles or 'grammar' of cell biology. The following elaborates on the utility of generative AI in single-cell biology:

- *Data augmentation*: scRNA-seq datasets often have limited sample sizes due to the high cost and complexity of the technology. Generative AI can be used to augment a dataset by generating additional data that resemble collected data. This augmented dataset can then be used to improve the accuracy and robustness of downstream analysis, such as cell-type classification and trajectory inference.
- *Data imputation*: scRNA-seq data often contain missing values due to technical limitations or low RNA capture rates. Generative AI models can predict these missing values by generating synthetic data that align with existing patterns in the dataset. This imputation process helps preserve more information during analysis, potentially improving the quality and reliability of subsequent analytical steps.

- *Generating synthetic data for benchmarking*: In single-cell biology, generating ground truth datasets can be challenging and time consuming. Generative AI can create synthetic datasets with known properties, allowing researchers to benchmark and validate their algorithms, pipelines, and analytical methods more efficiently.

In summary, generative AI, particularly GANs, can be a valuable tool in single-cell biology for data augmentation, imputation, and generating synthetic datasets. These techniques enable researchers to overcome limitations in sample sizes, handle missing data, and benchmark their analytical methods more effectively, ultimately advancing our understanding of the complex biological processes at the single-cell level.

EMPOWERING CLINICIANS TO MAKE BETTER DECISIONS THROUGH AI

Clinicians are increasingly becoming involved in the field of scRNA-seq due to the profound impact this technology has on advancing our understanding of disease mechanisms, diagnosis, and personalised treatments [Zhang et al., 2020a; Kim et al., 2021; Jovic et al., 2022]. scRNA-seq allows clinicians to analyse the gene expression profiles of individual cells in healthy and diseased tissues at the single-cell level. They can identify specific cell types, gene expression changes, and potential disease-related biomarkers, which can guide treatment decisions by identifying the most appropriate therapies based on the patient's molecular profile.

Clinicians can also harness the power of AI to enhance their decision-making process, improve diagnostic accuracy, enhance patient care, speed up drug development, and ultimately deliver more efficient and effective healthcare [Basu et al., 2020; Giordano et al., 2021; Briganti, 2023]. However, the successful integration of AI in clinical data and treatment requires careful validation, regulatory bodies to ensure the technology is used responsibly, and collaboration between healthcare professionals and AI experts to ensure that patient safety, data privacy, and ethical considerations are addressed.

The role of AI in clinical data and treatment is becoming increasingly significant and transformative in healthcare. AI technologies have created a paradigm shift in the way medical data are analysed, interpreted, and used to support clinical decision-making and patient treatment. A diagram of how scRNA-seq paired with AI can help diagnose and characterise a disease, and create a personalised treatment plan, is shown in Figure 13.5. Here are some examples of how AI can be used by clinicians:

- *Data integration and interoperability*: AI algorithms can efficiently analyse large volumes of clinical data, from various sources, including electronic health records, medical imaging, lab results, and genetic data. AI can identify patterns, trends, and correlations that may be challenging for human clinicians to detect, leading to faster and more accurate diagnoses.
- *Clinical decision support systems (CDSS)*: AI-powered CDSS provides real-time evidence-based recommendations to clinicians during the patient care process. It can assist in diagnosis, treatment planning, medication selection, and using personalised medicine by considering a patient's unique characteristics and medical history.
- *Risk prediction and prognostication*: AI can predict patient outcomes and assess the risk of disease progression. By analysing historical patient data, AI models can identify high-risk individuals who may benefit from early interventions or more intensive monitoring.
- *Drug discovery and development*: AI is playing a crucial role in drug discovery and development by analysing vast amounts of biomedical data to identify potential drug candidates, predict drug efficacy, and optimise drug properties.
- *Image analysis and medical imaging*: AI algorithms can interpret medical images, such as X-rays, MRIs, and CT scans, with high accuracy. This aids in early detection of diseases, reduces diagnostic errors, and enhances treatment planning.

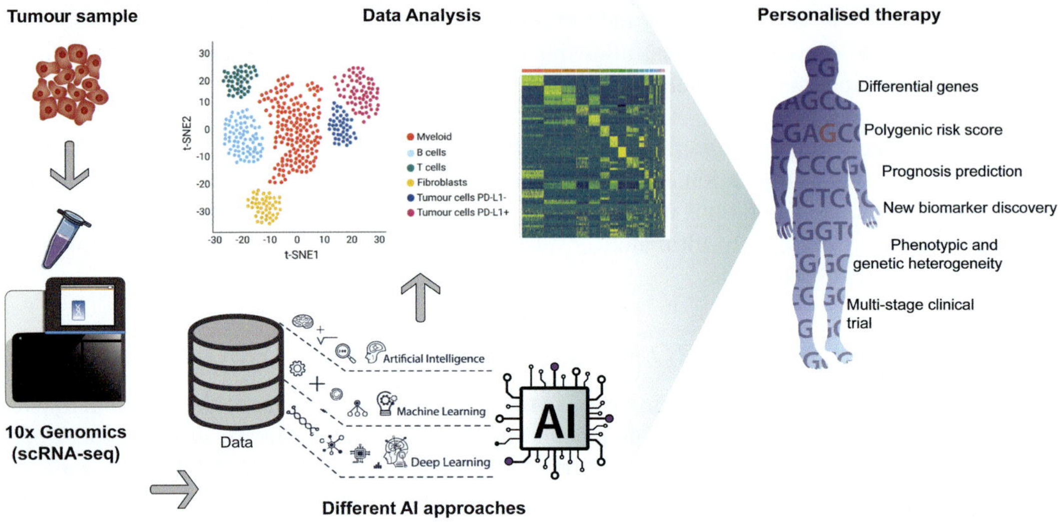

FIGURE 13.5 Applying scRNA-seq and AI to diagnose and characterise a disease, and create a personalised treatment plan in a clinical environment.

- *Natural language processing (NLP)*: NLP enables AI systems to understand and extract valuable information from unstructured clinical notes, research papers, and patient reports. This helps towards better data organisation and knowledge extraction.
- *Personalised medicine*: AI enables precision medicine by considering individual patient characteristics, genetics, and lifestyle factors to tailor treatment plans for better outcomes and reduced adverse effects.
- *Remote patient monitoring*: AI-powered wearable devices and remote monitoring systems can continuously collect patient data, allowing real-time tracking of vital signs and health parameters. This can help detect early signs of deterioration and prevent hospital readmissions.
- *Clinical trials optimisation*: AI can enhance the efficiency of clinical trials by identifying suitable patient populations, improving trial designs, and predicting patient responses to treatments.
- *Feedback and collaboration*: AI developers can encourage clinicians to provide feedback on performance and usability. Developers can collaborate with clinicians to continuously improve AI algorithms based on real-world use cases.
- *Data privacy and security*: AI developers need to ensure strict adherence to data privacy and security regulations. Clinicians need to trust that patient data used by AI tools are protected and handled responsibly.
- *Healthcare operations and resource management*: AI can optimise hospital workflows, patient scheduling, and resource allocation, leading to better patient care, reduced wait times, and cost savings.
- *Ethical considerations*: AI developers need to address ethical concerns associated with it, such as bias in algorithms, data privacy, and the role of AI in clinical decision-making. Ethical guidelines should be established to guide AI implementation.

However, in terms of transparency and explainability, AI systems should be transparent and provide explanations for their recommendations; clinicians need to understand how AI arrives at a particular decision to trust and validate its suggestions.

CONFLICT OF INTEREST STATEMENT

The authors declare no competing interests.

REFERENCES

Adil, A., Kumar, V., Jan, A. T. and Asger, M. (2021). Single-cell transcriptomics: Current methods and challenges in data acquisition and analysis. *Frontiers in Neuroscience*, 15:591122.

Albrecht, S., Andreani, T., Andrade-Navarro, M. A. and Fontaine, J. F. (2022). Single-cell specific and interpretable machine learning models for sparse scChIP-seq data imputation. *PLoS One*, 17(7):e0270043.

Alquicira-Hernandez, J., Sathe, A., Ji, H. P., Nguyen, Q. and Powell, J. E. (2019). ScPred: Accurate supervised method for cell-type classification from single-cell RNA-seq data. *Genome Biology*, 20(1):1–17.

Arisdakessian, C., Poirion, O., Yunits, B., Zhu, X. and Garmire, L. X. (2019). Deepimpute: An accurate, fast, and scalable deep neural network method to impute single-cell RNA-seq data. *Genome Biology*, 20(1):1–14.

Ashburner, M., Ball, C. A., Blake, J. A., Botstein, D., Butler, H., Cherry, J. M., Davis, A. P., Dolinski, K., Dwight, S. S., Eppig, J. T., et al. (2000). Gene ontology: Tool for the unification of biology. *Nature Genetics*, 25(1):25–29.

Barkas, N., Petukhov, V., Kharchenko, P. and Biederstedt, E. (2021). pagoda2: Single cell analysis and differential expression. *R Package Version*, 102.

Basu, K., Sinha, R., Ong, A. and Basu, T. (2020). Artificial intelligence: How is it changing medical sciences and its future? *Indian Journal of Dermatology*, 65(5):365.

Bergen, V., Lange, M., Peidli, S., Wolf, F. A. and Theis, F. J. (2020). Generalizing RNA velocity to transient cell states through dynamical modeling. *Nature Biotechnology*, 38(12):1408–1414.

Bishara, I., Chen, J., Griffiths, J. I., Bild, A. H. and Nath, A. (2022). A machine learning framework for scRNA-seq UMI threshold optimization and accurate classification of cell types. *Frontiers in Genetics*, 13:982019.

Brendel, M., Su, C., Bai, Z., Zhang, H., Elemento, O. and Wang, F. (2022). Application of deep learning on single-cell RNA sequencing data analysis: A review. *Genomics, Proteomics & Bioinformatics*, 20(5):814–835.

Briganti, G. (2023). Artificial intelligence: An introduction for clinicians. *Revue Des Maladies Respiratoires*, 40(4):S0761–8425.

Buettner, M., Ostner, J., Mueller, C. L., Theis, F. J. and Schubert, B. (2021). scCODA is a Bayesian model for compositional single-cell data analysis. *Nature Communications*, 12(1):6876.

Butler, A., Hoffman, P., Smibert, P., Papalexi, E. and Satija, R. (2018). Integrating single-cell transcriptomic data across different conditions, technologies, and species. *Nature Biotechnology*, 36(5):411–420.

Cao, Y., Lin, Y., Ormerod, J. T., Yang, P., Yang, J. Y. and Lo, K. K. (2019). scDC: Single cell differential composition analysis. *BMC Bioinformatics*, 20(19):1–12.

Chen, J., Cheong, C., Lan, L., Zhou, X., Liu, J., Lyu, A., Cheung, W. K. and Zhang, L. (2021). Deepdrim: A deep neural network to reconstruct cell-type-specific gene regulatory network using single-cell RNA-seq data. *Briefings in Bioinformatics*, 22(6):bbab325.

Chen, Y., Zhang, H. and Sun, X. (2023). Improving the performance of single-cell RNA-seq data mining based on relative expression orderings. *Briefings in Bioinformatics*, 24(1).):bbac556.

Choi, S. R. and Lee, M. (2023). Transformer architecture and attention mechanisms in genome data analysis: A comprehensive review. *Biology*, 12(7):1033.

Clayton, E. A., Pujol, T. A., McDonald, J. F. and Qiu, P. (2020). Leveraging TCGA gene expression data to build predictive models for cancer drug response. *BMC Bioinformatics*, 21:1–11.

Cui, H., Wang, C., Maan, H., Pang, K., Luo, F. and Wang, B. (2023). scGPT: Towards building a foundation model for single-cell multi-omics using generative ai. Nat Methods 21, 1470–1480 (2024). https://doi.org/10.1038/s41592-024-02201-0

Dann, E., Henderson, N. C., Teichmann, S. A., Morgan, M. D. and Marioni, J. C. (2022). Differential abundance testing on single-cell data using k-nearest neighbor graphs. *Nature Biotechnology*, 40(2):245–253.

De Silva, H. M. and Perera, A. S. (2017). Evolutionary k-nearest neighbor imputation algorithm for gene expression data. *The International Journal on Advances in ICT for Emerging Regions*, 10(1):1–8.

Ding, J., Condon, A. and Shah, S. P. (2018). Interpretable dimensionality reduction of single cell transcriptome data with deep generative models. *Nature Communications*, 9(1):2002.

Ding, J. and Regev, A. (2021). Deep generative model embedding of single-cell RNA-seq profiles on hyperspheres and hyperbolic spaces. *Nature Communications*, 12(1):2554.

Du, J.-H., Gao, M. and Wang, J. (2020). Model-based trajectory inference for single-cell RNA sequencing using deep learning with a mixture prior. *bioRxiv*, pages 2020–12.

Eraslan, G., Simon, L. M., Mircea, M., Mueller, N. S. and Theis, F. J. (2019). Single-cell RNA-seq denoising using a deep count autoencoder. *Nature Communications*, 10(1):390.

Esposito, S., Carputo, D., Cardi, T. and Tripodi, P. (2019). Applications and trends of machine learning in genomics and phenomics for next-generation breeding. *Plants*, 9(1):34.

Finak, G., McDavid, A., Yajima, M., Deng, J., Gersuk, V., Shalek, A. K., Slichter, C. K., Miller, H. W., McElrath, M. J., Prlic, M., et al. (2015). MAST: A flexible statistical framework for assessing transcriptional changes and characterizing heterogeneity in single-cell RNA sequencing data. *Genome Biology*, 16(1):1–13.

Flores, M., Liu, Z., Zhang, T., Hasib, M. M., Chiu, Y.-C., Ye, Z., Paniagua, K., Jo, S., Zhang, J., Gao, S.-J., et al. (2022). Deep learning tackles single-cell analysis – a survey of deep learning for scRNA-seq analysis. *Briefings in Bioinformatics*, 23(1):bbab531.

Forrow, A. and Schiebinger, G. (2021). Lineageot is a unified framework for lineage tracing and trajectory inference. *Nature Communications*, 12(1):4940.

Garcia-Alonso, L., Holland, C. H., Ibrahim, M. M., Turei, D. and Saez-Rodriguez, J. (2019). Benchmark and integration of resources for the estimation of human transcription factor activities. *Genome Research*, 29(8):1363–1375.

Gierahn, T. M., Wadsworth, M. H., Hughes, T. K., Bryson, B. D., Butler, A., Satija, R., Fortune, S., Love, J. C. and Shalek, A. K. (2017). Seq-well: Portable, low-cost RNA sequencing of single cells at high throughput. *Nature Methods*, 14(4):395–398.

Gillespie, M., Jassal, B., Stephan, R., Milacic, M., Rothfels, K., Senff-Ribeiro, A., Griss, J., Sevilla, C., Matthews, L., Gong, C., et al. (2022). The reactome pathway knowledgebase 2022. *Nucleic Acids Research*, 50(D1):D687–D692.

Giordano, C., Brennan, M., Mohamed, B., Rashidi, P., Modave, F. and Tighe, P. (2021). Accessing artificial intelligence for clinical decision-making. *Frontiers in Digital Health*, 3:645232.

Grün, D., Kester, L. and Van Oudenaarden, A. (2014). Validation of noise models for single-cell transcriptomics. *Nature Methods*, 11(6):637–640.

Haghverdi, L., Lun, A. T., Morgan, M. D. and Marioni, J. C. (2018). Batch effects in single-cell RNA-sequencing data are corrected by matching mutual nearest neighbors. *Nature Biotechnology*, 36(5):421–427.

Hao, J., Zou, J., Zhang, J., Chen, K., Wu, D., Cao, W., Shang, G., Yang, J. Y., Wong-Lin, K., Sun, H., et al. (2023). scSTAR reveals hidden heterogeneity with a real-virtual cell pair structure across conditions in single-cell RNA sequencing data. *Briefings in Bioinformatics*, 24(2):bbad062.

Haque, A., Engel, J., Teichmann, S. A. and Lönnberg, T. (2017). A practical guide to single-cell RNA-sequencing for biomedical research and clinical applications. *Genome Medicine*, 9(1):1–12.

Hashimshony, T., Senderovich, N., Avital, G., Klochendler, A., De Leeuw, Y., Anavy, L., Gennert, D., Li, S., Livak, K. J., Rozenblatt-Rosen, O., et al. (2016). Cel-seq2: Sensitive highly-multiplexed single-cell RNA-seq. *Genome Biology*, 17:1–7.

Hashimshony, T., Wagner, F., Sher, N. and Yanai, I. (2012). Cel-seq: Single-cell RNA-seq by multiplexed linear amplification. *Cell Reports*, 2(3):666–673.

Herring, C. A., Banerjee, A., McKinley, E. T., Simmons, A. J., Ping, J., Roland, J. T., Franklin, J. L., Liu, Q., Gerdes, M. J., Coffey, R. J., et al. (2018). Unsupervised trajectory analysis of single-cell RNA-seq and imaging data reveals alternative tuft cell origins in the gut. *Cell Systems*, 6(1):37–51.

Hie, B., Bryson, B. and Berger, B. (2019). Efficient integration of heterogeneous single-cell transcriptomes using scanorama. *Nature Biotechnology*, 37(6):685–691.

Hu, Q. and Greene, C. S. (2018). Parameter tuning is a key part of dimensionality reduction via deep variational autoencoders for single cell RNA transcriptomics. In *BIOCOMPUTING 2019: Proceedings of the Pacific Symposium*, 362–373. World Scientific.

Huang, Z., Wang, J., Lu, X., Mohd Zain, A. and Yu, G. (2023). ScGGAN: Single-cell RNA-seq imputation by graph-based generative adversarial network. *Briefings in Bioinformatics*, 24(2).):bbad040.

Islam, S., Zeisel, A., Joost, S., La Manno, G., Zajac, P., Kasper, M., Lönnerberg, P. and Linnarsson, S. (2014). Quantitative single-cell RNA-seq with unique molecular identifiers. *Nature Methods*, 11(2): 163–166.

Jaitin, D. A., Kenigsberg, E., Keren-Shaul, H., Elefant, N., Paul, F., Zaretsky, I., Mildner, A., Cohen, N., Jung, S., Tanay, A., et al. (2014). Massively parallel single-cell RNA-seq for marker-free decomposition of tissues into cell types. *Science*, 343(6172):776–779.

Jia, S., Lysenko, A., Boroevich, K., Sharma, A. and Tsunoda, T. (2023). scDeepInsight: A supervised cell-type identification method for scRNA-seq data with deep learning. *bioRxiv*, pages 2023–03.

Jiang, J., Xu, J., Liu, Y., Song, B., Guo, X., Zeng, X. and Zou, Q. (2023). Dimensionality reduction and visualization of single-cell RNA-seq data with an improved deep variational autoencoder. *Briefings in Bioinformatics*, 24(3).):bbad152.

Jiao, L., Ren, Y., Wang, L., Gao, C., Wang, S. and Song, T. (2023). MulCNN: An efficient and accurate deep learning method based on gene embedding for cell type identification in single-cell RNA-seq data. *Frontiers in Genetics*, 14:1179859.

Jolliffe, I. T. (2002). *Principal Component Analysis for Special Types of Data*. Springer.

Jovic, D., Liang, X., Zeng, H., Xu, L. and Luo, F. (2022). Single-cell RNA sequencing technologies and applications: A brief overview. *Clinical and Translational Medicine*, 12(3):e694.

Junttila, S., Smolander, J. and Elo, L. L. (2022). Benchmarking methods for detecting differential states between conditions from multi-subject single-cell RNA-seq data. *Briefings in Bioinformatics*, 23(5):bbac286.

Kanehisa, M., Furumichi, M., Tanabe, M., Sato, Y. and Morishima, K. (2017). KEGG: New perspectives on genomes, pathways, diseases and drugs. *Nucleic Acids Research*, 45(D1):D353–D361.

Kashima, Y., Sakamoto, Y., Kaneko, K., Seki, M., Suzuki, Y. and Suzuki, A. (2020). Single-cell sequencing techniques from individual to multiomics analyses. *Experimental & Molecular Medicine*, 52(9): 1419–1427.

Kharchenko, P. V. (2021). The triumphs and limitations of computational methods for scRNA-seq. *Nature Methods*, 18(7):723–732.

Kim, N., Eum, H. H. and Lee, H.-O. (2021). Clinical perspectives of single-cell RNA sequencing. *Biomolecules*, 11(8):1161.

Kiselev, V. Y., Andrews, T. S. and Hemberg, M. (2019). Challenges in unsupervised clustering of single-cell RNA-seq data. *Nature Reviews Genetics*, 20(5):273–282.

Kivioja, T., Vähärautio, A., Karlsson, K., Bonke, M., Enge, M., Linnarsson, S. and Taipale, J. (2012). Counting absolute numbers of molecules using unique molecular identifiers. *Nature Methods*, 9(1):72–74.

Klein, A. M., Mazutis, L., Akartuna, I., Tallapragada, N., Veres, A., Li, V., Peshkin, L., Weitz, D. A. and Kirschner, M. W. (2015). Droplet barcoding for single-cell transcriptomics applied to embryonic stem cells. *Cell*, 161(5):1187–1201.

Korsunsky, I., Millard, N., Fan, J., Slowikowski, K., Zhang, F., Wei, K., Baglaenko, Y., Brenner, M., Loh, P.-r and Raychaudhuri, S. (2019). Fast, sensitive and accurate integration of single-cell data with harmony. *Nature Methods*, 16(12):1289–1296.

La Manno, G., Soldatov, R., Zeisel, A., Braun, E., Hochgerner, H., Petukhov, V., Lidschreiber, K., Kastriti, M. E., Lönnerberg, P., Furlan, A., et al. (2018). RNA velocity of single cells. *Nature*, 560(7719):494–498.

Lakkis, J., Schroeder, A., Su, K., Lee, M. Y., Bashore, A. C., Reilly, M. P. and Li, M. (2022). A multi-use deep learning method for cite-seq and single-cell RNA-seq data integration with cell surface protein prediction and imputation. *Nature Machine Intelligence*, 4(11):940–952.

Lange, M., Bergen, V., Klein, M., Setty, M., Reuter, B., Bakhti, M., Lickert, H., Ansari, M., Schniering, J., Schiller, H. B., et al. (2022). Cellrank for directed single-cell fate mapping. *Nature Methods*, 19(2):159–170.

Le, H., Peng, B., Uy, J., Carrillo, D., Zhang, Y., Aevermann, B. D. and Scheuermann, R. H. (2022). Machine learning for cell type classification from single nucleus RNA sequencing data. *PLoS One*, 17(9):e0275070.

Lee, M. Y. and Li, M. (2023). Integration of multi-modal single-cell data. *Nature Biotechnology*, 1–2.

Li, X., Wang, K., Lyu, Y., Pan, H., Zhang, J., Stambolian, D., Susztak, K., Reilly, M. P., Hu, G. and Li, M. (2020). Deep learning enables accurate clustering with batch effect removal in single-cell RNA-seq analysis. *Nature Communications*, 11(1):2338.

Liberzon, A., Subramanian, A., Pinchback, R., Thorvaldsdóttir, H., Tamayo, P. and Mesirov, J. P. (2011). Molecular signatures database (MSigDB) 3.0. *Bioinformatics*, 27(12):1739–1740.

Liebermeister, W. (2002). Linear modes of gene expression determined by independent component analysis. *Bioinformatics*, 18(1):51–60.

Lin, X., Gao, L., Whitener, N. and Ahmed, A. (2022). A model-based constrained deep learning clustering approach for spatially resolved single-cell data. *Genome Research*, 32(10):1906–1917.

Liu, J., Fan, Z., Zhao, W. and Zhou, X. (2021). Machine intelligence in single-cell data analysis: Advances and new challenges. *Frontiers in Genetics*, 12:655536.

Lotfollahi, M., Wolf, F. A. and Theis, F. J. (2019). scGen predicts single-cell perturbation responses. *Nature Methods*, 16(8):715–721.

Love, M. I., Huber, W. and Anders, S. (2014). Moderated estimation of fold change and dispersion for RNA-seq data with deseq2. *Genome Biology*, 15(12):1–21.

Luo, Q., Yu, Y. and Lan, X. (2022). Signet: Single-cell RNA-seq-based gene regulatory network prediction using multiple-layer perceptron bagging. *Briefings in Bioinformatics*, 23(1):bbab547.

Luo, Z., Xu, C. and Zhang, Z. (2021). A topology-preserving dimensionality reduction method for single-cell RNA-seq data using graph autoencoder. *Scientific Reports*, 11(1):20028.

Ma, Q. and Xu, D. (2022). Deep learning shapes single-cell data analysis. *Nature Reviews Molecular Cell Biology*, 23(5):303–304.

Macosko, E. Z., Basu, A., Satija, R., Nemesh, J., Shekhar, K., Goldman, M., Tirosh, I., Bialas, A. R., Kamitaki, N., Martersteck, E. M., et al. (2015). Highly parallel genome-wide expression profiling of individual cells using nanoliter droplets. *Cell*, 161(5):1202–1214.

Mancuso, C. A., Canfield, J. L., Singla, D. and Krishnan, A. (2020). A flexible, interpretable, and accurate approach for imputing the expression of unmeasured genes. *Nucleic Acids Research*, 48(21):e125–e125.

Marouf, M., Machart, P., Bansal, V., Kilian, C., Magruder, D. S., Krebs, C. F. and Bonn, S. (2020). Realistic *in silico* generation and augmentation of single-cell RNA-seq data using generative adversarial networks. *Nature Communications*, 11(1):166.

Martorell-Maruga´n, J., Tabik, S., Benhammou, Y., del Val, C., Zwir, I., Herrera, F. and Carmona-Sáez, P. (2019). Deep Learning in Omics Data Analysis and Precision Medicine. *Exon Publications*, pages 37–53.

McInnes, L., Healy, J. and Melville (2018). UMAP: Uniform manifold approximation and projection for dimension reduction. *arXiv preprint arXiv:1802.03426*.

Minsky, Marvin L., and Seymour A. Papert. (1988). Perceptrons: expanded edition.

Moerman, T., Aibar Santos, S., Bravo González-Blas, C., Simm, J., Moreau, Y., Aerts, J. and Aerts, S. (2019). Grnboost2 and arboreto: Efficient and scalable inference of gene regulatory networks. *Bioinformatics*, 35(12):2159–2161.

Nasteski, V. (2017). An overview of the supervised machine learning methods. *Horizons. b*, 4:51–62.

Ni, Y., Aghamirzaie, D., Elmarakeby, H., Collakova, E., Li, S., Grene, R. and Heath, L. S. (2016). A machine learning approach to predict gene regulatory networks in seed development in arabidopsis. *Frontiers in Plant Science*, 7:1936.

Ostner, J., Carcy, S. and Müller, C. L. (2021). TascCODA: Bayesian tree-aggregated analysis of compositional amplicon and single-cell data. *Frontiers in Genetics*, 12:766405.

Pasquini, G., Arias, J. E. R., Schäfer, P. and Busskamp, V. (2021). Automated methods for cell type annotation on scRNA-seq data. *Computational and Structural Biotechnology Journal*, 19:961–969.

Petegrosso, R., Li, Z. and Kuang, R. (2020). Machine learning and statistical methods for clustering single-cell RNA-sequencing data. *Briefings in Bioinformatics*, 21(4):1209–1223.

Picelli, S., Bjo¨rklund, ÅK., Faridani, O. R., Sagasser, S., Winberg, G. and Sandberg, R. (2013). Smart-seq2 for sensitive full-length transcriptome profiling in single cells. *Nature Methods*, 10(11):1096–1098.

Pollen, A. A., Nowakowski, T. J., Shuga, J., Wang, X., Leyrat, A. A., Lui, J. H., Li, N., Szpankowski, L., Fowler, B., Chen, P., et al. (2014). Low-coverage single-cell mRNA sequencing reveals cellular heterogeneity and activated signaling pathways in developing cerebral cortex. *Nature Biotechnology*, 32(10):1053–1058.

Ritchie, M. E., Phipson, B., Wu, D., Hu, Y., Law, C. W., Shi, W. and Smyth, G. K. (2015). Limma powers differential expression analyses for RNA-sequencing and microarray studies. *Nucleic Acids Research*, 43(7):e47–e47.

Robinson, M. D., McCarthy, D. J. and Smyth, G. K. (2010). Edger: A bioconductor package for differential expression analysis of digital gene expression data. *Bioinformatics*, 26(1):139–140.

Rosenberg, A. B., Roco, C. M., Muscat, R. A., Kuchina, A., Sample, P., Yao, Z., Graybuck, L. T., Peeler, D. J., Mukherjee, S., Chen, W., et al. (2018). Single-cell profiling of the developing mouse brain and spinal cord with split-pool barcoding. *Science*, 360(6385):176–182.

Saelens, W., Cannoodt, R., Todorov, H. and Saeys, Y. (2019). A comparison of single-cell trajectory inference methods. *Nature Biotechnology*, 37(5):547–554.

Schubert, M., Klinger, B., Klünemann, M., Sieber, A., Uhlitz, F., Sauer, S., Garnett, M. J., Blüthgen, N. and Saez-Rodriguez, J. (2018). Perturbation-response genes reveal signaling footprints in cancer gene expression. *Nature Communications*, 9(1):20.

Sheng, K., Cao, W., Niu, Y., Deng, Q. and Zong, C. (2017). Effective detection of variation in single-cell transcriptomes using MATQ-seq. *Nature Methods*, 14(3):267–270.

Song, T., Dai, H., Wang, S., Wang, G., Zhang, X., Zhang, Y. and Jiao, L. (2022). Transcluster: A cell-type identification method for single-cell RNA-seq data using deep learning based on transformer. *Frontiers in Genetics*, 13:1038919.

Street, K., Risso, D., Fletcher, R. B., Das, D., Ngai, J., Yosef, N., Purdom, E. and Dudoit, S. (2018). Slingshot: Cell lineage and pseudotime inference for single-cell transcriptomics. *BMC Genomics*, 19:1–16.

Sun, S., Zhu, J., Ma, Y. and Zhou, X. (2019). Accuracy, robustness and scalability of dimensionality reduction methods for single-cell RNA-seq analysis. *Genome Biology*, 20(1):1–21.

Thompson, N. C., Greenewald, K., Lee, K. and Manso, G. F. (2020). The computational limits of deep learning. *arXiv preprint arXiv:2007.05558*.

Torroja, C. and Sanchez-Cabo, F. (2019). DigitalDLsorter: Deep-learning on scRNA-seq to deconvolute gene expression data. *Frontiers in Genetics*, 10:978.

Traag, V. A., Waltman, L. and Van Eck, N. J. (2019). From Louvain to Leiden: Guaranteeing well-connected communities. *Scientific Reports*, 9(1):5233.

Tran, T. N. and Bader, G. D. (2020). Tempora: Cell trajectory inference using time-series single-cell RNA sequencing data. *PLoS Computational Biology*, 16(9):e1008205.

Turki, T. and Taguchi, Y. (2020). ScGRNs: Novel supervised inference of single-cell gene regulatory networks of complex diseases. *Computers in Biology and Medicine*, 118:103656.

Van der Maaten, L. and Hinton, G. (2008). Visualizing data using t-SNE. *Journal of Machine Learning Research*, 9(11).).

Vaswani, A., Shazeer, N., Parmar, N., Uszkoreit, J., Jones, L., Gomez, A. N., Kaiser, Ł and Polosukhin, I. (2017). Attention is all you need. *Advances in Neural Information Processing Systems*, 30.

Wagner, F. and Yanai, I. (2018). Moana: a robust and scalable cell type classification framework for single-cell RNA-seq data. *bioRxiv*, page 456129.

Wang, B., Zhu, J., Pierson, E., Ramazzotti, D. and Batzoglou, S. (2017). Visualization and analysis of single-cell RNA-seq data by kernel-based similarity learning. *Nature Methods*, 14(4):414–416.

Wang, D. and Gu, J. (2018). Vasc: Dimension reduction and visualization of single-cell RNA-seq data by deep variational autoencoder. *Genomics, Proteomics & Bioinformatics*, 16(5):320–331.

Wang, D., Hou, S., Zhang, L., Wang, X., Liu, B. and Zhang, Z. (2021). iMAP: Integration of multiple single-cell datasets by adversarial paired transfer networks. *Genome Biology*, 22(1):1–24.

Wang, L., Nie, R., Yu, Z., Xin, R., Zheng, C., Zhang, Z., Zhang, J. and Cai, J. (2020). An interpretable deep-learning architecture of capsule networks for identifying cell-type gene expression programs from single-cell RNA-sequencing data. *Nature Machine Intelligence*, 2(11):693–703.

Wang, S.-W., Herriges, M. J., Hurley, K., Kotton, D. N. and Klein, A. M. (2022). Cospar identifies early cell fate biases from single-cell transcriptomic and lineage information. *Nature Biotechnology*, 40(7):1066–1074.

Wang, T., Johnson, T. S., Shao, W., Lu, Z., Helm, B. R., Zhang, J. and Huang, K. (2019a). Bermuda: A novel deep transfer learning method for single-cell RNA sequencing batch correction reveals hidden high-resolution cellular subtypes. *Genome Biology*, 20(1):1–15.

Wang, T., Li, B., Nelson, C. E. and Nabavi, S. (2019b). Comparative analysis of differential gene expression analysis tools for single-cell RNA sequencing data. *BMC Bioinformatics*, 20(1):1–16.

Wolf, F. A., Hamey, F. K., Plass, M., Solana, J., Dahlin, J. S., Göttgens, B., Rajewsky, N., Simon, L. and Theis, F. J. (2019). Paga: Graph abstraction reconciles clustering with trajectory inference through a topology preserving map of single cells. *Genome Biology*, 20(1):9.

Xia, F., Allen, J., Balaprakash, P., Brettin, T., Garcia-Cardona, C., Clyde, A., Cohn, J., Doroshow, J., Duan, X., Dubinkina, V., et al. (2022). A cross-study analysis of drug response prediction in cancer cell lines. *Briefings in Bioinformatics*, 23(1):bbab356.

Xu, J., Mao, C., Hou, Y., Luo, Y., Binder, J. L., Zhou, Y., Bekris, L. M., Shin, J., Hu, M., Wang, F., et al. (2022). Interpretable deep learning translation of GWAS and multi-omics findings to identify pathobiology and drug repurposing in Alzheimer's disease. *Cell Reports*, 41(9):111717.

Xu, Y., Zhang, Z., You, L., Liu, J., Fan, Z. and Zhou, X. (2020). ScIGANs: Single-cell RNA-seq imputation using generative adversarial networks. *Nucleic Acids Research*, 48(15):e85–e85.

Yang, B., Bao, W., Chen, B. and Song, D. (2022). Single cell GRN: gene Regulatory network identification based on supervised learning method and single-cell RNA-seq data. *BioData Mining*, 15(1):1–18.

Yang, M. Q., Weissman, S. M., Yang, W., Zhang, J., Canaann, A. and Guan, R. (2018). Misc: Missing imputation for single-cell RNA sequencing data. *BMC Systems Biology*, 12:55–63.

Yu, B., Chen, C., Qi, R., Zheng, R., Skillman-Lawrence, P. J., Wang, X., Ma, A. and Gu, H. (2021). Scgmai: A Gaussian mixture model for clustering single-cell RNA-seq data based on deep autoencoder. *Briefings in Bioinformatics*, 22(4).):bbaa316.

Yuan, B., Shen, C., Luna, A., Korkut, A., Marks, D. S., Ingraham, J. and Sander, C. (2021). Cellbox: Interpretable machine learning for perturbation biology with application to the design of cancer combination therapy. *Cell Systems*, 12(2):128–140.

Zhang, J., Wang, W., Huang, J., Wang, X. and Zeng, Y. (2020a). How far is single-cell sequencing from clinical application?. *Clinical and Translational Medicine*, 10(3): p. e117.

Zhang, Y., Ma, Y., Huang, Y., Zhang, Y., Jiang, Q., Zhou, M. and Su, J. (2020b). Benchmarking algorithms for pathway activity transformation of single-cell RNA-seq data. *Computational and Structural Biotechnology Journal*, 18:2953–2961.

Zhao, J., Jaffe, A., Li, H., Lindenbaum, O., Sefik, E., Jackson, R., Cheng, X., Flavell, R. A. and Kluger, Y. (2021). Detection of differentially abundant cell subpopulations in scRNA-seq data. *Proceedings of the National Academy of Sciences*, 118(22):e2100293118.

Zheng, G. X., Terry, J. M., Belgrader, P., Ryvkin, P., Bent, Z. W., Wilson, R., Ziraldo, S. B., Wheeler, T. D., McDermott, G. P., Zhu, J., et al. (2017). Massively parallel digital transcriptional profiling of single cells. *Nature Communications*, 8(1):14049.

Zhu, M., Li, C., Lv, K., Guo, H., Hou, R., Tian, G. and Yang, J. (2023). MLSpatial: A machine-learning method to reconstruct the spatial distribution of cells from scRNA-seq by extracting spatial features. *Computers in Biology and Medicine*, 159:106873.

Ziegenhain, C., Vieth, B., Parekh, S., Reinius, B., Guillaumet-Adkins, A., Smets, M., Leonhardt, H., Heyn, H., Hellmann, I. and Enard, W. (2017). Comparative analysis of single-cell RNA sequencing methods. *Molecular Cell*, 65(4):631–643.

14 The Central Dogma of Single-Molecule Sequencing

From DNA to RNA to Proteins

Shri Hari Prasad and Adarsh Pooradan Prasannakumar

INTRODUCTION

Proteins are considered the workhorses of the living cell. They determine cellular architecture and activity, provide the mechanisms for cell communication, catalyse chemical reactions that support cell metabolism, and their structure can affect the function or dysfunction of a process. Consequently, proteins are often implicated in disease mechanisms while also serving as therapeutic targets. The term 'protein', derived from the Ancient Greek 'proteios', meaning 'primary' or 'of utmost importance', aptly reflects their significance. And so, it is also of great importance to determine their sequences.

The development of high-throughput, low-cost sequencing technologies has led to a great number of advancements in genomics and transcriptomics over the past decade, enabling parallel and rapid sequencing of DNA and RNA molecules (Gawad et al., 2016). These developments also facilitated the quantification of gene repeats and of the transcription levels of biomolecules (Bray et al., 2016).

The development of similar technology for sequencing-based identification and quantification of proteins is still in its infancy. This technological lag is highlighted by the concept of the 'missing proteome', which refers to the set of proteins predicted from genomic open reading frames but not yet experimentally observed or identified (Omenn, 2014). Identifying these missing proteins is one of the objectives of the Human Proteome Project, and to date, 93.01% of the human proteome has been experimentally confirmed (HUPO, 2023).

THE RISE OF TRANSCRIPTOMICS

The transcriptome comprises the complete set of RNA molecules in a cell, including regulatory non-coding RNAs, ribosomal RNAs (rRNAs), transfer RNAs (tRNAs) – crucial for ribosome function – and messenger RNAs (mRNA) that encode proteins. High-throughput sequencing technology, introduced in 2005 (Margulies et al., 2005), enabled detailed examination of the transcriptome at the nucleotide level. However, a limitation emerged with popular RNA sequencing (RNA-seq) techniques. These methods, while allowing comprehensive transcriptome analysis, often employed 'bulk' approaches. Consequently, physiologically significant cell-to-cell variations could be obscured, as the analysed RNA represented only an 'average gene expression value' across hundreds to millions of distinct cellular transcriptomes within a sample (Hwang et al., 2018). This averaging effect potentially masked important biological heterogeneity.

Advances in sequencing technologies have led to many significant discoveries. It has been found that protein-coding genes make up a mere 2% of the human genome, even though more than 80% of it could potentially be transcribed (Hangauer et al., 2013). Another degree of complexity is added by the fact that a single genomic locus can exhibit various and varied splicing patterns of pre-mRNA, leading to distinct mRNA isoforms, which may in turn have different transcriptional start sites or polyadenylation sites (Shalek et al., 2013). Additionally, research by Deng et al. (2014) indicates that gene expression can exhibit random monoallelic behaviour, where only one allele – either maternal or paternal – is expressed at a particular time point. As the transcriptome information in each cell

DOI: 10.1201/9781003354062-14

reflects the unique activity of only a subset of genes, it is crucial to understand the precise expression pattern of each individual cell, as a majority of transcripts are expressed in small numbers per cell, and most long noncoding RNAs (lncRNAs) are present in one or fewer copies per cell (Ma et al., 2019). Furthermore, transcript expression can be influenced by both deterministic factors, such as cell cycle stage, and stochastic elements. This can cause adjacent cells in the same microenvironment to express a transcript at different levels, a phenomenon known as 'transcriptional noise'. This random, burst-like fluctuation in expression levels is now understood to have transcriptome-wide effects and play a crucial role in determining cell fate (Chang et al., 2008).

A STEP TOWARDS SINGLE-CELL RNA SEQUENCING

One of the most reliable methods to investigate cell identification, condition, function, and response is to profile the gene expression activity in individual cells. The first conceptual, as well as technological, breakthrough for carrying out single-cell RNA-seq (scRNA-seq) was achieved by Tang et al. (2009).

Since then, a growing number of modified and improved scRNA-seq technologies have been created to bring about critical modifications and improvements in sample collection, single-cell capture, barcoded reverse transcription, complementary DNA (cDNA) amplification, library preparation, sequencing, and downstream analysis (Table 14.1). Even though several procedures and

TABLE 14.1
Principal Characteristics of scRNA-seq Approaches

Name	Isolation Strategy	Transcript Coverage	UMI	Amplification	Platform	Reaction Volume	Throughput (Number of Cells)	Typical Read Depth (per Cell)	References
Tang method	FACS	Nearly full-length	No	PCR					Tang et al., 2009, 2010
Smart-seq	FACS	Full-length	No	PCR	Plate-based	Microliter	10^2–10^3	106	Goetz and Trimarchi, 2012; Picelli et al., 2013
C1	Microfluidic	5′-only	Yes	PCR		Nanoliter	10^2–10^3	106	Xin et al., 2016
CEL-seq2	FACS	3′-only	Yes	IVT	Plate-based	Nanoliter	10^2–10^3	10^4–10^5	Hashimshony et al., 2012
MARS-seq	FACS	3′-only	Yes	IVT	Plate-based	Microliter	10^2–10^3	10^4–10^5	Jaitin et al., 2014
Drop-seq	Microdroplets	3′-only	Yes	PCR	Droplet	Nanoliter	10^3–10^4	10^4–10^5	Macosko et al., 2015
InDrop	Microdroplets	3′-only	Yes	IVT	Droplet	Nanoliter	10^3–10^4	10^4–10^5	Klein et al., 2015
MATQ-seq	FACS	Full-length	Yes	PCR	Plate-based	Microliter	10^2–10^3	106	Sheng et al., 2017
Chromium	Microdroplets	3′-only	Yes	PCR	Droplet	Nanoliter	10^3–10^4	10^4–10^5	Zheng et al., 2017
SEQ-well	Microfluidic	3′-only	Yes	PCR	Nanowell array	Nanoliter	10^3–10^4	10^4–10^5	Gierahn et al., 2017
SPLIT-seq		3′-only	Yes		Plate-based	Microliter	10^3–10^5	104	Rosenberg et al., 2017

methods for scRNA-seq have been created and improved, all of these protocols adhere to the same fundamental concepts, with some variances.

Prior to processing individual cells, they must first be separated from one another. Enzymatic dissociation of tissue into a liquid cell suspension is frequently required for this. Following isolation, the cells are lysed, and the RNA molecules inside each cell are isolated and reverse-transcribed to produce cDNA. In many cases, the cDNA molecules are also given cell-specific DNA barcodes at this stage. After that, the cDNA is amplified. This is followed by the creation of scRNA-seq libraries. The libraries are then sequenced, and the data that are gathered are evaluated to, among other analyses, visualise and categorise the gene expression landscape.

CONSTRAINTS IN SINGLE-CELL RNA SEQUENCING

Despite recent advancements in 'sensitive' scRNA-seq platforms, which can detect low abundance transcripts and capture a broader range of the cellular transcriptome, the data obtained remain limited due to frequent dropout events, where specific transcripts fail to be detected. The issue is compounded by the digital nature of gene expression at the single-cell level and the phenomenon of transcriptional bursting. Also known as 'transcriptional pulsing', this process involves brief periods of intense transcriptional activity followed by extended inactive phases, leading to temporal fluctuations in transcript levels (Suter et al., 2011). These factors contribute to the high frequency of 'zero counts' for individual genes across cells in scRNA-seq data, significantly increasing the likelihood of missing weakly or even moderately expressed genes (Haque et al., 2017).

Another significant limitation is that current approaches rely on converting RNA to cDNA through reverse transcription, followed by amplification using either polymerase chain reaction (PCR) or *in vitro* transcription/translation (IVTT), before sequencing. As direct RNA sequencing is not yet possible with second-generation sequencing methods, these intermediary steps introduce unavoidable biases and losses. Despite the recent introduction of higher-capacity short-read sequencing platforms, these technological constraints remain substantial (Picelli, 2017).

Advances in long-read sequencing technologies, such as single-molecule real-time (SMRT) and Ion Torrent sequencers, have enabled RNA-seq with a median read length of 1.5 kb. However, this approach still relies on converting RNA into cDNA using Moloney Murine Leukemia Virus (MMLV)-based reverse transcription (RT) enzymes. Consequently, it suffers from the same sensitivity issues and loss of rare transcripts as other methods. Additionally, long-read sequencing technology has higher error rates than short-read technology (Sharon et al., 2013), and there are only a few computational pipelines available to handle the raw datasets of scRNA-seq. There are commercial companies that offer software tools, but this field is still in its infancy, and there are no gold-standard solutions at this time (Hwang et al., 2018). Thus, there is plenty of room for improvement in scRNA-seq.

CURRENT LIMITATIONS IN PROTEOMICS

Relying solely on genomic or transcriptomic sequences for protein analysis is inadequate (Timp and Timp, 2020). While transcriptomics studies serve as a bridge by indicating which genes are actively expressed, they do not directly reflect protein levels. This persistent gap arises from two key factors: the variable translational efficiency of different mRNAs (Boersma et al., 2019) and the discrepancies between mRNA and protein half-lives (Haider and Pal, 2013). Also, posttranslational modifications (PTMs) further influence the function and structure of proteins. Proteome analysis is, therefore, key to understanding biological processes and their dynamic nature, because proteins dictate most biological functions and are directly related to the phenotype of a cell (Pandey and Mann, 2000; Tyers and Mann, 2003). So, while genomics provides a blueprint of an organism's potential, proteomics reveals its operational state at a given time.

HISTORICAL METHODS

Frederick Sanger laid the foundation for sequencing technologies. His protein sequencing work developed from a method for end-group determination in proteins. In 1953, he completed sequencing an entire protein, bovine insulin (Sanger and Thompson, 1953). The method involved the use of 1-fluoro-2,4-dinitrobenzene (FDNB), now known as Sanger's reagent. The process involves treating a protein sample with FDNB, which then undergoes acid hydrolysis to form a coloured dinitrophenol derivative, which is then separated and identified by chromatographic methods (Sanger, 1949; Sanger and Tuppy, 1951; Sanger and Thompson, 1953).

During the same period, Pehr Edman developed an elegant protocol that used phenylisothiocyanate as the reagent, which facilitated the stepwise degradation of protein (Edman, 1950). 'Edman degradation' became more popular than the Sanger method due to its improved sensitivity, reliability, and potential for automation. His method facilitated the identification of the exact N-terminal amino acid and is useful for the N-terminal sequencing of proteins in mixtures. Even though Edman degradation is a valuable tool for sequencing, it is generally limited to the analysis of purified peptides that are shorter than ~50 amino acids. The method is not suitable for the analysis of complex protein mixtures, such as those present in most biological samples. Additionally, each degradation cycle typically takes 30-45 minutes, making the process time consuming (Li and Geraerts, 1997). N-terminus modifications can also interfere with the process. For example, acetylation (a common PTM) of the N-terminus of the peptide can significantly hinder the reaction, potentially making sequencing challenging (Restrepo-Pérez et al., 2018).

MASS SPECTROMETRY (MS) FOR PROTEIN SEQUENCING

MS is a commonly used, high-throughput tool that can be used to study proteins (Steen and Mann, 2004). The procedure of MS-based protein identification involves digesting proteins into peptides, which are then separated, fragmented, ionised, and captured by mass spectrometers (Figure 14.1). Peptides are then identified from the peaks of the captured mass spectra using computational methods, where each peak theoretically represents a peptide fragment ion (Domon and Aebersold, 2006; Restrepo-Pérez et al., 2018). Overlapping peptide sequences are then used to assemble the complete protein sequence (Sinitcyn et al., 2018).

Decades ago, the applicability of MS to larger biomolecules like proteins was enhanced by combining protein digestion and matrix-assisted laser desorption ionisation (MALDI), followed by screening a protein database. This approach significantly reduced protein identification time (Pappin et al., 1993). The method improved sensitivity, allowing detection of genomic material in femtomole (10^{-15} mole) ranges. The use of time-of-flight spectrometers reduced mass errors, and the technique's high tolerance to impurities decreased sample purification time (Standing, 2003). Consequently, MALDI emerged as a ready-to-use, inexpensive, and reliable tool for peptide mass fingerprinting (Pappin, 2003).

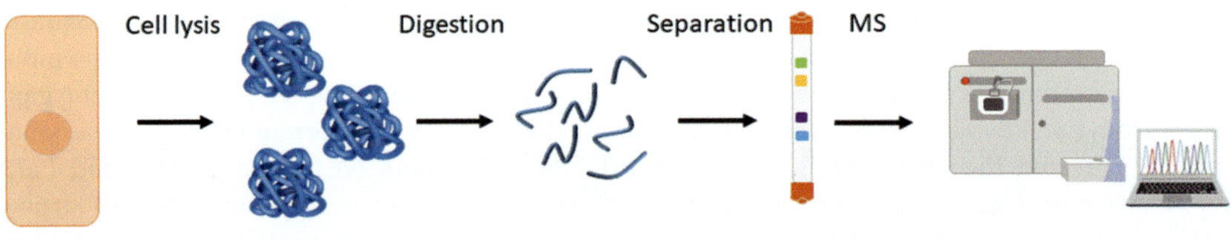

FIGURE 14.1 Single-cell MS method: To be compatible with single-cell samples, MS involves numerous stages between cell lysis and analysis that must be optimised, which include protein digestion, separation, fragmentation, and ionisation.

Theoretically, MS has attomole (10^{-18} mole) detection sensitivity (Tsedilin et al., 2015). However, because there is a need for in-depth analysis to resolve peptides with similar masses (Medzihradszky and Chalkley, 2015), practical protein identification often requires picomole (10^{-12} mole) amounts of purified protein samples (Scheffler et al., 2018). Another challenge is when proteins are present in low copy numbers (less than 1,000 molecules per cell, which is in the zeptomole or 10^{-21} mole range per cell) (Huang et al., 2007). In such a case, millions of cells would be required just to reach the typical MS detection limit of 0.1 to 10 femtomoles (10^{-16} to 10^{-14} moles). Sequence information analysed using MS technologies, *viz.* KASPar and Sequenom, which is beyond the scope of this review (Domon and Aebersold, 2010; Zubarev, 2013; Pagel et al., 2015).

While other technologies, such as the SEQUENOM MassArray system (which combines PCR and MALDI-TOF), are valuable tools in genetics, particularly for SNP genotyping and targeted gene expression analysis (Cullinan and Cantor, 2008), they fall short in the comprehensive, high-sensitivity, and multiplexed protein analysis required for in-depth single-cell proteomics. Although proven effective in analysing gene expression of multiple genes and multiple samples, these technologies do not provide a comprehensive solution for the challenges facing single-cell protein analysis.

CHALLENGES IN LONG-READ DNA SEQUENCING

Next-generation sequencing (NGS) techniques that process short reads have many advantages in terms of cost, accuracy, a low error rate, and the availability of a wide range of benchmarked tools for data analysis. However, short-read sequencing (SRS) in NGS uses sequences typically between 100 to 500 bp, with the ideal size varying, depending on the sequencing platform and experiment goals. With this technique, the short reads are assembled, and complex repetitive regions are aligned, but SRS can struggle with detecting certain types of structural variants, especially in complex or repetitive regions (Amarasinghe et al., 2020). Furthermore, there could be complications such as PCR bias and depth of coverage issues associated with SRS.

Long-read sequencing (LRS) technologies emerged in 2010 with the promise of resolving these problems and outperforming SRS. Also known as 'third-generation sequencing', LRS is able to sequence DNA of lengths between Kbps and Mbps (Giani et al., 2020), and it is the first technology to extract genomic information in its entirety with high accuracy. Oxford Nanopore Technologies (ONT), which performs LRS by using nanopores and changes in electrical current, is a key player. Pacific Biosciences (PacBio), which has a fluorescence-based SMRT sequencer, is equally competitive (Giani et al., 2020; Wohlers et al., 2023).

LRS has significantly improved *de novo* assembly, accurate read alignment, and identification of transcript isoforms and structural variants. Importantly, it avoids PCR amplification, e.g., with nanopore sequencing (Goodwin et al., 2015; Amarasinghe et al., 2021). LRS has made it easier to discover DNA overlaps, an advancement that has improved genome assembly and the generation of long, continuous consensus sequences (Wohlers et al., 2023). Despite these advancements, LRS technologies are still evolving and have several more challenges to face.

Those challenges include input material requirements, cost of sequencing, availability of dedicated tools, accuracy of base calling and error correction, information technology (IT) infrastructure, and pipeline benchmarking. Compared to SRS, LRS technologies are still costly, which increases with genome and sample size. Among the LRS technologies, ONT competes with SMRT in terms of readability, quality, portability, and cost. At one point in time, ONT readings did exhibit sequencing biases that were challenging to fix (Istace et al., 2017).

Another challenge faced by LRS technologies is base-calling accuracy and error correction (Amarasinghe et al., 2020; Giani et al., 2020; Wohlers et al., 2023). Base-calling accuracy for SMRT long reads is significantly lower than that for Illumina short reads. Studies published in 2012 show that SMRT sequencing experiences approximately one error every 10 nucleotides (Carneiro et al., 2012), while Illumina sequencing reportedly achieves a much higher accuracy with only one error occurring every 1,000 nucleotides (Liu et al., 2012). With ONT, a median single-pass accuracy of 95% and a median consensus accuracy of 98% are attained utilising 1D and 1D2 sequencing protocols, respectively (Clive Brown, 2018. However, with recent improvements, ONT Q20+ and PacBio High Fidelity (HiFi) devices can generate reads with an accuracy of greater than 99% for lengths ranging from Kbps to Mbps (Wohlers et al., 2023).

Since the data generated by SMRT and nanopore technology differ from those produced by short-read techniques, tools specific to those technologies are needed. But, unlike NGS tools, many are not yet standardised, and the benchmarking of pipelines is still ongoing. To increase overall data quality, on which research depends, improvements must be made to the error-correction tools that are available. Some of those used for long-read data processing are given in Table 14.2. Additionally,

TABLE 14.2
Tools Used for Long-Read Data Processing

Base-Calling Tools	
Tools	**References**
Guppy	(Oxford Nanopore Technologies, 2013/2014)
Chiron	(Teng et al., 2018)
CCS	(PacificBiosciences/Ccs: CCS: Generate Highly Accurate Single-Molecule Consensus Reads (HiFi Reads), n.d.)
Bonito	(Nanoporetech/Bonito: A PyTorch Basecaller for Oxford Nanopore Reads, n.d.)
Genome Assembly Tools	
Tools	**Reference**
Canu	(Jiang et al., 2020)
FALCON	(Chin et al., 2013)
ABrujin	(Lin et al., 2016)
Miniasm	(Li, 2016)
HINGE	(Kamath et al., 2017)
Marvel	(Nowoshilow et al., 2018)
Wtdbg2	(Ruan and Li, 2019)
Flye	(Kolmogorov et al., 2019)
Peregrine	(C.-S. Chin and Khalak, 2019)
Shasta	(Shafin et al., 2019)
Alignment Tools	
Tools	**References**
Minimap2	(Li, 2018)
NGMLR	(Sedlazeck et al., 2018)
BLASR	(Chaisson and Tesler, 2012)
Ira	(Ren and Chaisson, 2021)
pbmm2	(Frith and Kawaguchi, 2015)
GraphMap	(Sović et al., 2016)
MashMap	(Jain et al., 2018)
Variant-Calling Tools	
Tools	**References**
Nanocaller	(Ahsan et al., 2021)
Clair3	(Zheng et al., 2022)
Longshot	(Edge and Bansal, 2019)
Medaka	(Nanoporetech/Medaka: Sequence Correction Provided by ONT Research, n.d.)
Clair3-trio	(Su et al., 2022)
Nanopanel2	(Popitsch et al., 2021)

(Continued)

TABLE 14.2 ***(Continued)***
Tools Used for Long-Read Data Processing

Structural Variant-Calling Tools

Tools	References
Pbsv	(PacificBiosciences/Pbsv: Pbsv – PacBio Structural Variant (SV) Calling and Analysis Tools, n.d.)
Sniffles	(Sedlazeck et al., 2018)
SVIM	(Heller and Vingron, 2019)
cuteSV	(Jiang et al., 2020)
nplnv	(Shao et al., 2018)
NextSV	(Fang et al., 2018)
NanoVar	(Tham et al., 2020)

Modified Base Identification Tools

Tools	References
Tombo	(Stoiber et al., 2017)
NanoMod	(Liu et al., 2019)
mCaller	(McIntyre et al., 2019)
DeepSignal	(Ni et al., 2019)
DeepMod	(Ni et al., 2019)

Amarasinghe et al. have built an online database containing a comprehensive collection of tools that can be used for long-read data analysis (Amarasinghe et al., 2020 & 2021). However, a standardised methodology and choice of tools are still under development.

A significant advantage of LRS over SRS is its superior ability to identify structural variants, primarily due to its extended read length. The NGS chemistry used in SRS is less effective at detecting larger SVs. Despite the availability of various tools designed for LRS data, evaluating the performance of these SV-calling tools remains challenging. This difficulty arises because most benchmarking datasets are derived from short reads, which inherently lack comprehensive SV annotations (Ho et al., 2019; Mahmoud et al., 2019). Consequently, there is a pressing need for the development of robust benchmarking methods specifically tailored for long-read SV detection tools.

THE NANOPORE MENAGERIE

The world of nanopores offers an assortment of tiny molecular gateways, each with unique characteristics and potential uses. This menagerie includes biological nanopores like α-hemolysin and MspA, solid-state nanopores fabricated from synthetic materials, hybrid nanopores combining biological and artificial components, and even DNA origami nanopores.

Nanopores typically range in size from 1 to 100 nanometres in diameter and can be either biological (formed by proteins or cellular structures) or artificially produced. They function by measuring changes in ionic current flow when a voltage is applied across them. While nanopores have diverse applications, their development was largely driven by the need for improved methods to characterise protein properties and activities (Varongchayakul et al., 2018).

Nanopore technology has shown significant potential for DNA sequencing (Deamer et al., 2016) due to its ability to detect molecules at extremely low concentrations. This method excels in identifying minor molecular variations and can provide detailed information about biomolecule conformation without the need for labelling (Acharya et al., 2015; Robertson and Reiner, 2018; Varongchayakul et al., 2018). These advantages position nanopore technology as a promising and potentially superior approach for protein characterisation. Research laboratories have begun to

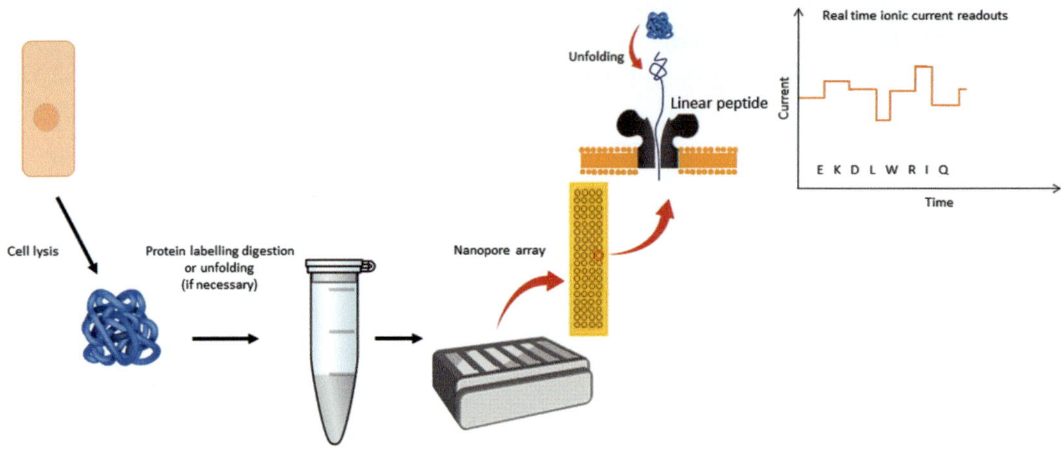

FIGURE 14.2 Nanopore protein sequencing method.

explore nanopore analysis of proteins more extensively, and manufacturers are investigating its commercial viability (Movileanu, 2009; Oukhaled et al., 2012; Restrepo-Pérez et al., 2018; Wei et al., 2023).

Recent protein nanopore experiments can be broadly classified into three main categories: (i) those detecting protein transport and protein unfolding, (ii) those interrogating protein–pore, protein–protein, and protein–ligand interactions, and (iii) those detecting native proteins using covalently bonded aptamers (Nivala et al., 2014). In the third category, a notable recent development is the experimental detection of the 20 standard, proteinogenic amino acids using nanopore technology. This advancement represents a significant step towards protein sequencing at the single-molecule level.

Employing nanopore technology for single-molecule protein sequencing requires overcoming a number of challenges unique to proteins. For example, the approaches require the distinct identification of 20 amino acid residues, which is fivefold greater than the four bases in DNA. Also, amino acid residues vary widely in charge distribution, unlike DNA, which is essentially uniformly charged. Electrophoresis-driven unidirectional translocation of polypeptides through nanopores thus cannot be simply employed. Furthermore, proteins generally exist in complex conformations and are folded in their native state. Therefore, disrupting their secondary, tertiary, and sometimes quaternary structures is necessary (Restrepo-Pérez et al., 2018). A nanopore workflow for protein sequencing is given in Figure 14.2.

TOWARDS NANOPORE SEQUENCING OF PROTEINS

Early protein nanopore experiments primarily focused on the translocation of short polypeptides, typically 20-30 amino acids in length, through biological nanopores such as α-haemolysin and aerolysin (Sutherland et al., 2004; Movileanu et al., 2005; Stefureac et al., 2006). These short peptides were preferred due to their ability to translocate through nanopores without denaturing agents and because of their lack of a stable tertiary structure (Goodrich et al., 2007; Mohammad and Movileanu, 2008). By studying peptides with specific motifs like α-helices and β-hairpins, researchers gained valuable insights into peptide–nanopore interactions, protein capturing, and partitioning during molecular passage (Movileanu et al., 2005; Goodrich et al., 2007). This single-molecule approach, combined with the analysis of both the translocating polypeptide and the pore lumen architecture, opened new avenues for high-resolution studies of polypeptide–pore interactions.

While peptide translocation studies provide valuable insights into the fundamental dynamics of protein movement through nanopores, the development of nanopore-based protein sequencers for whole proteins requires an additional step: protein denaturation. This unfolding process is necessary to enable the linear passage of proteins through the nanopore, allowing for potential sequence readout. To achieve this, researchers have proposed and investigated various protein unfolding methods, encompassing both physical and chemical approaches, each with its own advantages and challenges in the context of nanopore analysis.

Researches have successfully demonstrated the denaturation and transport of unfolded proteins through nanopores using various chemical denaturants such as guanidine hydrochloride (Li et al., 2014), sodium dodecyl sulphate (SDS) (Restrepo-Pérez et al., 2017), and urea (Sauciuc et al., 2023). Physical approaches, including elevated temperatures, have been employed in both solid-state and biological nanopores (Freedman et al., 2011; Payet et al., 2012). Additionally, studies by Talaga and Li (2009), Oukhaled et al. (2011), Cressiot et al. (2012), and Freedman et al. (2013) have shown that electrical forces under appropriate potentials can aid in protein unfolding. However, the application of voltages around 0.4 V can cause electroporation of the lipid bilayer, limiting this approach's applicability in biological nanopore systems.

Among the chemical denaturants, SDS offers distinct advantages. It not only effectively unfolds proteins but also imparts a uniform negative charge to the denatured polypeptide chain. This uniform charge is particularly beneficial because amino acid residues naturally have non-uniform charge distributions, which can impede smooth transport through nanopores. By coating the protein with a consistent negative charge, SDS facilitates more efficient and predictable peptide translocation through the nanopore (Kennedy et al., 2016; Dong et al., 2017; Restrepo-Pérez et al., 2017).

The translocation of proteins through nanopores typically occurs in less than 1 millisecond, which is too rapid for accurate sequencing (Plesa et al., 2013). Controlling this speed is crucial to allow sufficient time for nanopores to accurately read the amino acid sequence. Nivala et al. (2013) developed a promising approach using the AAA+ unfoldase ClpX to unfold proteins and translocate them through the α-hemolysin nanopore. This method achieves a slower translocation rate of 80 amino acids per second, suitable for sequencing, while providing enough force to unfold proteins (Aubin-Tam et al., 2011).

A follow-up study by Nivala et al. (2014) introduces a machine learning (ML) algorithm capable of distinguishing between different protein domains and their variants, including mutations or truncations. This differentiation is achieved using three parameters: dwell time, average current amplitude (the magnitude of the ionic current flowing through the nanopore), and standard deviation of the current amplitude. The ClpX method successfully addressed two critical requirements for nanopore protein sequencing: effective protein unfolding and controlled substrate translocation. While other approaches have been proposed, such as double nanopore systems (Sampath, 2015) and perpendicular nanochannels (Boynton and Di Ventra, 2016), these lack experimental validation, unlike the ClpX method, which has been demonstrated in practice.

Wilson et al. (2016) conducted a molecular dynamics simulation to explore the potential of solid two-dimensional graphene nanopores for protein sequencing. The study demonstrated that peptides translocate through the nanopore in a stepwise manner, adhering to the pore under transmembrane bias or pressure gradient. By adjusting the peptide's charge density and hydrophobicity, the researchers were able to control the translocation speed. Crucially, the peptide passage caused discrete changes in ionic current, corresponding to the amino acid group in the nanopore. This study addressed many challenges in nanopore protein sequencing. Additionally, the study examined electroosmotic flow, wherein the net ion-charge buildup inside the nanopore drives water flow through the pore, further facilitating controlled peptide translocation (Wilson et al., 2016; Mehrafrooz and Aksimentiev, 2023; Sauciuc et al., 2023).

Recent years have seen many promising advances in nanopore proteomics for single-molecule protein sequencing. Piguet et al. (2018) utilised a wild-type aerolysin nanopore to discriminate between six arginine homopeptides, varying in length from 5 to 10 residues. The same method was also effective in identifying and discriminating between arginine and lysine homopeptides and arginine/lysine heteropeptides in an equimolar mixture.

Di Muccio et al. (2019) employed all-atom molecular dynamics (MD) simulations to evaluate the α-hemolysin nanopore's capacity to distinguish between standard amino acids. Recognising that peptide folding influences ionic current blockades, they developed a theoretical model to calculate nanopore resistance. The researchers initially tested their model using MD simulations with four representative amino acids (A, Q, F, and W), revealing a correlation between the estimator of nanopore steric exclusion (pore clogging) and ionic current blockage. This correlation allowed them to extend their analysis to all 20 standard amino acids. Their findings demonstrated that amino acid volume, charge, and hydrophobicity are the primary determinants of nanopore steric exclusion, providing a comprehensive understanding of how different amino acids interact with the α-hemolysin nanopore.

Ouldali et al. (2020) attempted to identify the 20 standard amino acids using the aerolysin nanopore. The experiment involved the use of an XR_7 peptide construct, where the carboxyl terminus of a cationic carrier of seven arginine amino acids (R_7) is chemically bound to the eighth amino acid, X, to produce 20 XR_7 peptides. To achieve high sensitivity, individual XR_7 peptides were passed through the nanopore, and 13 amino acids (R, W, F, K, L, N, T, P, D, A, C, S, G) were able to be identified, whereas the remaining seven fell into two groups: (Q, M, H, I, Y) and (E, V). However, those amino acids produced current blockades too similar to be differentiated. The researchers were also able to identify modified amino acids from those that were unmodified. Additionally, it was observed that by increasing the time peptides resided in the nanopore, the likelihood of correctly identifying amino acids could reach more than 90% for 16 of the 20 amino acids. The researchers were able to identify the rest of the amino acids by modifying sample preparations and through nanopore engineering.

Even though the results described above are promising, significant challenges remain in the field of nanopore-based protein sequencing. For instance, a study by Nivala et al., (2014) used ClpX protein unfoldase to draw a peptide through a nanopore. Although their signal data could differentiate between various peptides, interpreting the readings proved challenging, partly due to the erratic stepping behaviour of the ClpX protein. This example illustrates some of the complexities involved in translating nanopore signals into accurate protein sequences.

A nanopore-based single-molecule peptide reader was developed by Brinkerhoff et al. (2021), where a DNA-translocating motor was used to precisely pull the peptide through the nanopore. This method is able to recognise single amino acid alterations and acquire high-fidelity signals by repeatedly scanning the same individual molecule. Yan et al. (2021) employed a related approach, utilising peptide–oligonucleotide conjugates and nanopore-induced phase-shift sequencing for their measurements.

In summary, three significant challenges must be addressed to develop a comprehensive peptide sequencing system that uses nanopores. Firstly, a mechanism is needed for precise control of peptide translocation through the nanopore. Secondly, an enzyme capable of repeatedly cleaving the peptide one amino acid at a time needs to be engineered. Finally, a method to accurately distinguish between all 20 standard amino acids based on their ionic current signatures must be developed (Liu et al., 2023). While progress has been made in each of these areas, none of these key requirements have been fully met, highlighting the ongoing nature of research in this field.

CONCLUSIONS

In this chapter, we have summarised the evolution of sequencing approaches and the recent advancements in single-molecule protein sequencing. Protein-nanopore experiments are progressing rapidly, and a nanopore-based protein sequencer is likely to be available in the near future.

Recent research has primarily concentrated on identifying large posttranslational modifications and distinguishing between phosphorylated proteins. While numerous proposed methods for single-molecule protein sequencing appear promising in theory, many still lack robust experimental validation. It is important to note that differentiating between peptides with single-residue substitutions is significantly different from achieving complete peptide sequencing, which necessitates the sequential identification of each residue in a strand. To realise this goal, present-day nanopore systems will likely require substantial reengineering. This may involve implementing controlled, unidirectional protein motion via application of a pressure differential across the pore. Furthermore, utilising both blockade amplitude and duration could potentially allow for dual identification of peptide residues, enhancing the accuracy of the sequencing process. The development of smaller, subnanometre-scale reading heads, analogous to those employed in DNA sequencing, may also prove beneficial in detecting individual amino acids in translocating peptides.

The development of a single-molecule protein sequencer is a formidable challenge. However, when developed, such a system will transform existing proteomic research by expediting the identification of low-abundance proteins. It will also facilitate the following: detection and study of post-translational modifications, precise detection of disease markers, identification of drug targets, and investigation of protein expression changes in single cells, in response to various stimuli.

ACKNOWLEDGEMENTS

We would like to express our sincere gratitude to Dr. Deepu Mathew and Dr. Prashanth N. Suravajhala for their unwavering support, invaluable guidance, and continuous motivation throughout the writing of this chapter. Their expertise and mentorship were instrumental in the successful completion of this work.

REFERENCES

Acharya, S., Edwards, S., & Schmidt, J. (2015). Research highlights: Nanopore protein detection and analysis. *Lab on a Chip*, *15*(17), 3424–3427.

Ahsan, M. U., Liu, Q., Fang, L., & Wang, K. (2021). NanoCaller for accurate detection of SNPs and indels in difficult-to-map regions from long-read sequencing by haplotype-aware deep neural networks. *Genome Biology*, *22*(1), 1–33. https://doi.org/10.1186/s13059-021-02472-2

Amarasinghe, S. L., Ritchie, M. E., & Gouil, Q. (2021). Long-read-tools.org: An interactive catalogue of analysis methods for long-read sequencing data. *GigaScience*, *10*(2), giab003. https://doi.org/10.1093/GIGASCIENCE/GIAB003

Amarasinghe, S. L., Su, S., Dong, X., Zappia, L., Ritchie, M. E., & Gouil, Q. (2020). Opportunities and challenges in long-read sequencing data analysis. *Genome Biology*, *21*(1), 1–16. https://doi.org/10.1186/S13059-020-1935-5

Aubin-Tam, M. E., Olivares, A. O., Sauer, R. T., Baker, T. A., & Lang, M. J. (2011). Single-molecule protein unfolding and translocation by an ATP-fueled proteolytic machine. *Cell*, *145*(2), 257–267.

Boersma, S., Khuperkar, D., Verhagen, B. M., Sonneveld, S., Grimm, J. B., Lavis, L. D., & Tanenbaum, M. E. (2019). Multi-color single-molecule imaging uncovers extensive heterogeneity in mRNA decoding. *Cell*, *178*(2), 458–472.

Boynton, P., & Di Ventra, M. (2016). Sequencing proteins with transverse ionic transport in nanochannels. *Scientific Reports*, *6*(1), 1–10.

Bray, N. L., Pimentel, H., Melsted, P., & Pachter, L. (2016). Near-optimal probabilistic RNA-seq quantification. *Nature Biotechnology*, *34*(5), 525–527.

Brinkerhoff, H., Kang, A. S., Liu, J., Aksimentiev, A., & Dekker, C. (2021). Multiple rereads of single proteins at single-amino acid resolution using nanopores. *Science*, *374*(6574), 1509–1513.

Carneiro, M. O., Russ, C., Ross, M. G., Gabriel, S. B., Nusbaum, C., & DePristo, M. A. (2012). Pacific biosciences sequencing technology for genotyping and variation discovery in human data. *BMC Genomics*, *13*(1), 1–7. https://doi.org/10.1186/1471-2164-13-375

Chaisson, M. J., & Tesler, G. (2012). Mapping single molecule sequencing reads using basic local alignment with successive refinement (BLASR): Application and theory. *BMC Bioinformatics*, *13*(1), 1–18. https://doi.org/10.1186/1471-2105-13-238

Chang, H. H., Hemberg, M., Barahona, M., Ingber, D. E., & Huang, S. (2008). Transcriptome-wide noise controls lineage choice in mammalian progenitor cells. *Nature*, *453*(7194), 544–547.

Chin, C. S., Alexander, D. H., Marks, P., Klammer, A. A., Drake, J., Heiner, C., Clum, A., Copeland, A., Huddleston, J., Eichler, E. E., Turner, S. W., & Korlach, J. (2013). Nonhybrid, finished microbial genome assemblies from long-read SMRT sequencing data. *Nature Methods, 10*(6), 563–569. https://doi.org/10.1038/nmeth.2474

Chin, C.-S., & Khalak, A. (2019). Human genome assembly in 100 minutes. *BioRxiv*, 705616. https://doi.org/10.1101/705616

Clive Brown. (2018). ONT Nanopore Technologies. Clive Brown's keynote at nanopore community meeting. Retrieved August 4, 2023, from https://nanoporetech.com/resource-centre/clive-brown-ncm-2018

Cressiot, B., Oukhaled, A., Patriarche, G., Pastoriza-Gallego, M., Betton, J. M., Auvray, L., Muthukumar, M., Bacri, L., & Pelta, J. (2012). Protein transport through a narrow solid-state nanopore at high voltage: Experiments and theory. *ACS Nano*, *6*(7), 6236–6243.

Cullinan, A. and Cantor, C. (2008). 'Sequenom, Inc.', *Pharmacogenomics*, *9*(9), pp. 1211–1215. doi: 10.2217/14622416.9.9.1211.

Deamer, D., Akeson, M. and Branton, D. (2016). Three decades of nanopore sequencing. *Nature biotechnology*, *34*(5), 518–524.

Deng, Q., Ramsköld, D., Reinius, B., & Sandberg, R. (2014). Single-cell RNA-seq reveals dynamic, random monoallelic gene expression in mammalian cells. *Science*, *343*(6167), 193–196.

Di Muccio, G., Rossini, A. E., Di Marino, D., Zollo, G., & Chinappi, M. (2019). Insights into protein sequencing with an α-hemolysin nanopore by atomistic simulations. *Scientific Reports*, *9*(1), 6440.

Domon, B., & Aebersold, R. (2006). Mass spectrometry and protein analysis. *Science*, 312(5771), 212–217. doi:10.1126/science.1124619

Domon, B., & Aebersold, R. (2010). Options and considerations when selecting a quantitative proteomics strategy. *Nature Biotechnology*, 28(7), 710–721. doi.org/10.1038/nbt.1661

Dong, Z., Kennedy, E., Hokmabadi, M., & Timp, G. (2017). Discriminating residue substitutions in a single protein molecule using a sub-nanopore. *ACS Nano*, *11*(6), 5440–5452.

Edge, P., & Bansal, V. (2019). Longshot enables accurate variant calling in diploid genomes from single-molecule long read sequencing. *Nature Communications, 10*(1), 1–10. https://doi.org/10.1038/s41467-019-12493-y

Edman, P. (1950). Method for the determination of the amino acid sequence in peptides. *Acta Chemica Scandinavica*, *4*, 283–293.

Fang, L., Hu, J., Wang, D., & Wang, K. (2018). NextSV: A meta-caller for structural variants from low-coverage long-read sequencing data. *BMC Bioinformatics*, *19*(1), 1–11. https://doi.org/10.1186/S12859-018-2207-1

Freedman, K. J., Haq, S. R., Edel, J. B., Jemth, P., & Kim, M. J. (2013). Single molecule unfolding and stretching of protein domains inside a solid-state nanopore by electric field. *Scientific Reports*, *3*(1), 1–8.

Freedman, K. J., Jürgens, M., Prabhu, A., Ahn, C. W., Jemth, P., Edel, J. B., & Kim, M. J. (2011). Chemical, thermal, and electric field induced unfolding of single protein molecules studied using nanopores. *Analytical Chemistry*, *83*(13), 5137–5144.

Frith, M. C., & Kawaguchi, R. (2015). Split-alignment of genomes finds orthologies more accurately. *Genome Biology*, *16*(106). https://doi.org/10.1186/S13059-015-0670-9

Gawad, C., Koh, W., & Quake, S. R. (2016). Single-cell genome sequencing: Current state of the science. *Nature Reviews Genetics*, *17*(3), 175.

Giani, A. M., Gallo, G. R., Gianfranceschi, L., & Formenti, G. (2020). Long walk to genomics: History and current approaches to genome sequencing and assembly. *Computational and Structural Biotechnology Journal*, *18*, 9–19. https://doi.org/10.1016/J.CSBJ.2019.11.002

Gierahn, T. M., Wadsworth, M. H., Hughes, T. K., Bryson, B. D., Butler, A., Satija, R., Fortune, S., Love, J. C., & Shalek, A. K. (2017). Seq-well: Portable, low-cost RNA sequencing of single cells at high throughput. *Nature Methods*, *14*(4), 395–398.

Goetz, J. J., & Trimarchi, J. M. (2012). Transcriptome sequencing of single cells with smart-seq. *Nature Biotechnology*, *30*(8), 763–765.

Goodrich, C. P., Kirmizialtin, S., Huyghues-Despointes, B. M., Zhu, A., Scholtz, J. M., Makarov, D. E., & Movileanu, L. (2007). Single-molecule electrophoresis of β-hairpin peptides by electrical recordings and Langevin dynamics simulations. *The Journal of Physical Chemistry B*, *111*(13), 3332–3335.

Goodwin, S., Gurtowski, J., Ethe-Sayers, S., Deshpande, P., Schatz, M. C., & McCombie, W. R. (2015). Oxford Nanopore sequencing, hybrid error correction, and *de novo* assembly of a eukaryotic genome. *Genome Research*, *25*(11), 1750–1756. https://doi.org/10.1101/GR.191395.115

Haider, S., & Pal, R. (2013). Integrated analysis of transcriptomic and proteomic data. *Curr. Genomics*, *14*(2), 91–110.

Hangauer, M. J., Vaughn, I. W., & McManus, M. T. (2013). Pervasive transcription of the human genome produces thousands of previously unidentified long intergenic noncoding RNAs. *PloS Genetics*, *9*(6), e1003569.

Haque, A., Engel, J., Teichmann, S. A., & Lönnberg, T. (2017). A practical guide to single-cell RNA-sequencing for biomedical research and clinical applications. *Genome Medicine*, *9*(1), 1–2.

Hashimshony, T., Wagner, F., Sher, N., & Yanai, I. (2012). CEL-seq: Single-cell RNA-seq by multiplexed linear amplification. *Cell Reports*, *2*(3), 666–673.

Heller, D., & Vingron, M. (2019). SVIM: Structural variant identification using mapped long reads. *Bioinformatics*, *35*(17), 2907–2915. https://doi.org/10.1093/BIOINFORMATICS/BTZ041

Ho, S. S., Urban, A. E., & Mills, R. E. (2019). Structural variation in the sequencing era. *Nature Reviews Genetics*, *21*(3), 171–189. https://doi.org/10.1038/s41576-019-0180-9

Huang, B., Wu, H., Bhaya, D., Grossman, A., Granier, S., Kobilka, B. K. and Zare, R. N. (2007). Counting low-copy number proteins in a single cell. *Science*, *315*(5808), 81–84.

Hwang, B., Lee, J. H., & Bang, D. (2018). Single-cell RNA sequencing technologies and bioinformatics pipelines. *Experimental & Molecular Medicine*, *50*(8), 1–4.

Istace, B., Friedrich, A., d'Agata, L., Faye, S., Payen, E., Beluche, O., Caradec, C., Davidas, S., Cruaud, C., Liti, G., Lemainque, A., Engelen, S., Wincker, P., Schacherer, J., & Aury, J. M. (2017). *De novo* assembly and population genomic survey of natural yeast isolates with the Oxford nanopore MinION sequencer. *GigaScience*, *6*(2), 1–13. https://doi.org/10.1093/GIGASCIENCE/GIW018

Jain, C., Koren, S., Dilthey, A., Phillippy, A. M., & Aluru, S. (2018). A fast adaptive algorithm for computing whole-genome homology maps. *Bioinformatics*, *34*(17), i748–i756. https://doi.org/10.1093/BIOINFORMATICS/BTY597

Jaitin, D. A., Kenigsberg, E., Keren-Shaul, H., Elefant, N., Paul, F., Zaretsky, I., Mildner, A., Cohen, N., Jung, S., Tanay, A., & Amit, I. (2014). Massively parallel single-cell RNA-seq for marker-free decomposition of tissues into cell types. *Science*, *343*(6172), 776––779.

Jiang, T., Liu, Y., Jiang, Y., Li, J., Gao, Y., Cui, Z., Liu, Y., Liu, B., & Wang, Y. (2020). Long-read-based human genomic structural variation detection with cuteSV. *Genome Biology*, *21*(1), 1–24. https://doi.org/10.1186/S13059-020-02107-Y

Kamath, G. M., Shomorony, I., Xia, F., Courtade, T. A., & Tse, D. N. (2017). HINGE: Long-read assembly achieves optimal repeat Resolution. *Genome Research*, *27*(5), 747–756. https://doi.org/10.1101/GR.216465.116

Kennedy, E., Dong, Z., Tennant, C., & Timp, G. (2016). Reading The primary structure of a protein with 0.07 nm 3 Resolution using a subnanometre-diameter Pore. *Nature Nanotechnology*, *11*(11), 968–976.

Klein, A. M., Mazutis, L., Akartuna, I., Tallapragada, N., Veres, A., Li, V., Peshkin, L., Weitz, D. A., & Kirschner, M. W. (2015). Droplet barcoding for single-cell transcriptomics applied to embryonic stem cells. *Cell*, *161*(5), 1187–1201.

Kolmogorov, M., Yuan, J., Lin, Y., & Pevzner, P. A. (2019). Assembly of long, error-prone reads using repeat graphs. *Nature Biotechnology*, *37*(5), 540–546. https://doi.org/10.1038/s41587-019-0072-8

Li, H. (2016). Minimap and miniasm: Fast mapping and *de novo* assembly for noisy long sequences. *Bioinformatics*, *32*(14), 2103–2110. https://doi.org/10.1093/BIOINFORMATICS/BTW152

Li, H. (2018). Minimap2: Pairwise alignment for nucleotide sequences. *Bioinformatics*, *34*(18), 3094–3100. https://doi.org/10.1093/BIOINFORMATICS/BTY191

Li, J., Fologea, D., Rollings, R., & Ledden, B. (2014). Characterization of protein unfolding with solid-state nanopores. *Protein and Peptide Letters*, *21*(3), 256–265.

Li, K. W., & Geraerts, W. P. (1997). Amino acid sequencing of neuropeptides. *Neuropeptide Protocols*, 17–25. doi:10.1385/0-89603-399-6:17

Lin, Y., Yuan, J., Kolmogorov, M., Shen, M. W., Chaisson, M., & Pevzner, P. A. (2016). Assembly of long error-prone reads using De Bruijn graphs. *Proceedings of the National Academy of Sciences of the United States of America*, *113*(52), E8396–E8405. https://doi.org/10.1073/PNAS.1604560113/SUPPL_FILE/PNAS.1604560113.SAPP.PDF

Liu, L., Li, Y., Li, S., Hu, N., He, Y., Pong, R., Lin, D., Lu, L., & Law, M. (2012). Comparison of next-generation sequencing systems. *Journal of Biomedicine and Biotechnology*, *2012*. https://doi.org/10.1155/2012/251364

Liu, M., Li, J., & Tan, C. S. (2023). Unlocking the power of nanopores: Recent advances in biosensing applications and analog front-end. *Biosensors*, *13*(6), 598.

Liu, Q., Georgieva, D. C., Egli, D., & Wang, K. (2019). NanoMod: A computational tool to detect DNA modifications using nanopore long-read sequencing data. *BMC Genomics*, *20*(1), 31–42. https://doi.org/10.1186/S12864-018-5372-8/FIGURES/6

Ma, L., Cao, J., Liu, L., Du, Q., Li, Z., Zou, D., Bajic, V. B., Zhang, Z. (2019). LncBook: A curated knowledgebase of human long non-coding RNAs. *Nucleic Acids Research*, 47(D1), D128–D134.

Macosko, E. Z., Basu, A., Satija, R., Nemesh, J., Shekhar, K., Goldman, M., Tirosh, I., Bialas, A. R., Kamitaki, N., Martersteck, E. M., & Trombetta, J. J. (2015). Highly parallel genome-wide expression profiling of individual cells using nanoliter droplets. *Cell*, *161*(5), 1202–1214.

Mahmoud, M., Gobet, N., Cruz-Dávalos, D. I., Mounier, N., Dessimoz, C., & Sedlazeck, F. J. (2019). Structural variant calling: The long and the short of it. *Genome Biology*, *20*(1), 1–14. https://doi.org/10.1186/S13059-019-1828-7/TABLES/2

Margulies, M., Egholm, M., Altman, W. E., Attiya, S., Bader, J. S., Bemben, L. A., Berka, J., Braverman, M. S., Chen, Y. J., Chen, Z., & Dewell, S. B. (2005). Correction: Corrigendum: Genome sequencing in microfabricated high-density picolitre reactors. *Nature*, *441*(7089), 120.

McIntyre, A. B. R., Alexander, N., Grigorev, K., Bezdan, D., Sichtig, H., Chiu, C. Y., & Mason, C. E. (2019). Single-molecule sequencing detection of N6-methyladenine in microbial reference materials. *Nature Communications, 10*(1), 1–11. https://doi.org/10.1038/s41467-019-08289-9

Medzihradszky, K. F., & Chalkley, R. J. (2015). Lessons in *de novo* peptide sequencing by tandem mass spectrometry. *Mass Spectrometry Reviews*, *34*(1), 43–63.

Mehrafrooz, B., & Aksimentiev, A. (2023). Engineering electro-osmotic flow through biological channels for nanopore protein sequencing. *Biophysical Journal*, *122*(3), 551a.

Mohammad, M. M., & Movileanu, L. (2008). Excursion of a single polypeptide into a protein pore: Simple physics, but complicated biology. *European Biophysics Journal*, *37*(6), 913–925.

Movileanu, L., Schmittschmitt, J. P., Scholtz, J. M., & Bayley, H. (2005). Interactions of peptides with a protein pore. *Biophysical Journal*, *89*(2), 1030–1045.

Movileanu, L. (2009). Interrogating single proteins through nanopores: challenges and opportunities. *Trends in Biotechnology*, *27*(6), 333–341.

nanoporetech/bonito: A PyTorch Basecaller for Oxford Nanopore Reads. (n.d.). Retrieved August 5, 2023, from https://github.com/nanoporetech/bonito

nanoporetech/medaka: Sequence correction provided by ONT Research. (n.d.). Retrieved August 8, 2023, from https://github.com/nanoporetech/medaka

Ni, P., Huang, N., Zhang, Z., Wang, D. P., Liang, F., Miao, Y., Xiao, C., Le, Luo, F., & Wang, J. (2019). DeepSignal: Detecting DNA methylation state from nanopore sequencing reads using deep-learning. *Bioinformatics*, *35*(22), 4586–4595. https://doi.org/10.1093/BIOINFORMATICS/BTZ276

Nivala, J., Marks, D. B., & Akeson, M. (2013). Unfoldase-mediated protein translocation through an α-hemolysin nanopore. *Nature Biotechnology*, *31*(3), 247–250.

Nivala, J., Mulroney, L., Li, G., Schreiber, J., & Akeson, M. (2014). Discrimination among protein variants using an unfoldase-coupled nanopore. *ACS Nano*, *8*(12), 12365–12375.

Nowoshilow, S., Schloissnig, S., Fei, J. F., Dahl, A., Pang, A. W. C., Pippel, M., Winkler, S., Hastie, A. R., Young, G., Roscito, J. G., Falcon, F., Knapp, D., Powell, S., Cruz, A., Cao, H., Habermann, B., Hiller, M., Tanaka, E. M., & Myers, E. W. (2018). The axolotl genome and the evolution of key tissue formation regulators. *Nature, 554*(7690), 50–55. https://doi.org/10.1038/nature25458

Omenn, G. S. (2014). The strategy, organization, and progress of the HUPO human proteome project. *Journal of Proteomics*, *100*, 3–7.

Oukhaled, A., Bacri, L., Pastoriza-Gallego, M., Betton, J. M. and Pelta, J. (2012). Sensing proteins through nanopores: Fundamental to applications. *ACS chemical biology*, *7*(12), 1935–1949.

Oukhaled, A., Cressiot, B., Bacri, L., Pastoriza-Gallego, M., Betton, J. M., Bourhis, E., Jede, R., Gierak, J., Auvray, L., & Pelta, J. (2011). Dynamics of completely unfolded and native proteins through solid-state nanopores as a function of electric driving force. *ACS Nano*, *5*(5), 3628–3638.

Ouldali, H., Sarthak, K., Ensslen, T., Piguet, F., Manivet, P., Pelta, J., Behrends, J. C., Aksimentiev, A., & Oukhaled, A. (2020). Electrical recognition of the twenty proteinogenic amino acids using an aerolysin nanopore. *Nature Biotechnology*, *38*(2), 176–181.

Oxford Nanopore Technologies. (n.d.). Retrieved August 5, 2023, from https://github.com/nanoporetech

PacificBiosciences/ccs: CCS: Generate Highly Accurate Single-Molecule Consensus Reads (HiFi Reads). (n.d.). Retrieved August 5, 2023, from https://github.com/PacificBiosciences/ccs

PacificBiosciences/pbsv: pbsv - PacBio structural variant (SV) calling and analysis tools. (n.d.). Retrieved August 8, 2023, from https://github.com/PacificBiosciences/pbsv

Pagel, O., Loroch, S., Sickmann, A., & Zahedi, R. P. (2015). Current strategies and findings in clinically relevant post-translational modification-specific proteomics. *Expert Review of Proteomics*, 12(3), 235–253. https://doi.org/10.1586/14789450.2015.1042867

Pandey, A., & Mann, M., (2000). Proteomics to study genes and genomes. *Nature*, 405(6788), 837–846. doi:10.1038/35015709

Pappin, D. J., (2003). Peptide mass fingerprinting using MALDI-TOF mass spectrometry. In Protein Sequencing Protocols (pp. 211–219). Humana Press.

Pappin, D. J., Hojrup, P., & Bleasby, A. J. (1993). Rapid identification of proteins by peptide-mass fingerprinting. *Current Biology*, *3*(6), 327–332.

Payet, L., Martinho, M., Pastoriza-Gallego, M., Betton, J. M., Auvray, L., Pelta, J., & Mathé, J. (2012). Thermal unfolding of proteins probed at the single molecule level using nanopores. *Analytical Chemistry*, *84*(9), 4071–4076.

Picelli, S. (2017). Single-cell RNA-sequencing: The future of genome biology is now. *RNA Biology*, *14*(5), 637–650.

Picelli, S., Björklund, Å. K., Faridani, O. R., Sagasser, S., Winberg, G., & Sandberg, R. (2013). Smart-seq2 for sensitive full-length transcriptome profiling in single cells. *Nature Methods*, *10*(11), 1096–1098.

Piguet, F., Ouldali, H., Pastoriza-Gallego, M., Manivet, P., Pelta, J., & Oukhaled, A. (2018). Identification of single amino acid differences in uniformly charged homopolymeric peptides with aerolysin nanopore. *Nature Communications*, *9*(1), 1–13.

Plesa, C., Kowalczyk, S. W., Zinsmeester, R., Grosberg, A. Y., Rabin, Y., & Dekker, C. (2013). Fast translocation of proteins through solid state nanopores. *Nano Letters*, *13*(2), 658–663.

Popitsch, N., Preuner, S., & Lion, T. (2021). Nanopanel2 calls phased low-frequency variants in nanopore panel sequencing data. *Bioinformatics*, *37*(24), 4620–4625. https://doi.org/10.1093/BIOINFORMATICS/BTAB526

Ren, J., & Chaisson, M. J. P. (2021). LRA: A long read aligner for sequences and contigs. *PLOS Computational Biology*, *17*(6), e1009078. https://doi.org/10.1371/journal.pcbi.1009078

Restrepo-Pérez, L., John, S., Aksimentiev, A., Joo, C., & Dekker, C. (2017). SDS-assisted protein transport through solid-state nanopores. *Nanoscale*, *9*(32), 11685–11693.

Restrepo-Pérez, L., Joo, C., & Dekker, C. (2018). Paving the way to single-molecule protein sequencing. *Nature Nanotechnology*, *13*(9), 786–796. doi: 10.1038/s41565-018-0236-6

Robertson, J. W., & Reiner, J. E. (2018). The utility of nanopore technology for protein and peptide sensing. *Proteomics*, *18*(18), 1800026.

Rosenberg, A. B., Roco, C. M., Muscat, R. A., Kuchina, A., Mukherjee, S., Chen, W., Peeler, D. J., Yao, Z., Tasic, B., Sellers, D. L., & Pun, S. H. (2017). Scaling single cell transcriptomics through split pool barcoding. BioRxiv. 105163.

Ruan, J., & Li, H. (2019). Fast and accurate long-read assembly with wtdbg2. *Nature Methods, 17*(2), 155–158. https://doi.org/10.1038/s41592-019-0669-3

Sampath, G. (2015). Amino acid discrimination in a nanopore and the feasibility of sequencing peptides with a tandem cell and exopeptidase. *RSC Advances*, *5*(39), 30694–30700.

Sanger, F. (1949). The terminal peptides of insulin. *Biochemical Journal*, *45*(5), p.563.

Sanger, F. and Thompson, E.O.P. (1953). The amino-acid sequence in the glycyl chain of insulin. 1. The identification of lower peptides from partial hydrolysates. *Biochemical Journal*, *53*(3), p.353.

Sanger, F. and Tuppy, H. (1951). The amino-acid sequence in the phenylalanyl chain of insulin. 1. The identification of lower peptides from partial hydrolysates. *Biochemical journal*, *49*(4), p.463.

Sauciuc, A., Morozzo, B., Tadema, M., Chinappi, M., & Maglia, G. (2023). An engineered electroosmotic flow transports unravelled proteins across nanopores. *bioRxiv*. https://doi.org/10.1101/2023.02.17.528930

Scheffler, K., Viner, R., & Damoc, E. (2018). High resolution top-down experimental strategies on the orbitrap platform. *Journal of Proteomics*, *175*, 42–55.

Sedlazeck, F. J., Rescheneder, P., Smolka, M., Fang, H., Nattestad, M., Von Haeseler, A., & Schatz, M. C. (2018). Accurate detection of complex structural variations using single-molecule sequencing. *Nature Methods, 15*(6), 461–468. https://doi.org/10.1038/s41592-018-0001-7

Shafin, K., Pesout, T., Lorig-Roach, R., Haukness, M., Olsen, H. E., Bosworth, C., Armstrong, J., Tigyi, K., Maurer, N., Koren, S., Sedlazeck, F. J., Marschall, T., Mayes, S., Costa, V., Zook, J. M., Liu, K. J., Kilburn, D., Sorensen, M., Munson, K. M., & Paten, B. (2019). Efficient *de novo* assembly of eleven human genomes using PromethION sequencing and a novel nanopore toolkit. *BioRxiv*, 715722. https://doi.org/10.1101/715722

Shalek, A. K., Satija, R., Adiconis, X., Gertner, R. S., Gaublomme, J. T., Raychowdhury, R., Schwartz, S., Yosef, N., Malboeuf, C., Lu, D., & Trombetta, J. J. (2013). Single-cell transcriptomics reveals bimodality in expression and splicing in immune cells. *Nature*, *498*(7453), 236–240.

Shao, H., Ganesamoorthy, D., Duarte, T., Cao, M. D., Hoggart, C. J., & Coin, L. J. M. (2018). npInv: Accurate detection and genotyping of inversions using long read sub-alignment. *BMC Bioinformatics*, *19*(1), 1–13. https://doi.org/10.1186/S12859-018-2252-9

Sharon, D., Tilgner, H., Grubert, F., & Snyder, M. (2013). A single-molecule long-read survey of the human transcriptome. *Nature Biotechnology*, *31*(11), 1009–1014.

Sheng, K., Cao, W., Niu, Y., Deng, Q., & Zong, C. (2017). Effective detection of variation in single-cell transcriptomes using MATQ-seq. *Nature Methods*, *14*(3), 267–270.

Sinitcyn, P., Rudolph, J. D., & Cox, J. (2018). Computational methods for understanding mass spectrometry–based shotgun proteomics data. *Annual Review of Biomedical Data Science*, *1*, 207–234.

Sović, I., Šikić, M., Wilm, A., Fenlon, S. N., Chen, S., & Nagarajan, N. (2016). Fast and sensitive mapping of nanopore sequencing reads with GraphMap. *Nature Communications*, *7*(1), 1–11. https://doi.org/10.1038/ncomms11307

Standing, K. G. (2003). Peptide and protein *de novo* sequencing by mass spectrometry. *Current Opinion in Structural Biology*, *13*(5), 595–601.

Steen, H., & Mann, M. (2004). The ABC's (and XYZ's) of peptide sequencing. *Nature Reviews Molecular Cell Biology*, 5(9), 699–711. doi:10.1038/nrm1468

Stefureac, R., Long, Y. T., Kraatz, H. B., Howard, P., & Lee, J. S. (2006). Transport of α-helical peptides through α-hemolysin and aerolysin pores. *Biochemistry*, *45*(30), 9172–9179.

Stoiber, M., Quick, J., Egan, R., Lee, J. E., Celniker, S., Neely, R. K., Loman, N., Pennacchio, L. A., & Brown, J. (2017). *De novo* identification of DNA modifications enabled by genome-guided nanopore signal processing. *BioRxiv*, 094672. https://doi.org/10.1101/094672

Suter, D. M., Molina, N., Gatfield, D., Schneider, K., Schibler, U., & Naef, F. (2011). Mammalian genes are transcribed with widely different bursting kinetics. *Science*, *332*(6028), 472–4.

Sutherland, T. C., Long, Y. T., Stefureac, R. I., Bediako-Amoa, I., Kraatz, H. B., & Lee, J. S. (2004). Structure of peptides investigated by nanopore analysis. *Nano Letters*, *4*(7), 1273–1277.

Su, J., Zheng, Z., Ahmed, S. S., Lam, T. W., & Luo, R. (2022). Clair3-trio: High-performance nanopore long-read variant calling in family trios with trio-to-trio deep neural networks. *Briefings in Bioinformatics*, *23*(5), 1–12. https://doi.org/10.1093/BIB/BBAC301

Talaga, D. S., & Li, J. (2009). Single-molecule protein unfolding in solid state nanopores. *Journal of the American Chemical Society*, *131*(26), 9287–9297.

Tang, F., Barbacioru, C., Nordman, E., Li, B., Xu, N., Bashkirov, V. I., Lao, K., & Surani, M. A. (2010). RNA-seq analysis to capture the transcriptome landscape of a single cell. *Nature Protocols*, *5*(3), 516–535.

Tang, F., Barbacioru, C., Wang, Y., Nordman, E., Lee, C., Xu, N., Wang, X., Bodeau, J., Tuch, B. B., Siddiqui, A., & Lao, K. (2009). mRNA-seq whole-transcriptome analysis of a single cell. *Nature Methods*, *6*(5), 377–382.

Teng, H., Cao, M. D., Hall, M. B., Duarte, T., Wang, S., & Coin, L. J. M. (2018). Chiron: Translating nanopore raw signal directly into nucleotide sequence using deep learning. *GigaScience*, *7*(5), 1–9. https://doi.org/10.1093/GIGASCIENCE/GIY037

Tham, C. Y., Tirado-Magallanes, R., Goh, Y., Fullwood, M. J., Koh, B. T. H., Wang, W., Ng, C. H., Chng, W. J., Thiery, A., Tenen, D. G., & Benoukraf, T. (2020). NanoVar: Accurate characterization of patients' genomic structural variants using low-depth nanopore sequencing. *Genome Biology*, *21*(1), 1–15. https://doi.org/10.1186/S13059-020-01968-7

Timp, W., & Timp, G. (2020). Beyond mass spectrometry, the next step in proteomics. *Science Advances*, *6*(2), eaax8978.

Tsedilin, A. M., Fakhrutdinov, A. N., Eremin, D. B., Zalesskiy, S. S., Chizhov, A. O., Kolotyrkina, N. G., & Ananikov, V. P. (2015). How sensitive and accurate are routine NMR and MS measurements? *Mendeleev Communications*, *25*(6), 454–456.

Tyers, M., & Mann, M. 2003. From genomics to proteomics. *Nature*, 422(6928), 193–197. doi:10.1038/nature01510

Varongchayakul, N., Song, J., Meller, A., & Grinstaff, M. W. (2018). Single-molecule protein sensing in A nanopore: A tutorial. *Chemical Society Reviews*, *47*(23), 8512–8524.

Wei, X., Penkauskas, T., Reiner, J.E., Kennard, C., Uline, M.J., Wang, Q., Li, S., Aksimentiev, A., Robertson, J.W. and Liu, C. (2023). Engineering biological nanopore approaches toward protein sequencing. *ACS nano*, *17*(17), 16369–16395.

Wilhelm, M., Schlegl, J., Hahne, H., Gholami, A. M., Lieberenz, M., Savitski, M. M., Ziegler, E., Butzmann, L., Gessulat, S., Marx, H., & Mathieson, T. (2014). Mass-spectrometry-based draft of the human proteome. *Nature*, *509*(7502), 582–587.

Wilson, J., Sloman, L., He, Z., & Aksimentiev, A. (2016). Graphene nanopores for protein sequencing. *Advanced Functional Materials*, *26*(27), 4830–4838.

Wohlers, I., Garg, S., & Hehir-Kwa, J. Y. (2023). Editorial: Long-read sequencing – Pitfalls, benefits and success stories. *Frontiers in Genetics*, *13*, 1114542. https://doi.org/10.3389/FGENE.2022.1114542

Xin, Y., Kim, J., Ni, M., Wei, Y., Okamoto, H., Lee, J., Adler, C., Cavino, K., Murphy, A. J., Yancopoulos, G. D., & Lin, H. C. (2016). Use of the Fluidigm C1 platform for RNA sequencing of single mouse pancreatic islet cells. *Proceedings of the National Academy of Sciences*, *113*(12), 3293–3298.

Yan, S., Zhang, J., Wang, Y., Guo, W., Zhang, S., Liu, Y., Cao, J., Wang, Y., Wang, L., Ma, F., & Zhang, P. (2021). Single molecule ratcheting motion of peptides in a mycobacterium smegmatis porin a (MspA) nanopore. *Nano Letters*, *21*(15), 6703–6710.

Zheng, Z., Li, S., Su, J., Leung, A., Lam, S., & Luo, T. W. (2022). Symphonizing pileup and full-alignment for deep learning-based long-read variant calling. *Nature Computational Science, 2*(12), 797–803. https://doi.org/10.1038/s43588-022-00387-x

Zheng, G. X., Terry, J. M., Belgrader, P., Ryvkin, P., Bent, Z. W., Wilson, R., Ziraldo, S. B., Wheeler, T. D., McDermott, G. P., Zhu, J., & Gregory, M. T. (2017). Massively parallel digital transcriptional profiling of single cells. *Nature Communications*, *8*(1), 1–2.

Zubarev, R. A. (2013). The challenge of the proteome dynamic range and its implications for in-depth proteomics. *Proteomics*, 13(5), 723–726. doi:10.1002/pmic.201200451

15 Standard Operating Procedures for Effective Galaxy Workflows

Sona Charles, Kiran K. Telukunta, Darshana Joshi, Shalini Rajagopal, and Harpreet Singh

AN INTRODUCTION TO GALAXY

The past two decades have seen significant advances in 'omics' technologies, opening up new data-driven opportunities in biomedical sciences. However, biological data currently face the 'three V' challenges of big data: volume, velocity, and variety [1]. The rapid influx of such data has made its analysis computationally intensive, requiring advanced skills and competencies for executing complex algorithms. This has created a significant bottleneck in the field, with growing demand for efficient resources for data storage, handling, and analysis outpacing supply. As researchers increasingly rely on computational methods to extract meaningful biological insights from large-scale sequence data, those with little programming expertise often find themselves overwhelmed. They struggle with multiple aspects of data analysis: identifying appropriate tools, managing software installation, and selecting optimal parameters. These challenges frequently prevent the successful integration of various analytical tools into a single, cohesive analysis pipeline.

Since its inception in 2005, The Galaxy Project has been developing an open, web-based platform called 'Galaxy' for analysing high-throughput biomedical datasets [2]. The platform comprises an integrated collection of tools, user-friendly applications for exploratory data analysis, and installations of applications for research. The Galaxy Community Hub (https://galaxyproject.org, last accessed 13 December 2024) is a website that hosts the project and a training network. The project also has a community of developers with diverse competencies and skills who assist with platform development [3]. The project focuses on transparent, accessible, and reproducible data analysis, which is needed in the genomic sciences [4]. The Galaxy platform has been developed with three goals in mind: (i) as a bridge for experimental scientists to easily access computation-driven tools and algorithms; (ii) to develop user-friendly workflows connecting various tools for analysis of their own data, which may be saved or reused later; and (iii) to improve the reproducibility of bioinformatics analysis by sharing and publishing workflows adapted for any study. These workflows may be inspected, edited, and expanded by other users [4].

The Galaxy Project develops open-source software that enables the creation of publicly accessible and computationally rich bioinformatics servers. The main public Galaxy server, launched in 2007 and maintained by The Galaxy Project at https://usegalaxy.org (United States) (last accessed 13 December 2024), houses numerous tools for various uses, such as next-generation sequencing (NGS) data analysis, machine learning (ML), geographic information systems (GIS), climate analysis, and computational chemistry. This server also possesses a collection of ready-to-use reference genomes. Independent instances of the Galaxy platform are maintained by other institutions, with two major public servers launched in March 2018: https://usegalaxy.eu (last accessed 13 December 2024), operated by the University of Freiburg (Europe), and https://usegalaxy.org.au (last accessed 13 December 2024), operated by Australian research organisations. Several other servers are currently available and are listed in Table 15.1. Servers for specific research domains are available with preset workflows and tools that may require additional logins and have limited data processing

DOI: 10.1201/9781003354062-15

TABLE 15.1
List of Galaxy Servers and Public Instances Specific for Countries or a Domain

No.	Instance	Description	URL
1	Main Galaxy Server	The main free public server of the Galaxy Project, biomedical research	https://galaxyproject.org/use/usegalaxy-org/
2	Galaxy Europe	The European free public Galaxy server	https://galaxyproject.org/use/usegalaxy-eu/
3	Galaxy Australia	The Australian free public Galaxy server	https://galaxyproject.org/use/usegalaxy-au/
4	Galaxy Belgium	ELIXIR Belgium free public Galaxy Server	https://galaxyproject.org/use/usegalaxy-be/
5	Galaxy France	French free public instance of the Galaxy server	https://galaxyproject.org/use/usegalaxy-fr/
7	Galaxy Norway	ELIXIR Norway Galaxy Server, free for Norwegian researchers and affiliates	https://galaxyproject.org/use/usegalaxy-no/
8	Galaxy India	Free public instance of the Galaxy server hosted on EU instance dedicated to the Indian community	https://galaxyproject.org/use/galaxy-india/
9	Galaxy Africa	Free public instance of the Galaxy server hosted on EU instance dedicated to the African Continent	https://galaxyproject.org/use/galaxy-africa/

All URLs were last accessed 13 December 2024.

quotas, such as the Galaxy-P multi-omics informatics platform, plus servers for metagenomics, ML, etc. It is also possible to set up a custom Galaxy server as a local instance or on the cloud [2]. The Galaxy community is a vital group of members that support the activities of these servers. The project also maintains the Galaxy ToolShed [5], which consists of over 8,500 tools that may be accessed with a web browser. The increasing number of resources demands comprehensive training in the respective disciplines, so the Galaxy Training Network [6] has evolved into a global community of trainers who develop materials and provide in-house or online training, available at https://training.galaxyproject.org (last accessed 13 December 2024). The network supports the findable, accessible, interoperable, reusable (FAIR) policy for data sharing [7].

The Galaxy platform, being a web-based scientific analysis tool, is used by tens of thousands of scientists worldwide to analyse large biomedical datasets found in genomics, proteomics, metabolomics, transcriptomics, variant analysis, metagenomics, and imaging. Over the last few years, the project has expanded both the quantity and quality of the tools available in the Galaxy ToolShed [5]. The project's main public server is maintained by the Texas Advanced Computing Center (TACC) and provides common tools for the analysis of large datasets, generous disk space, and CPU usage. In this chapter, we focus on how the platform has revolutionised handling and analysing high-throughput sequencing data despite the inherent computational challenges.

NGS IN GALAXY

Recent advancements in NGS technology have overcome the bottleneck of raw data generation in many genome-scale studies. However, the sheer volume of data creates technological hurdles for individual researchers, particularly in terms of analysis, collaboration, and reproducibility.

The Galaxy platform's interface offers a gateway to explore various types of analyses, including NGS. Typically, NGS analysis on the platform commences with raw sequencing data in the standard FASTQ format. To accommodate supplementary data for further analysis, the platform enables users to upload data in other common file formats. The existing pipeline leverages wrappers to perform tasks such as quality filtering and trimming of reads (Trimmomatic, Cutadapt), generating sequence quality reports (FastQC, MultiQC), mapping reads to a reference genome (BWA-MEM, Bowtie, TopHat), and downstream functional annotation (VCFtools, SnpSift, InterProScan, DESeq, etc.). These pipelines on the platform can be easily tailored to meet the user's specific requirements [8].

WORKFLOWS IN GALAXY

Workflows are structured sequences of analytical tasks that connect multiple tools. They are becoming an essential component of the Galaxy platform and enable users to orchestrate tools in a predefined order with a set of parameters that may be tailor-made according to the requirements. A typical workflow, also known as a pipeline, receives output from one tool and uses it as input for another [9]. Galaxy workflows can be created from scratch or automatically generated from the step-by-step history of a completed analysis. They enhance the reproducibility of analyses by capturing and standardising each processing step and can be modified to accommodate different research requirements. These workflows are currently available for complex analyses such as sequence analysis, whole genome sequencing data assembly, RNA sequencing (RNA-seq), single-cell RNA sequencing (scRNA-seq), genome-wide alternative splicing analysis, ChIPseq data analysis, etc. Standardised workflows ensure a well-documented protocol that reduces the inclusion of human errors in an analysis procedure.

The Galaxy Workflow Editor is a powerful tool that is capable of handling thousands of samples through a user-friendly interface. Reports are generated in support of running workflows and are useful for gaining meaningful insights from an analysis and for developing visualisations. Created workflows may be stored in the Galaxy ToolShed.

NGS WORKFLOWS IN GALAXY

NGS analyses require input data to be in standardised formats that are compatible with the Galaxy platform's tools, workflows, and tutorials. NGS tools can be found in the Galaxy ToolShed. Pre-configured workflows for common NGS analysis protocols are available through the public Galaxy servers and can be shared within the Galaxy community. A list specific to NGS data analysis appears in Table 15.2.

TABLE 15.2
Reproducible Workflows Related to NGS from WorkflowHub

Title	Creators	Teams	Licence
Stacks RAD-seq reference-guided workflow	Anna Syme	Galaxy Australia, Australian BioCommons	Apache Software License 2.0
Partial *de novo* workflow: c-s-g-pops only	Anna Syme	Galaxy Australia, Australian BioCommons	Apache Software License 2.0
Partial ref-guided workflow – gstacks and pops	Anna Syme	Galaxy Australia, Australian BioCommons	Apache Software License 2.0
Partial ref-guided workflow – bwa mem only	Anna Syme	Galaxy Australia, Australian BioCommons	Apache Software License 2.0
Stacks RAD-seq *de novo* workflow	Anna Syme	Galaxy Australia, Australian BioCommons	Apache Software License 2.0
StringTie workflow	Cristóbal Gallardo	Galaxy Training Network	Creative Commons Attribution, NonCommercial 4.0
BAM to FASTQ + QC v1.0	Gareth Price	Australian BioCommons, Galaxy Australia	GNU General Public License 3.0
Purge duplicates from hifiasm assembly v1.0	Gareth Price	Galaxy Australia, Australian BioCommons	GNU General Public License 3.0
RNAseq_UMG_SDumont_v1	Kary Ocaña	ParslRNA-seq: an efficient and scalable RNA-seq analysis workflow for studies of differentiated gene expression	Creative Commons Attribution 4.0
PacBio HiFi genome assembly using hifiasm v2.1	Gareth Price and Katherine Farquharson	Galaxy Australia, Australian BioCommons	GNU General Public License 3.0

(Continued)

TABLE 15.2 ***(Continued)***
Reproducible Workflows Related to NGS from WorkflowHub

Title	Creators	Teams	Licence
Combined workflows for large genome assembly	Anna Syme	Galaxy Australia, Australian BioCommons	Apache Software License 2.0
Data QC	Anna Syme	Galaxy Australia, Australian BioCommons	Apache Software License 2.0
Assembly with Flye	Anna Syme	Galaxy Australia, Australian BioCommons	Apache Software License 2.0
Genomics – read preprocessing	Dannon Baker et al.	GalaxyProject SARS-CoV-2	MIT License
Genomics – read preprocessing without downloading from SRA	Dannon Baker et al.	GalaxyProject SARS-CoV-2	MIT License
Assembly of the genome sequence	Dannon Baker et al.	GalaxyProject SARS-CoV-2	MIT License
Genomics – MRCA analysis	Dannon Baker et al.	GalaxyProject SARS-CoV-2	MIT License
Assembly using Tophat2 and annotation	None	GalaxyProject SARS-CoV-2	Apache Software License 2.0
Genome-wide alternative splicing analysis	Cristóbal Gallardo	Galaxy Training Network	Creative Commons Attribution, NonCommercial 4.0
Genome-wide alternative splicing analysis v.2	Cristóbal Gallardo	Galaxy Training Network	Creative Commons Attribution, NonCommercial 4.0
Trim and filter reads – fastp	Anna Syme	Galaxy Australia, Australian BioCommons	Apache Software License 2.0
Assess genome quality	Anna Syme	Galaxy Australia, Australian BioCommons	Apache Software License 2.0
Assembly polishing	Anna Syme	Galaxy Australia, Australian BioCommons	Apache Software License 2.0
scRNA-seq single-sample processing (counts matrix)	Sarah Williams et al.	QCIF Bioinformatics	Other (not open)
scRNA-seq single-sample processing STARSolo	Sarah Williams et al.	QCIF Bioinformatics	Creative Commons Attribution 4.0
scRNAseq_QCtoBasicProcessing	Sarah Williams et al.	QCIF Bioinformatics	Creative Commons Attribution 4.0
Generic variation analysis on WGS PE data	Wolfgang Maier	usegalaxy-eu	Creative Commons Attribution 4.0
SARS-CoV-2 Illumina Amplicon pipeline – SANBI, v. 1.2	Peter van Heusden and Bradley W. Langhorst	SANBI Pathogen Bioinformatics	MIT License
sars-cov-2-pe-illumina-wgs-variant-calling/ COVID-19-PE-WGS-ILLUMINA	Wolfgang Maier	Iwc	MIT License
scRNAseq_CellQC	Sarah Williams et al.	QCIF Bioinformatics	Creative Commons Attribution 4.0
chipseq-pe/main	Lucille Delisle	Iwc	MIT License

GALAXY WORKFLOWS ON WORKFLOWHUB

WorkflowHub (https://workflowhub.eu, last accessed 13 December 2024) is a centralised repository where researchers can search, share, and collaborate on scientific workflows across multiple disciplines and computational platforms, including the Galaxy platform. These workflows automate and simplify challenging tasks, with a focus on data analysis. Of particular interest to bioinformaticians, the repository includes workflows optimised for various sequencing platforms, such as Illumina or

Oxford Nanopore, and different library preparation strategies, including ampliconic or metatranscriptomic approaches.

Researchers can import these workflows directly into a public Galaxy server, where they can be customised to suit specific experimental designs and research questions. The Galaxy interface supports FAIR [7] data analysis principles and make it straightforward to adjust parameters, inputs, and outputs while abstracting the underlying computational complexities. This allows researchers to focus on their analysis results rather than technical implementation.

Galaxy workflows available through WorkflowHub include pipelines for quality control, read alignment, variant calling, and annotation. After completing their analyses, users can contribute their modified workflows back to WorkflowHub, along with documentation and best practices. This creates a virtuous cycle of community improvement, supported by discussions and collaborative projects across both platforms. Through this integration of WorkflowHub and the Galaxy platform, researchers can effectively share expertise and continuously refine their analytical approaches.

RECENT TRENDS IN GALAXY WORKFLOWS

Galaxy's capacity to create reproducible and shared workflows has established it as a popular tool within the bioinformatics and genomics communities. As the community's use of Galaxy instances, such as the public servers, expands, so does the accessibility of various workflows and tools. A growing trend is the increased interoperability of Galaxy workflows with other tools and platforms, enabling seamless integration with external data sources, databases, and analysis pipelines. This enhances the overall capabilities of the workflows.

As mentioned earlier, the Galaxy community plays an active role in the development and improvement of workflows. Members collaboratively create, refine, and share workflows to tackle common challenges and tasks, ensuring the quality and relevance of the available workflows. Galaxy workflows also serve educational purposes, providing practical examples for students and researchers to learn specific analyses and understand underlying processes.

Galaxy workflows often incorporate visualisation and reporting steps, enabling researchers to generate graphical representations of their findings. This not only enhances data interpretability but also facilitates effective communication of results.

With the advent of technologies like containerisation and cloud computing, there is a rising trend in creating Galaxy workflows that can operate in containerised environments and leverage cloud resources for scalable analyses. Containers ensure consistent software performance across different computing environments, minimising software compatibility and version control issues. The Galaxy Docker project (https://galaxyproject.org/use/galaxy-docker/, last accessed 13 December 2024) provides Docker container images of Galaxy, enabling users to run their own Galaxy instances with pre-configured tools in portable, reproducible environments. The containers include all of the necessary tools and dependencies, eliminating the need for complex installations and configurations.

Galaxy workflows can be integrated with data repositories and databases for seamless access to reference datasets, annotations, and other analysis resources. The Galaxy community caters to various groups, including systems administrators, developers, tool authors, trainers, and scientists. These individuals can use resources with user credentials, install as an administrator, or contribute to the community by developing tools.

For trainers, the Galaxy Training Network [6], a key element of the Galaxy Project, provides updated and systematic training materials and tools for efficient use of the Galaxy platform. For further learning, scientists can find relevant topics on the Galaxy Community Hub and utilise tutorials in combination with the Galaxy Project's YouTube channel (https://www.youtube.com/c/GalaxyProject, last accessed 13 December 2024).

In essence, Galaxy fosters a collaborative environment where users can discover, use, and contribute to the development of workflows, promoting a cycle of continuous learning and improvement.

CASE STUDY: A GENERAL WORKFLOW FOR COMPLETE DOMAIN ANALYSIS OF COVID-19 DATA

COVID-19 was the first pandemic to emerge in the age of big sequence data. During this period, researchers used *in silico* methods to investigate SARS-CoV-2. NGS was crucial in tracking viral evolution and investigating potential origins and intermediate hosts of the virus. Hundreds of coronavirus and SARS-CoV-2 genomes, sequenced using NGS, were made publicly available for researchers to investigate the virus' lineage. Researchers then used the high-quality sequencing datasets and various Galaxy workflows to detect allelic variants. A basic workflow for detecting single nucleotide polymorphisms (SNPs) and insertions or deletions (indels) in COVID-19 data includes (i) trimming the data, (ii) checking its quality, (iii) mapping the reads, (iv) checking for duplicates, (v) extracting indels using GATK4 Mutect2, and (vi) annotating variants using SnpEff with the SARS-CoV-2 genome database. During the COVID-19 outbreak, ready-made workflows were of great assistance in analysing sequenced datasets and are typical examples where pertinent preconstructed workflows came to aid.

The process, datasets, and documentation are all available at https://covid19.galaxyproject.org (last accessed 13 December 2024). Later, the Galaxy Project started an initiative called BY-COVID (Beyond COVID), which is a European Commission (EC)-funded project that intends to integrate well-established data resources and provide access to heterogeneous yet interconnected and ordered data across domains and jurisdictions. It aims to address the data obstacles that can hinder effective pandemic response. With the help of the Research Objects with DataCrate (RO-Crate) team, BY-COVID members have contributed to several RO-Crate-related tutorials, which have been added to the Galaxy Training Network under a brand new FAIR topic, where RO-Crate is a collaborative initiative to create a lightweight method of packaging research data with information.

MAKING TOOLS WORKFLOW-READY FOR GALAXY

The exponential growth in biological data generation, particularly from high-throughput technologies, has created new challenges for data analysis. This has driven the development of specialised tools and software platforms to support research workflows and data processing at scale. When developing new tools, there are some important criteria to consider. To begin with, a tool should be simple to download and be accessible in terms of pre-built packages and containers. When a tool is ready to be made publicly available, sufficient documentation is needed, which should contain sections on requirements, installation, and usage. A tool needs to be flexible in that it should allow options to be set or changed prior to execution. Its inputs and outputs must be specifically named, and their behaviour must be documented. A tool should be maintained periodically for compatibility with modern systems, in which case version management (e.g., git) should be used. To prevent conflicts between concurrent instances (e.g., instances writing to the same file), the tool developer should make sure that they are independent. For example, a tool could use separate temporary folders for each run. The fundamental goal for any scientific tool developer should be reproducible results. This will ultimately require regular testing with multiple systems and test data. It is crucial to make sure the tool's outputs contain both understandable data (such as an HTML report) to guide user decisions and structured data (such as in JSON format) that can be used as input for subsequent steps in a workflow. Because tool development is typically an iterative process, it is not necessary to implement all of these criteria at once [7].

HOW IS GALAXY DIFFERENT?

The Galaxy platform runs on Linux/Unix-based servers and provides a browser-based user interface for running bioinformatics tools and constructing simple workflows. While a competing workflow environment named Anvaya requires local installation to run workflows [10], the Galaxy platform offers multiple deployment options. Users can either run workflows through public Galaxy servers (such as https://usegalaxy.org, last accessed 13 December 2024) or install and run the platform locally on their own systems (https://galaxyproject.org/admin/get-galaxy/, last accessed 13 December 2024).

The Taverna workflow suite (officially retired by Apache in 2020) was, like the Galaxy platform, designed for users to run complex analysis pipelines. Taverna workflows could run on local desktop computers or be executed through other clients or web interfaces using the Taverna Server [11]. However, while Taverna could connect to various external services and tools, Galaxy integrates tools directly into its platform through the ToolShed system.

SUMMARY OF GALAXY CONTRIBUTION

The Galaxy Project has served thousands of users with a framework useful for accessible, transparent, and reproducible data analysis. The Galaxy community has been expanding rapidly and making great use of the platform's tools and services to publish impactful research. Moreover, the community maintains an annotated list on the free and open-source Zotero service of publications that reference the Galaxy platform [4]. As of 26 January 2025, there are 12,973 published articles on research that used, referenced, implemented, and/or extended the platform, according to the library maintained there (https://www.zotero.org/groups/1732893/galaxy, last accessed 26 January 2025). The yearly growth in publications involving Galaxy is depicted in Figure 15.1, while ToolShed statistics are shown in Figure 15.2.

There has been an increase in the amount of software and workflows available in the ToolShed, and they may be used on any Galaxy server. Figure 15.2 depicts the yearly growth in the number of tools in the Galaxy ToolShed, which can be seen increasing from the year 2018.

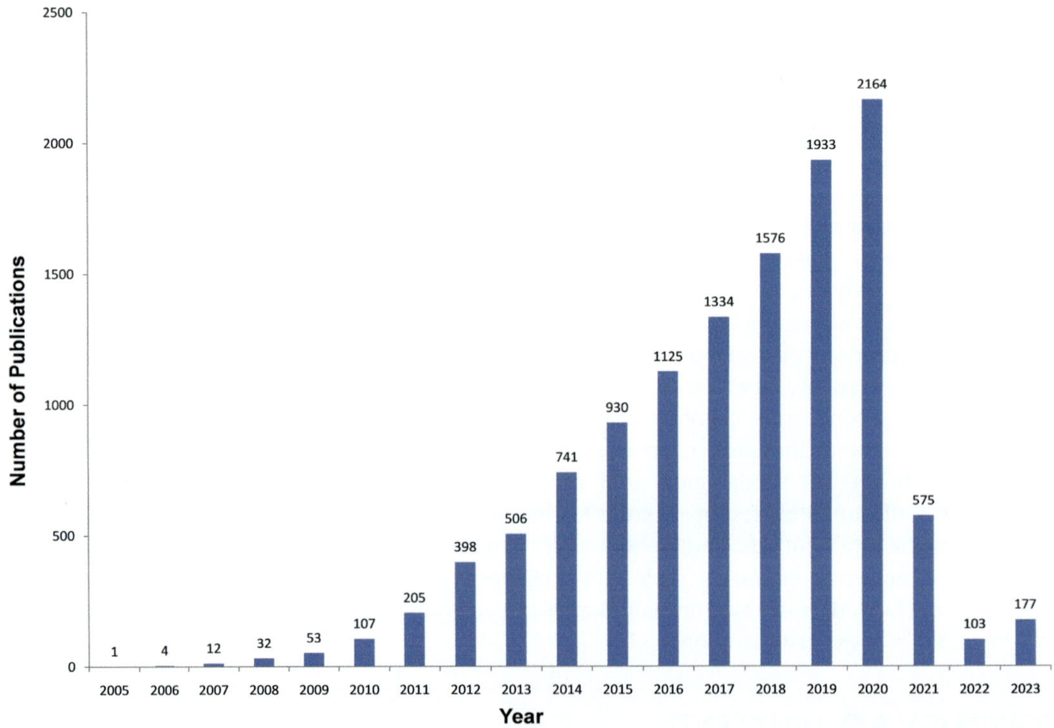

FIGURE 15.1 The yearly growth in publications involving (use, reference, extend, or implement) the Galaxy platform or a server, including journal articles, books, theses, and conference papers. The publications for the year 2023 are until 16 August 2023.

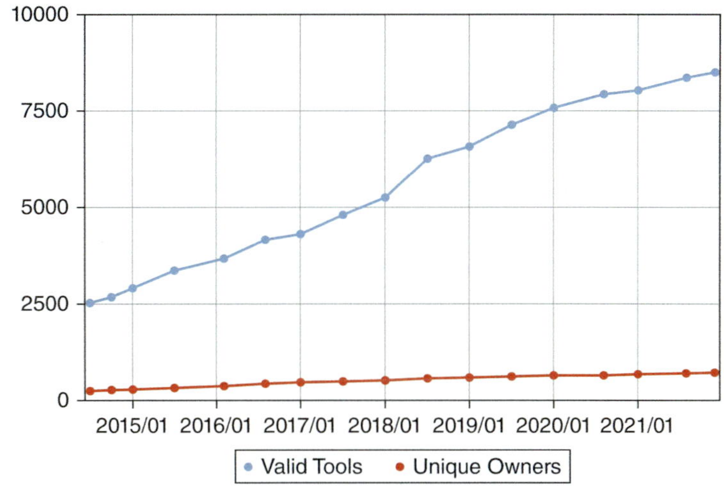

FIGURE 15.2 Yearly growth in tools in Galaxy ToolShed. (Source: Galaxy Project Statistics, https://galaxyproject.org/galaxy-project/statistics/, last accessed 13 December 2024.)

CONCLUSIONS AND CHALLENGES

The Galaxy platform has revolutionised genomics research by pioneering a new generation of interactive tools for large-scale genome analysis. It has also democratised such analyses by making the tools accessible to users who would have previously needed substantial programming experience and database management skills. Over the past few years, the Galaxy community has significantly enhanced the platform's core framework, user interface, tools, and training materials. This has led to the creation of numerous high-quality, curated tutorials focusing on common types of bioinformatics analyses.

Despite these advancements, scalability remains a significant impediment, especially as genomic datasets continue to grow. However, the platform is designed to scale both frontend and backend in a multiuser environment. In some instances, major cloud services like Amazon Web Services (AWS), Jetstream, and Microsoft Azure can bolster backend infrastructure. Yet, the public Galaxy servers often rely on public funding. With adequate funding, the Galaxy Project and its community can scale to resolve these constraints.

The Galaxy community remains committed to addressing three key barriers in data-driven biomedical science: (i) ensuring analyses are accessible to researchers regardless of computational expertise, (ii) guaranteeing reproducible results through standardised workflows and tool versioning, and (iii) enabling effective sharing of complete analytical procedures. As the Galaxy Project evolves, these principles continue to guide the development of new features and workflows to serve its growing user community.

ACKNOWLEDGEMENTS

We sincerely acknowledge Bioclues.org and Galaxy India for giving us this opportunity to make our fellow researchers aware of various global efforts being made by the Galaxy community for the analysis of vast amounts of NGS data using simplified workflows.

REFERENCES

[1] Mrozek, D. (2014). High-Performance Computational Solutions in Protein Bioinformatics. SpringerBriefs in Computer Science. Cham: Springer International Publishing. https://doi.org/10.1007/978-3-319-06971-5

[2] Goecks, J., Nekrutenko, A., Taylor, J. *et al.* Galaxy: A comprehensive approach for supporting accessible, reproducible, and transparent computational research in the life sciences. *Genome Biol*, ***11***, R86 (2010). https://doi.org/10.1186/gb-2010-11-8-r86

[3] Afgan, E., Nekrutenko, A., Grüning, B.A., Blankenberg, D., Goecks, J., Schatz, M.C., Ostrovsky, A.E., Mahmoud, A., Lonie, A.J., Syme, A., Fouilloux, A., Bretaudeau, A., Nekrutenko, A., Kumar, A., Eschenlauer, A.C., Desanto, A.D., Guerler, A., Serrano-Solano, B., Batut, B., Briggs, P.J. The galaxy platform for accessible, reproducible and collaborative biomedical analyses: 2022 update. *Nucleic Acids Research, 50*(W1), W345 (2022). https://doi.org/10.1093/NAR/GKAC247

[4] Afgan, E., Baker, D., van den Beek, M., Blankenberg, D., Bouvier, D., Čech, M., Chilton, J., Clements, D., Coraor, N., Eberhard, C., Grüning, B., Guerler, A., Hillman-Jackson, J., Von Kuster, G., Rasche, E., Soranzo, N., Turaga, N., Taylor, J., Nekrutenko, A., Goecks, J. The galaxy platform for accessible, reproducible and collaborative biomedical analyses: 2016 update. *Nucleic Acids Res, 44*(W1), W3–W10 (2016). https://doi.org/10.1093/nar/gkw343. Epub 2016 May 2. PMID: 27137889; PMCID: PMC4987906.

[5] Blankenberg, D., Von Kuster, G., Bouvier, E., Baker, D., Afgan, E., Stoler, N., Taylor, J., Nekrutenko, A., the Galaxy Team. Dissemination of scientific software with Galaxy ToolShed. *Genome Biology, 15*, 403 (2014). https://doi.org/10.1186/gb4161

[6] Hiltemann, S., Rasche, H., Gladman, S., Hotz, H.-R., Larivière, D., Blankenberg, D., Jagtap, P.D., Wollmann, T., Bretaudeau, A., Goué, N., Griffin, T.J., Royaux, C., Le Bras, Y., Mehta, S., Syme, A., Coppens, F., Droesbeke, B., Soranzo, N., Bacon, W., Batut, B. Galaxy training: A powerful framework for teaching! *PLoS Comput Biol, 19*(1), e1010752 (2023). https://doi.org/10.1371/journal.pcbi.1010752

[7] Boeckhout, M., Zielhuis, G.A., Bredenoord, A.L. The fair guiding principles for data stewardship: Fair enough? *Eur J Hum Genet, 26*(7), 931–936 (2018). https://doi.org/10.1038/s41431-018-0160-0. Epub 2018 May 17. PMID: 29777206; PMCID: PMC6018669.

[8] Griffiths, S.M., Fox, G., Briggs, P.J. *et al.* A galaxy-based bioinformatics pipeline for optimized, streamlined microsatellite development from illumina next-generation sequencing data. *Conserv Genet Resour, 8*, 481–486 (2016). https://doi.org/10.1007/s12686-016-0570-7

[9] Brack, P., Crowther, P., Soiland-Reyes, S., Owen, S., Lowe, D., Williams, R., Groom, A., Dillen, Q., Coppens, M., Gruning, F., Eguinoa, B., Ewels, I., Goble, C. Ten simple rules for making a software tool workflow-ready. *PLoS Comput Biol, 18*(3) (2022). https://doi.org/10.1371/JOURNAL.PCBI.1009823

[10] Limaye, B., Banerjee, R., Datta, A., Inamdar, H., Vats, P., Dahale, S., Bhandari, A., Ramakrishnan, E.P., Tupakula, R., Malviya, S., Bayaskar, A., Gadhari, R., Jain, S., Gavane, V., Mahajan, R., Sunitha, K., Joshi, R. Anvaya: A workflows environment for automated genome analysis. *J Bioinform Comput Biol, 10*(4), 1250006 (2012). https://doi.org/10.1142/S0219720012500060. Epub 2012 Jun 22. PMID: 22809419.

[11] Wolstencroft, K., Haines, R., Fellows, D., Williams, A., Withers, D., Owen, S., Soiland-Reyes, S., Dunlop, I., Nenadic, A., Fisher, P., Bhagat, J., Belhajjame, K., Bacall, F., Hardisty, A., Nieva de la Hidalga, A., Balcazar Vargas, M.P., Sufi, S., Goble, C. The taverna workflow suite: Designing and executing workflows of web services on the desktop, web or in the cloud. *Nucleic Acids Res, 41*, W557–W561 (2013). https://doi.org/10.1093/nar/gkt328. PMID: 23640334; PMCID: PMC3692062.

16 Motif Prediction Using ChIP-seq Data Analysis with Galaxy

Vrushabh Anil Nikhade, Kalyan Ram Uppaluri, Hima Jyothi Challa, Kalyani Palasamudram, and K. Sri Manjari

INTRODUCTION

Multiple myeloma (MM) is a haematological cancer characterised by the abnormal proliferation of plasma cells, and bortezomib serves as a primary treatment by inhibiting the proteasome pathway in cancer cells. However, many patients develop resistance to bortezomib therapy, which recent studies attribute to the dysregulation of various cellular processes, including proteasome subunit expression, apoptosis, autophagy, drug efflux, epithelial–mesenchymal transition, and DNA repair (Di Lernia et al., 2020). To better understand the molecular mechanisms underlying this drug resistance, researchers employ motif prediction—a bioinformatics approach that identifies short, conserved sequences, or 'motifs', of nucleotides or amino acids across different biological molecules such as DNA, RNA, and proteins. These conserved sequences often serve as binding sites for regulatory proteins or indicate functional regions crucial for biological processes like gene regulation, protein–protein interactions, and signal transduction. By identifying these motifs in MM cells, researchers can potentially discover new drug targets or predict how therapeutic compounds like bortezomib interact with specific molecular targets, ultimately leading to more effective treatments for resistant cases (Bailey et al., 2007).

Recent studies have shown that specific sequence properties of proteins, such as hydrophobicity, solvent accessibility, and structural features, can help identify druggable motifs and binding regions. Li and Lai (2007) demonstrated that proteins with accessible and stable binding motifs are promising drug targets, as these characteristics often correlate with critical functional sites. This understanding has practical applications in drug development—when researchers identify disease-associated genetic variations that affect these druggable motifs, they can design molecules that specifically target these altered sites. This targeted approach reduces the likelihood of off-target effects and enables more personalised treatments based on a patient's genetic profile (Agamah et al., 2020).

MM is a kind of malignant plasma cell neoplasm characterised by clonal plasma cell growth in the bone marrow. Despite recent advances in treatment, the condition remains fatal (Fairfield et al., 2016). Several genes and pathways have been implicated in the pathogenesis of MM, including those involved in the regulation of the NFκB pathway (e.g., NFKB1, NFKB2), cell cycle regulation (e.g., CCND1, CCND2), DNA damage response (e.g., ATM, ATR), protein degradation (e.g., PSMA1, PSMB1), and apoptosis (e.g., BCL2, BAX) (Cardona-Benavides et al., 2021). Understanding the function of these genes and pathways is crucial for developing effective therapies and personalised treatment strategies for MM patients. Table 16.1 details various MM genes and their association with bortezomib.

Chromatin immunoprecipitation sequencing (ChIP-seq) is a powerful tool that identifies DNA-binding sites of transcription factors and other proteins in the genome (Park, 2009). With advancements in next-generation sequencing (NGS) techniques, ChIP-seq analysis has become increasingly accessible and allows for identifying genome-wide protein–DNA interactions. Using ChIP-seq analysis, it is possible to identify the binding motifs of bortezomib-responsive proteins in MM cells (Narlikar and Jothi, 2012).

DOI: 10.1201/9781003354062-16

TABLE 16.1
Multiple Myeloma Genes and Their Association with Bortezomib

S. No	Gene	Interaction with Bortezomib Drug	Evidence PMID
1	*PSMB5*	Bortezomib binds to the PSMB5 subunit of proteasome and inhibits its activity, accumulating unwanted proteins and causing cell death.	36495869 (Louros et al., 2023)
2	*NFκB*	NFκB is a transcription factor that regulates diverse cellular functions, and Bortezomib has demonstrated the ability to inhibit its activity.	37095322 (Yao et al., 2023)
3	*BIRC5*	BIRC5 encodes a protein that regulates cell division and prevents cell death. Bortezomib decreases its expression.	26537080 (Liapis et al., 2015)
4	*MYC*	The protein encoded by MYC is a transcription factor that controls cell proliferation. Bortezomib has been shown to reduce its expression.	36623719 (Arena et al., 2023)
5	*LEF1*	Bortezomib has been shown to inhibit Wnt/β-catenin signalling by decreasing the levels of LEF1 and β-catenin in MM cells.	24394665 (Gupta et al., 2014)
6	*ERK*	Bortezomib has demonstrated specific inhibition of ERK1/2 phosphorylation, resulting in the induction of apoptosis.	25375375 (Kao et al., 2014)

MM cell lines, such as MM.1S, H929, and RPMI-8226, are commonly used to study the role of HDAC3 in MM pathogenesis and treatment (Ho et al., 2020). Bortezomib is a proteasome inhibitor that the FDA has approved for the treatment of MM. It acts as a reversible inhibitor of the 26S proteasome complex in mammalian cells. It leads to the accumulation of misfolded and ubiquitinated proteins, which induces apoptosis in tumour cells. Bortezomib is effective against primary and refractory myeloma, alone or combined with other chemotherapeutic agents (Chen et al., 2011). Its efficacy in combination with dexamethasone has been demonstrated in several clinical trials, leading to its approval by the FDA in 2003 for treating MM (Field-Smith et al., 2006). Bortezomib has shown promising therapeutic potential in additional haematological malignancies, including mantle cell lymphoma (Robak and Robak, 2019).

The efficacy of a drug in addressing mutated cells, as seen in MM, relies on its capacity to bind to particular target sequences. Identifying these binding sites is crucial to understanding drug function and resistance mechanisms. In this study, we aimed to identify the core binding sites in the MM cell line genome to uncover key regulatory sequences involved in drug response. This information could aid in developing new therapies and personalised treatment strategies for MM patients.

METHODS

Data Retrieval

ChIP-seq data from MM cells treated with bortezomib were retrieved from the short read archive (SRA) database using the ID GSM5788461. Two samples, SRR17510890 and SRR17510891, were selected for further analysis. The European Nucleotide Archive (ENA) browser obtained the data in FASTQ format (https://www.ebi.ac.uk/ena/browser/home, last accessed on 7 November 2024). The data were then uploaded to the Galaxy server for analysis and annotation (https://usegalaxy.org, last accessed on 7 November 2024).

Several tools were employed on the Galaxy web platform for ChIP-seq data analysis and annotation, including FastQC, MultiQC, Map with BWA, Filter SAM and BAM, MACS2 callpeak, and ChIPseeker. The FastQC and MultiQC tools were used for quality control and assessment of the sequencing data. The BWA tool aligned the reads to the human genome assembly. The Filter SAM and BAM tool was then used to remove any reads that did not meet the desired criteria (Barnett et al., 2011). This process ensures that the retained reads are high quality and appropriate for subsequent analyses. The MACS2 callpeak tool was employed for peak calling, and ChIPseeker was used to annotate the peaks (Feng et al., 2012).

For motif discovery within the identified peaks, we used Multiple Expectation maximisations for Motif Elicitation (MEME), a widely used tool that detects recurring patterns in DNA sequences.

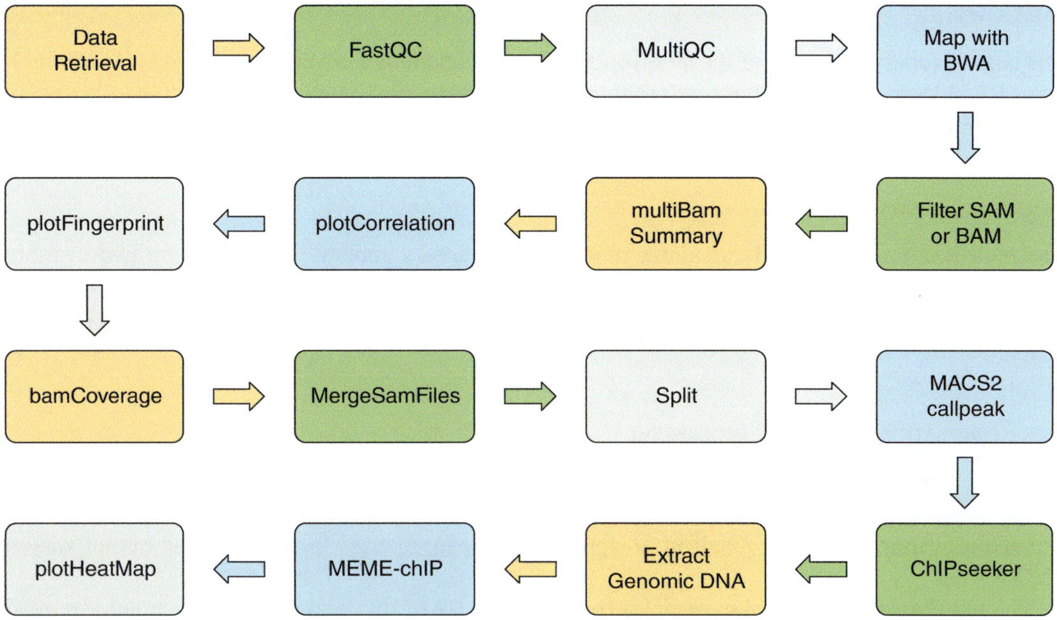

FIGURE 16.1 Workflow of the methodology used for ChiP-seq data analysis and motif identification.

Note: **This workflow demonstrates the phases in data preprocessing, tools for genomic alignment, motif prediction, and motif functional enrichment.**

The complete ChIP-seq analysis pipeline, from initial quality control through motif discovery, is illustrated in Figure 16.1.

MultiQC

To ensure the quality of the raw sequencing data, we used FastQC, a modular tool designed to identify any potential issues that require attention before further analysis. Raw sequence data from samples SRR17510890 and SRR17510891 in '`fasta.qz`' format were used as input for the FastQC analysis, with the resulting outputs provided in both HTML and '.txt' format. FastQC generates several statistical values that can be used to detect any necessary filtering or preprocessing steps.

MultiQC is a tool that compiles quality control data from several samples into a single report for easy comparison. The FastQC outputs in '.txt' format were used as inputs for the MultiQC tool, while the output plots generated by MultiQC were in HTML format (Ewels et al., 2016). The statistical values generated include per-base sequence quality, sequence length distribution, guanine-cytosine (GC) content, sequence duplication levels, and overrepresented sequences. This method efficiently and thoroughly analyses considerable datasets, identifying potential sequencing data flaws.

Map with BWA

The Map with BWA tool aligns short sequencing reads to the hg19 reference genome (Human Genome version 19, released in 2009) to determine the degree of similarity between read and genomic sequences. This version of the human reference genome, though superseded by newer assemblies, remains widely used in research for consistency with existing datasets and annotations. The 'fasta.qz' files SRR17510890 and SRR17510891 were used as inputs for Map with BWA. The output was in BAM format, which may be visualised in genome browsers like UCSC, Ensembl, bam.iobio (https://bam.iobio.io, last accessed on 7 November 2024), IGB, and IGV.

bamCoverage

The bamCoverage tool (part of the deepTools package) generates a coverage file from a BAM file. It takes BAM format files as input (such as those produced by BWA mapping) and outputs a coverage file in either bigWig or bedGraph format (Ramírez et al., 2016).

multiBamSummary

The multiBamSummary tool calculates read coverage across genomic regions using two or more BAM files. The input was the Filter SAM or BAM tool results, and the output was in 'deeptools_coverage_matrix' format. This result enables quantitative comparison of genomic alignments between different samples.

plotCorrelation and plotFingerprint

The plotCorrelation tool generated a heatmap of correlation values between samples. It visualises sample correlations based on either multiBamSummary or multiBigwigSummary results. The input was the multiBamSummary result in 'deeptools_coverage_matrix' format, and the output was in PNG format.

The plotFingerprint tool takes samples from indexed BAM files and generates a cumulative read coverage profile for each file. The input was the results of the map with BWA and Filter SAM or BAM in BAM format, and the output was in PNG format. This plot assesses the quality of ChIP enrichment by showing the distribution of reads across the genome.

MergeSamFiles

The MergeSamFiles tool merges multiple SAM or BAM files into a single file. The input was the BAM files produced by Map with BWA and processed by Filter SAM or BAM. Both the input and the output used the BAM file format.

Split

The Split tool separates a merged BAM file back into individual BAM files based on read groups. Each new BAM file contains aligned reads corresponding to the initial datasets. The input was the merged BAM file produced by MergeSamFiles, and the output was separate BAM files for each sample.

MACS2 callpeak

ChIP-seq model-based analysis (MACS), is a widely-used method for identifying transcription factor binding sites by evaluating local enrichment of ChIP-seq reads. The MACS algorithm accounts for local genomic complexity and biases when determining the statistical significance of enriched regions. While initially designed for transcription factor binding site detection, MACS2 has been adapted for broader peak-calling applications, including histone modifications. In our workflow, the MACS2 callpeak tool processed the split BAM files to generate peak calls in BED format, producing both standard and narrow peak outputs for downstream analysis (Zhang et al., 2008).

ChIPseeker

The ChIPseeker tool was employed to annotate the ChIP-seq data. ChIPseeker provides functional annotation of ChIP peaks and visualisation capabilities, including chromosome-wide peak coverage and binding profiles around transcription start sites (TSS). The tool enables detailed peak annotation and facilitates comparisons between different ChIP-seq datasets. ChIPseeker's capabilities are enhanced by referencing data from over 17,000 BED files in the Gene Expression Omnibus (GEO) database, expanding its annotation options (Robinson et al., 2010; Yu et al., 2015).

A matching gene annotation file is required to use ChIPseeker for downstream ChIP-seq data analysis. For this purpose, the GTF annotation file for the specific version of the human genome was downloaded from the official Gencode website (https://www.gencodegenes.org/human/, last accessed 7 November 2024). Once downloaded, the GTF file can be uploaded to the Galaxy server.

The ChIPseeker tool was used with the inputs of MACS2 callpeak results in tabular format and a GTF file downloaded from Gencode. The results were graphs and Excel files listing all introns and exons and their chromosomal numbers.

Extract Genomic DNA

The Extract Genomic DNA tool retrieves DNA sequences from specified genomic regions within a reference genome. Using the peak regions identified by MACS2 (in BED format) and the reference genome (in FASTA format), the tool generates a new FASTA file containing the DNA sequences corresponding to the enriched binding regions. These extracted sequences serve as input for subsequent motif discovery analysis.

MEME-ChIP

MEME-ChIP is a comprehensive motif analysis suite designed specifically for ChIP-seq data. It combines multiple algorithms to discover and analyse sequence patterns within ChIP-enriched regions. Using the DNA sequences obtained from the Extract Genomic DNA tool, MEME-ChIP identified recurring sequence patterns that could represent protein binding motifs. The analysis results were generated in HTML format, providing detailed visualisations and statistical assessments of the discovered motifs (Machanick and Bailey, 2011).

Compute Matrix

The ComputeMatrix tool allows users to generate a matrix of read coverage values for a set of genomic regions, such as genes or enhancers, from aligned sequencing data. The input data can be in BAM or bigWig format, and the output is a matrix that can be visualised as a heatmap or line plot to analyse gene expression patterns.

Plot Heatmap

The Plot Heatmap tool generates comprehensive visualisations of genomic binding patterns across multiple regions. It creates a colour-coded heatmap that represents the intensity and distribution of binding scores across the identified genomic regions, enabling the comparison of binding patterns across different samples or conditions. The resulting heatmap provides an intuitive visualisation of predicted binding site locations and their relative strengths within the genome, helping to identify patterns and clusters of regulatory activity.

ANALYSIS AND APPLICATIONS

Before aligning ChIP-seq data to a reference genome, it is necessary to perform quality control, a standard step in NGS data analysis. This ensures the data's reliability for subsequent analytical steps. The low number of duplicate reads and consistent quality metrics in our data suggest that we can confidently proceed with downstream analysis tasks such as variant calling, differential

Sample Name	% Dups	% GC	M Seqs
SRR17510890	18.0%	41%	39.9
SRR17510891	11.0%	41%	30.9

FIGURE 16.2 The number of unique and duplicate reads in HDAC3 bortezomib (SRR17510890) and HDAC3 DMSO (SRR17510891) ChiP-seq files.

Note: **This comparison reveals the number of unique and duplicated reads in both samples, with unique reads being 33 M in SRR17510890 and 27 M in SRR17510891.**

expression analysis, and functional annotation. In processing NGS data, removing adapters, barcodes, and unique molecular identifiers (UMIs) is a common necessity. Adapters, which are crucial for polymerase chain reaction (PCR) amplification and sequencing in standard NGS protocols, may extend beyond the read end during sequencing. To address this, it is essential to identify and excise any adapter sequences present in the reads. While barcodes are intentionally incorporated for multiplexing in some experiments and aid in data normalisation and comparison, considering barcode sequences was unnecessary for our analysis since the provided data were already demultiplexed. UMIs, which are similar to barcodes but unique for each read, were introduced to handle high duplication levels in cross-linking immunoprecipitation (CLIP) experiments. The presence of UMIs allows the fusion of reads with the same UMI, effectively addressing PCR duplicates. This strategy improves the accuracy of downstream analyses by collapsing duplicated reads that originate from the same molecular entity.

In the event of suboptimal results, interventions such as trimming poor-quality reads become essential. This involves the removal of sequences with low-quality bases or other anomalies identified through quality control tools, FastQC and MultiQC. While tools like Trimmomatic, Cutadapt, and Galaxy Trimming Tools offer various functionalities for adapter removal and quality-based trimming, appropriate tool selection depends on specific data characteristics and quality issues. Integration into workflows allows for tailored parameter adjustments, typically involving the removal of low-quality bases and adapters. By implementing these corrective measures, researchers can enhance data quality and ensure the robustness of subsequent analyses. This adherence to quality control measures aligns with established bioinformatics practices, emphasising the importance of data integrity in ChIP-seq analysis.

Following the protocol outlined in Figure 16.1, Galaxy was used to analyse samples SRR17510890 and SRR17510891. Figure 16.2 displays the FastQC sequence count results. The comparison of the samples indicates that both files are suitable for additional analysis, considering factors such as the percentage of duplicate reads, percent GC content, and total number of raw reads (M Seqs). The quality metrics for both samples are shown in Figure 16.3. The data revealed that duplicate rates were relatively low (18.0% and 11.0%), with consistent GC content (41%) across both samples, and

Sample Name	% Dups	% GC	M Seqs
SRR17510890	18.0%	41%	39.9
SRR17510891	11.0%	41%	30.9

FIGURE 16.3 MultiQC results.

Note: **The figure displays the duplicate percentage, GC percentage, and total number of raw reads (M Seqs) for SRR17510890 and SRR17510891.**

high sequence depth (39.9 and 30.9 million sequences respectively). This indicated that the sequencing reads were high quality with satisfactory per-base sequence quality scores and minimal per-base N content (not shown), making the data suitable for downstream processing.

To identify the genomic origins of DNA fragments, aligning the sequenced reads to a reference genome is imperative. The alignment serves as a crucial foundation for broader applications. In genomics, high alignment rates indicate the data's reliability, paving the way for downstream analyses. This includes but is not limited to the exploration of genomic variations, assessment of gene expression patterns, and the functional annotation of genomic elements more precisely. One primary function of alignment is to precisely localise each sequencing read within the reference genome, providing a spatial context for genomic information. Furthermore, alignment is crucial for identifying genetic variations, including single nucleotide polymorphisms (SNPs), insertions, deletions, and structural variations. This is particularly pertinent in studies exploring genetic diversity, disease-associated mutations, or evolutionary patterns.

Alignment facilitates the annotation of genomic elements such as coding regions, non-coding RNAs, and regulatory regions. Understanding the functional significance of different genomic regions is pivotal for deciphering the complexities of gene regulation and function. Additionally, alignment enables comparative genomics, allowing researchers to compare genomic sequences across different individuals, species, or populations, offering insights into evolutionary relationships and adaptive changes.

Map with BWA produced results in BAM format, which were visualised using the bam.iobio genome browser, as seen in Figures 16.4 and 16.5. The results of the read sequence alignment with the hg19 reference genome are indicated. Figure 16.6 shows the alignment between the HDAC3 bortezomib sample and the reference human genome (SRR17510890 to hg19). Figure 16.7 shows the alignment of the HDAC3 vehicle control reads and the reference genome (SRR17510891 to hg19). Both samples were more than 95% aligned with the reference genome.

The input control and our ChIP experiment might be strongly correlated, meaning that the potential binding regions are not truly enriched compared to our control. Therefore, further checking the correlation between our control and the ChIP experiment is necessary. The plotCorrelation tool generated a heatmap of correlation scores between the samples. Figure 16.6 shows the correlation between the control and bortezomib.

The plotFingerprint tool was used to identify the signal strength of the ChIP sample. Figure 16.7 shows the comparison of HDAC3 bortezomib (SRR17510890, green and blue lines) and HDAC3

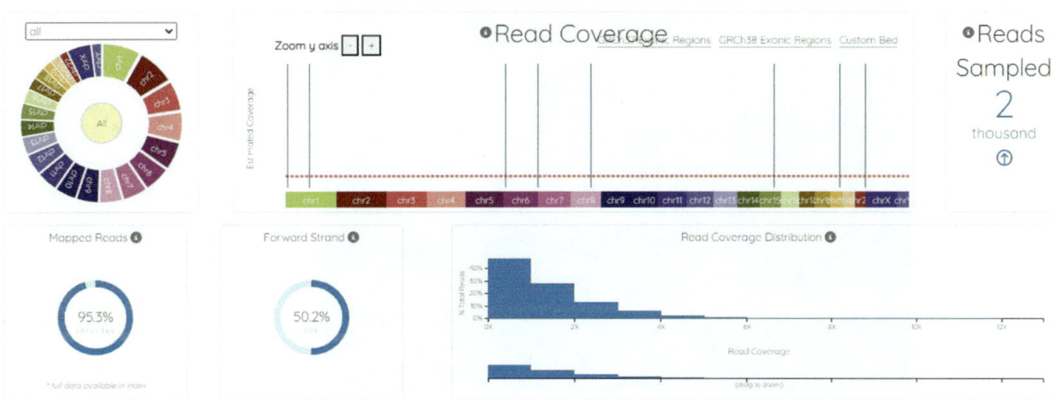

FIGURE 16.4 Visualisation of HDAC3 bortezomib (SRR17510890) genomic alignment to the reference genome hg19 using the bam.iobio browser. The interface displays multiple alignment metrics including chromosome-wise read coverage (centre), overall mapping rate (95.3%), strand distribution (50.2% forward strand), and read coverage distribution (bottom). The circular plot (left) provides a genome-wide view of read distribution across chromosomes.

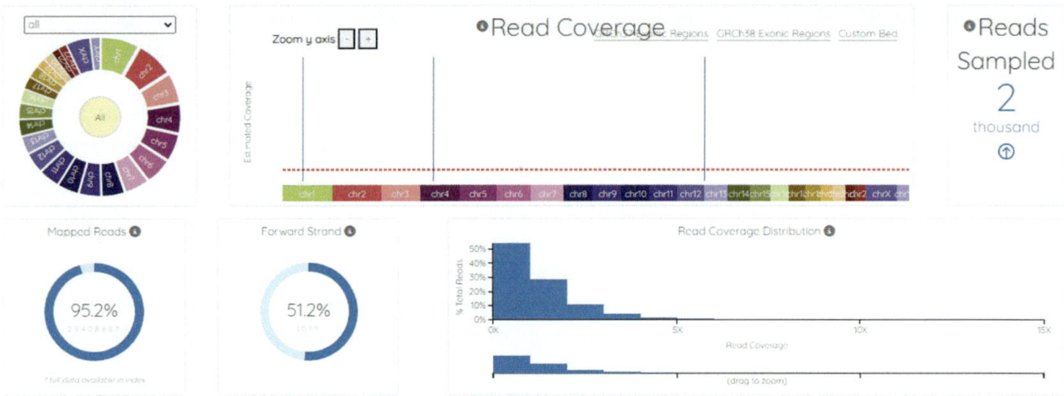

FIGURE 16.5 Visualisation of HDAC3 DMSO control (SRR17510891) genomic alignment to the reference genome hg19 using the bam.iobio browser. The interface shows strong alignment quality with 95.2% mapped reads and balanced strand distribution (51.2% forward strand). The chromosome-wise coverage plot (centre) and circular genome view (left) display read distribution across all chromosomes, while the coverage distribution histogram (right) indicates sequencing depth across the genome.

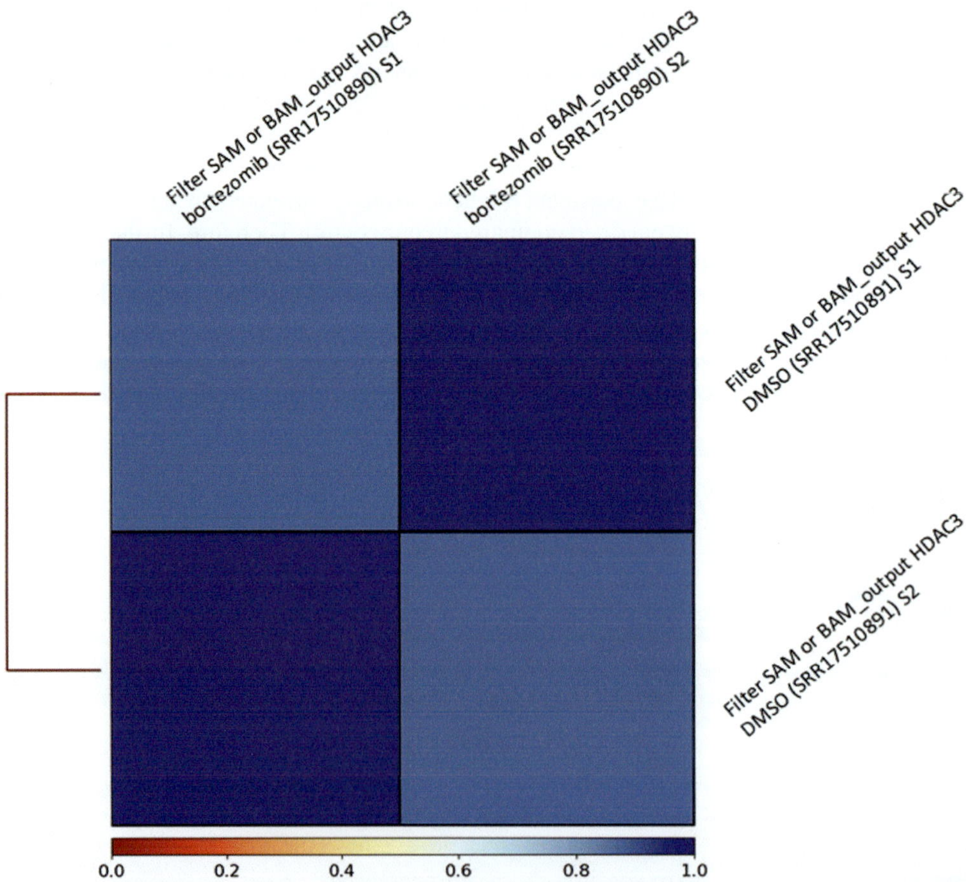

FIGURE 16.6 Heatmap correlation between aligned reads from the HDAC3 bortezomib (SRR17510890) and HDAC3 DMSO (SRR17510891) files.

Note: **This statistic compares the proportion of aligned reads in roughly the same sample file.**

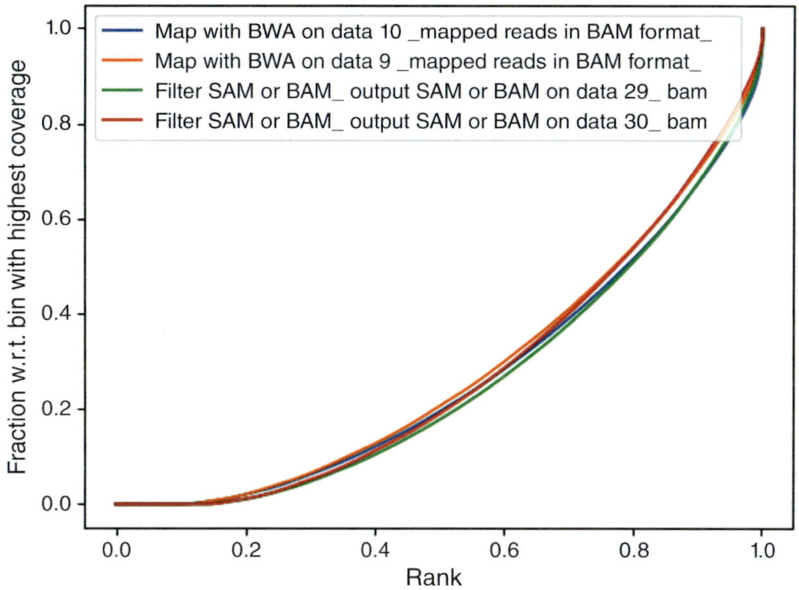

FIGURE 16.7 Fingerprint plots of both samples, generated with the plotFingerprint tool. The plots compare read distribution patterns before and after filtering for both samples. They show the cumulative read coverage distribution across genomic regions, with initial BWA mapping results (blue and orange lines) and post-filtering results (green and red lines). The similar curves indicate consistent enrichment patterns between samples both pre- and post-filtering, suggesting good quality ChIP-seq data. The x-axis represents the rank of genomic regions by read coverage, while the y-axis shows the fraction of reads that fell within these regions.

Note: **The plotFingerprint tool samples indexed BAM files and generates a cumulative read coverage profile for each. The plots demonstrate that the DMSO control sample (SRR17510891, red and orange line) has more read coverage.**

DMSO (SRR17510891, red and orange lines), which shows that there might be some variation in the regions that can be explored in further studies. The different coloured lines in the plot help distinguish between samples and processing stages. Similar to the functionality of plotFingerprint, multiBamSummary employs randomly sampling genomic regions (bins) with a specified length and aggregates per-base coverage overlapping with those regions. The resulting correlation values between these bins are then calculated for each pair of samples. An optimal plot is characterised by two distinct clusters: one for the biological replicates of the ChIP-seq experiment, and another for the replicates of the control. Conversely, an undesirable outcome occurs when the control and ChIP-seq experiment fall into a single cluster, indicating a close relationship and potential issues in the laboratory procedures.

Together, the plotCorrelation and plotFingerprint tools contribute to the optimisation of experimental design, guiding researchers in refining protocols and minimising biases. They offer a valuable benchmark for data quality, supporting informed decisions about downstream analyses.

Peak calling is a crucial stage in ChIP-seq data analysis, second only to mapping. The MACS2 package was used to perform peak calling, and the outcomes are displayed in Figure 16.8. This tool can identify the best parameters of the MACS2 peak calling.

MACS2 is instrumental in identifying peaks, representing regions of the genome enriched with sequencing reads. These peaks often correspond to regulatory elements like transcription factor binding sites and histone modifications. By pinpointing regions of increased read density, MACS2 aids researchers in characterising the epigenetic landscape of the genome. MACS2 facilitates the comparison of peak profiles between different biological states, such as normal and diseased tissues

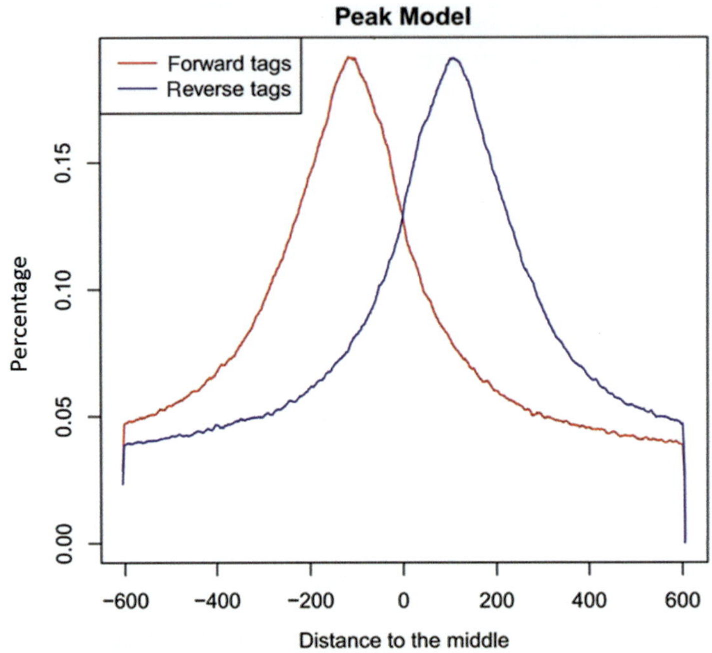

FIGURE 16.8 Peak model showing the distribution of DNA-Protein binding events across forward (red) and reverse (blue) strands. The x-axis represents the distance from the peak centre in base pairs, while the y-axis shows the relative enrichment of binding events. The complementary peaks offset by approximately 200 bp demonstrate the expected pattern of protein-DNA interactions on opposite strands, confirming the presence of genuine binding sites.

or cells, under different experimental conditions. This comparative analysis enables the identification of differential peaks, shedding light on genomic regions associated with specific phenotypes or responses to external stimuli. This information is valuable for understanding the molecular basis of diseases and designing targeted interventions. Peaks called by MACS2 serve as a foundation for downstream analyses, including motif analysis, pathway enrichment, and integration with other omics data, enhancing the depth and interpretability of genomics studies.

The result of the MACS2 callpeak analysis is shown in Figure 16.9, which displays the proportion of peaks detected in various genomic regions. The peaks are distributed across functional regions including promoters (45.11% total), exons (1.84%), introns (30.16%), and downstream regions (0.77%), with a substantial portion in distal intergenic regions (20.67%). The reference genome alignment demonstrates comprehensive coverage across these functional genomic regions. Visualising DNA–protein interaction percentages in both the forward and reverse strands, represented as peaks, is important in genomics research. These peaks represent active regulatory elements, including enhancers and promoters, where DNA–protein interactions are pivotal in gene regulation. The data obtained from these peaks can be integrated with other omics data, such as RNA-seq or ChIP-seq data, to understand the functional consequences of DNA–protein interactions comprehensively.

De novo motif identification, motif enrichment analysis, motif position analysis, and motif clustering were executed using MEME-ChIP to comprehensively identify enriched motifs and potential transcription factor binding sites (TFBS). This analysis is essential for understanding the regulatory code that governs gene expression. Motif enrichment analysis sheds light on the prevalence and significance of specific motifs, aiding in the annotation of functional regulatory elements, such as promoters, enhancers, and other cis-regulatory modules. Motif position analysis provides insights

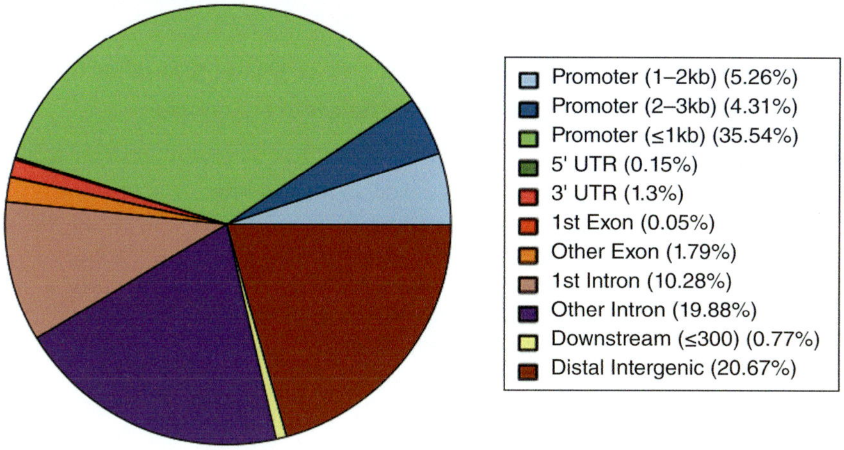

FIGURE 16.9 Distribution of HDAC3 binding sites in different genomic regions for the bortezomib-treated sample (SRR17510890). The pie chart shows the percentage of peaks in various genomic features, including promoter regions at different distances from transcription start sites, UTRs, exons, introns, and intergenic regions. The majority of binding occurs in promoter regions (≤1kb, 35.54%) and distal intergenic regions (20.67%).

Note: **ChIPSeeker was used to identify percentages across several genomic functional regions. It demonstrates that binding sites may be found in functional regions, such as the promoter and exon.**

into motif locations relative to genomic features, crucial for characterising the spatial organisation of regulatory elements. Motif clustering uncovers coordinated regulatory modules where multiple motifs work together to regulate gene expression, offering a holistic view of regulatory logic. The analysis employs two complementary *de novo* motif prediction approaches—MEME for matrix-based discovery and DREME for word-based discovery—to enhance both the accuracy and sensitivity of motif prediction.

The Extract Genomic DNA tool produced genomic sequences corresponding to the ChIP peaks. Figure 16.10 shows the motif that was predicted using the MEME-ChIP tool. The sequence motif plot visually represents the nucleotide positions of a motif. On the x-axis, each position is depicted, while the y-axis represents the information in bits at each position. This measurement reflects the

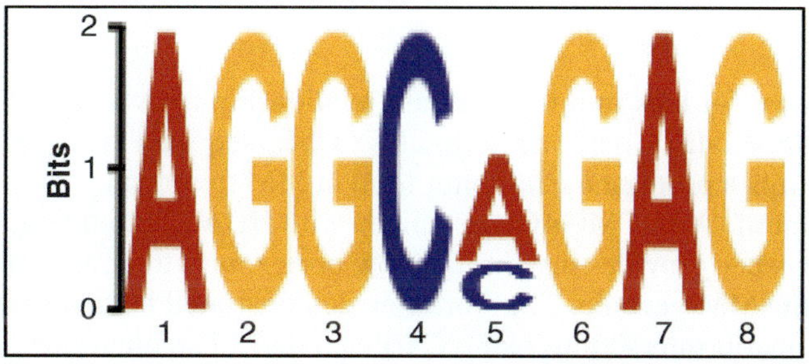

FIGURE 16.10 Sequence logo of the most enriched DNA motif (AGGCMGAG, 'M' represents either A or C at position 5) identified by MEME-ChIP analysis in HDAC3 binding regions from bortezomib-treated cells (SRR17510890). The motif was found in 12,478 peak sequences. Letter height indicates conservation at each position measured in bits (0-2), where taller letters show higher probability of finding that nucleotide at that position across all binding sites. This conserved sequence pattern suggests a specific DNA binding preference for HDAC3 in bortezomib-treated conditions.

TABLE 16.2
Identification of Motif AGGCMGAG, Which Was Identified in the 12,478 Positions in the Human Reference Genome Version hg19

Chromosome	Start	End	P-value	q-value
chr1	24710920	24710927	6.86E-05	0.245
chr1	169862951	169862958	3.43E-05	0.245
chr1	196621116	196621123	6.86E-05	0.245
chr2	201981160	201981167	6.86E-05	0.245
chr2	201982560	201982567	6.86E-05	0.245
chr2	202040725	202040732	6.86E-05	0.245
chr2	202040757	202040764	6.86E-05	0.245
chr2	202047446	202047453	1.64E-05	0.201
chr2	202645500	202645507	6.86E-05	0.245
chr7	2272725	2272732	6.86E-05	0.245
chr7	8215911	8215918	1.64E-05	0.201
chr7	8301042	8301049	6.86E-05	0.245
chr7	43519386	43519393	1.64E-05	0.201
chr7	91764144	91764151	3.43E-05	0.245
chr7	91874596	91874603	1.64E-05	0.201
chr7	120970650	120970657	6.86E-05	0.245
chr12	1059074	1059081	1.64E-05	0.201
chr12	9101935	9101942	6.86E-05	0.245
chr17	37026302	37026309	6.86E-05	0.245

probability of specific nucleotide (A, C, G, and T for DNA), with larger letters indicating higher probabilities. The motif plot offers insights into the positional preferences and probabilities of nucleotides within the represented sequence motif.

Table 16.2 shows the motif AGGCMGAG, which was identified in 12,478 positions in the human reference genome hg19. The table details the presence of predicted motifs in different chromosomes, along with their start and end positions. Due to space limitations, only a subset of these positions is shown, for illustrative purposes.

CONCLUSIONS

ChIP-seq offers a powerful lens into the intricate relationship between DNA and its regulators. This study utilised the Galaxy platform to analyse ChIP-seq data from HDAC3 in MM cells treated with bortezomib, a proteasome inhibitor used in MM treatment. The analysis successfully identified a binding motif (AGGCMGAG) associated with HDAC3 binding in bortezomib-treated conditions. This discovery provides insight into how bortezomib treatment affects HDAC3 binding patterns within the MM genome. The workflow demonstrated here can be applied to investigate other drug-chromatin interactions using ChIP-seq data. The identified binding motif offers potential applications in understanding drug response and resistance mechanisms. Future work should involve wet lab validation to explore the motif's function and implications for drug efficacy. This protocol provides a foundation for investigating drug-induced changes in transcription factor binding and their potential impact on treatment outcomes. Understanding such binding motifs may contribute to the development of more targeted therapeutic approaches, potentially advancing precision medicine strategies that could optimise treatment based on individual patient characteristics.

REFERENCES

Agamah FE, Mazandu GK, Hassan R, et al. Computational/in silico methods in drug target and lead prediction. *Brief Bioinform.* 2020;21(5):1663–1675. doi:10.1093/bib/bbz103

Arena A, Romeo MA, Benedetti R, Gilardini Montani MS, Cirone M. JQ-1/bortezomib combination strongly impairs MM and PEL survival by inhibiting c-Mmyc and mTOR despite the activation of prosurvival mechanisms. *Exp Hematol.* 2023;119–120:28–41. doi:10.1016/j.exphem.2022.12.007

Bailey TL. Discovering sequence motifs. *Methods Mol Biol.* 2007;395:271–292. doi:10.1007/978-1-59745-514-5_17

Barnett DW, Garrison EK, Quinlan AR, Strömberg MP, Marth GT. BamTools: A C++ API and toolkit for analyzing and managing BAM files. *Bioinformatics.* 2011;**27**(12):1691–1692. doi:10.1093/bioinformatics/btr174

Cardona-Benavides IJ, de Ramón C, Gutiérrez NC. Genetic abnormalities in multiple myeloma: Prognostic and therapeutic implications. *Cells.* 2021;**10**(2):336. Published 2021 Feb 5. doi:10.3390/cells10020336

Chen D, Frezza M, Schmitt S, Kanwar J, Dou QP. Bortezomib as the first proteasome inhibitor anticancer drug: Current status and future perspectives. *Curr Cancer Drug Targets.* 2011;**11**(3):239–253. doi:10.2174/156800911794519752

Di Lernia G, Leone P, Solimando AG, et al. Bortezomib treatment modulates autophagy in multiple myeloma. *J Clin Med.* 2020;9(2):552. Published 2020 Feb 18. doi:10.3390/jcm9020552

Ewels P, Magnusson M, Lundin S, Käller M. MultiQC: Summarize analysis results for multiple tools and samples in a single report. *Bioinformatics.* 2016;**32**(19):3047–3048. doi:10.1093/bioinformatics/btw354

Fairfield H, Falank C, Avery L, Reagan MR. Multiple myeloma in the marrow: Pathogenesis and treatments. *Ann N Y Acad Sci.* 2016;**1364**(1):32–51. doi:10.1111/nyas.13038

Feng J, Liu T, Qin B, Zhang Y, Liu XS. Identifying ChIP-seq enrichment using MACS. *Nat Protoc.* 2012;**7**(9): 1728–1740. doi:10.1038/nprot.2012.101

Field-Smith A, Morgan GJ, Davies FE. Bortezomib (Velcade trademark) in the treatment of multiple myeloma. *Ther Clin Risk Manag.* 2006;**2**(3):271–279. doi:10.2147/tcrm.2006.2.3.271

Gupta K, Kuznetsova I, Klimenkova O, et al. Bortezomib inhibits STAT5-dependent degradation of LEF-1, inducing granulocytic differentiation in congenital neutropenia CD34(+) cells. *Blood.* 2014;123(16): 2550–2561. doi:10.1182/blood-2012-09-456889

Ho M, Chen T, Liu J, et al. Targeting histone deacetylase 3 (HDAC3) in the bone marrow microenvironment inhibits multiple myeloma proliferation by modulating exosomes and IL-6 trans-signaling. *Leukemia.* 2020;34(1):196–209. doi:10.1038/s41375-019-0493-x

Kao C, Chao A, Tsai CL, et al. Bortezomib enhances cancer cell death by blocking the autophagic flux through stimulating ERK phosphorylation. *Cell Death Dis.* 2014;5(11):e1510. doi:10.1038/cddis.2014.468

Liapis K, Kastritis E, Bagratouni T, et al. Early tumor-cell gene expression changes may predict the response to first-line bortezomib-based therapy in patients with newly diagnosed multiple myeloma. *J BUON.* 2015;20(5):1314–1321.

Li Q, Lai L. Prediction of potential drug targets based on simple sequence properties. *BMC Bioinformatics.* 2007;8:353. doi:10.1186/1471-2105-8-353

Louros SR, Seo SS, Maio B, et al. Excessive proteostasis contributes to pathology in fragile x syndrome. *Neuron.* 2023;111(4):508–525.e7. doi:10.1016/j.neuron.2022.11.012

Machanick P, Bailey TL. MEME-ChIP: Motif analysis of large DNA datasets. *Bioinformatics.* 2011;**27**(12): 1696–1697. doi:10.1093/bioinformatics/btr189

Narlikar L, Jothi R. ChIP-seq data analysis: Identification of protein-DNA binding sites with SISSRs peakfinder. *Methods Mol Biol.* 2012;802:305–322. doi:10.1007/978-1-61779-400-1_20

Park PJ. ChIP-seq: Advantages and challenges of a maturing technology. *Nat Rev Genet.* 2009;**10**(10): 669–680. doi:10.1038/nrg2641

Ramírez F, Ryan DP, Grüning B, et al. deepTools2: A next generation web server for deep-sequencing data analysis. *Nucleic Acids Res.* 2016;44(W1):W160–W165. doi:10.1093/nar/gkw257

Robak P, Robak T. Bortezomib for the treatment of hematologic malignancies: 15 years later. *Drugs R D.* 2019;**19**(2):73–92. doi:10.1007/s40268-019-0269-9

Robinson MD, McCarthy DJ, Smyth GK. edgeR: A bioconductor package for differential expression analysis of digital Gene expression data. *Bioinformatics.* 2010;**26**(1):139–140. doi:10.1093/bioinformatics/btp616

Yao X, Hong JH, Nargund AM, et al. PBRM1-deficient PBAF complexes target aberrant genomic loci to activate the NF-κB pathway in clear cell renal cell carcinoma. *Nat Cell Biol.* 2023;25(5):765–777. doi:10.1038/s41556-023-01122-y

Yu G, Wang LG, He QY. ChIPseeker: An R/Bioconductor package for ChIP peak annotation, comparison and visualization. *Bioinformatics.* 2015;**31**(14):2382–2383. doi:10.1093/bioinformatics/btv145

Zhang Y, Liu T, Meyer CA, et al. Model-based analysis of ChIP-seq (MACS). *Genome Biol.* 2008;9(9):R137. doi:10.1186/gb-2008-9-9-r137

17 Developing a Whole Exome Consensus Variant-Calling Pipeline to Infer Causal Pathogenic Variants

Ujjwal Prathap Singh, Ranjana M Raju, Jeffrey W. Bizzaro, and Prashanth N. Suravajhala

INTRODUCTION

NEXT-GENERATION SEQUENCING (NGS)

NGS has transformed the field of genomics by enabling the rapid and cost-effective sequencing of large amounts of DNA. NGS has numerous applications, including genome sequencing, transcriptome analysis, and epigenetics. Emerging technologies in the field of genomics, such as long-read sequencing and single-cell sequencing, are advancing the capabilities of NGS and expanding its potential applications. The NGS technologies, particularly Illumina sequencing, rely on the generation of short DNA fragments, which are then sequenced in parallel. The resulting sequence reads are then aligned to a reference genome or assembled *de novo* to reconstruct the original sequence. The advantages of NGS, particularly for short-read sequencing data, include high throughput, high accuracy, and relatively low cost per base. These characteristics have made NGS a popular tool for a wide range of applications, including population genetics, cancer research, and microbiome analysis (Slatko et al., 2018). The transformative era of genomics in healthcare has been ushered in by NGS. Whole exome sequencing (WES), a tool that focuses on the protein-coding sections of the genome, has demonstrated efficacy in locating novel causative mutations for diseases with undetermined aetiology. Recent investigations have shown that its clinical utility has a successful detection rate for uncommon diseases that is nearing 25%. To enable accurate and secure transmission of results back to the bedside, WES data interpretation requires competence in genomic informatics and clinical medicine. A significant challenge is the large volume of sequencing data, which typically contains approximately 25,000 variants per sequenced individual when compared to the reference genome. Therefore, computational methodologies and meticulous filtering frameworks are required to identify potential disease-causing variants among common polymorphisms. The identified and prioritised variants require intensive biological scrutiny and judicious assessment alongside the clinical phenotype (Seaby et al., 2016).

Recently, the advent of third-generation sequencing, embodied by long-read sequencing, has fundamentally transformed high-throughput DNA sequencing capabilities. While, historically, NGS primarily referred to second-generation short-read methods, this innovative long-read technology represents a distinct advancement in sequencing capabilities. It can sequence significantly longer DNA fragments than short-read methods, facilitating a more comprehensive genome assembly and identification of structural variants, such as large insertions and deletions. Long-read sequencing proves particularly beneficial for sequencing repetitive regions of the genome, which pose challenges for accurate sequencing with short-read technologies. The two main approaches to long-read sequencing have been developed by Pacific Biosciences (PacBio), using single-molecule real-time

DOI: 10.1201/9781003354062-17

sequencing, and Oxford Nanopore Technologies (ONT), using nanopore sequencing technology. Both platforms are becoming increasingly popular in the genomics community.

Another emerging technology in the field of genomics is single-cell sequencing. This technology allows researchers to analyse individual cells, whether examining DNA sequences, RNA expression, or epigenetic modifications at a single-cell resolution. This provides unprecedented insights into cellular heterogeneity, gene expression patterns, and epigenetic regulation (Angerer et al., 2017). Single-cell sequencing is particularly useful for studying complex tissues, such as the brain, where cellular diversity plays a critical role in function and dysfunction. It has been used to study cancer, developmental biology, and neurodegenerative diseases. In addition to long-read sequencing and single-cell sequencing, other technological advances are expanding the applications of NGS. For example, spatial transcriptomics allows researchers to map gene expression patterns within tissues, providing spatial context to gene expression data (Wang et al., 2015). Additionally, metagenomics sequencing can be used to study the microbiome and its role in health and disease.

In summary, NGS technologies have transformed genomics and enabled numerous applications across a wide range of fields. Emerging technologies such as long-read sequencing and single-cell sequencing are expanding our analytical capabilities and providing new insights into the complexity of biological systems. As these technologies continue to develop, we can expect even greater advances in our understanding of genomics and its applications.

WES

WES enables the analysis of all protein-coding exons across the human genome, focusing on regions that harbour the majority of disease-causing mutations (Yang et al., 2013). This technology is particularly effective for identifying genetic variants within exons and offers a cost-effective, high-throughput alternative to whole genome sequencing. However, WES has certain limitations, such as reduced efficacy in detecting copy number variations (CNVs), even though it efficiently identifies single nucleotide variants (SNVs). To provide a more comprehensive approach, data from CNV and SNV analyses can be integrated. SNV variant-calling pipelines used in WES employ either machine learning (ML)–based combination methods or standalone solutions.

WES has become an indispensable tool in clinical diagnostics and treatment selection, especially in oncology. CNV detection tools analyse read depth across specific genomic regions to infer copy number alterations. Together, SNVs and CNVs facilitate the detection of mutations with potential therapeutic relevance. Additionally, microsatellite instability (MI) and tumour mutation burden, both detectable through WES, help identify patients who may benefit from immunotherapy. Understanding the limitations and advantages of WES is essential for its effective application in clinical settings. Key limitations include restricted CNV detection, lower coverage compared to targeted sequencing, lack of consensus on reference standards, and insufficiently defined minimal application criteria. Nevertheless, sequencing advancements have led to US Food and Drug Administration (FDA)-approved WES methods for cancer screening, detection, and follow-up. Consequently, WES has evolved into a technically mature and cost-effective tool for routine clinical use (Bartha and Győrffy, 2019).

Beyond oncology, WES has proven valuable for gene discovery in large cohorts with conditions such as autism, epilepsy, brain abnormalities, congenital heart disease, and neurodevelopmental disorders, leading to the identification of numerous novel disease genes and pathways. The diagnostic yield of WES in clinical series ranges between 22% to 46%, though success rates vary significantly by condition. For example, in clinical cases of ataxia, diagnostic yields have ranged from 12.8% to 44.1%. Furthermore, WES enables an unbiased and comprehensive analysis of disease-causing genes, allowing the identification of genetic conditions even when clinical presentations are ambiguous or complex (Retterer et al., 2016).

HETEROZYGOSITY IN VARIANT STUDIES

Heterozygosity, the presence of different alleles at a particular genomic locus in an individual's genome, is a fundamental measure in genetic variation studies. It provides valuable insights into the diversity and complexity of a population's genetic makeup, playing a crucial role in understanding population genetics, evolution, and conservation efforts. Heterozygosity reflects the occurrence of genetic variations, such as single nucleotide polymorphisms (SNPs), insertions, deletions, and structural variations within a population, serving as a measure of genetic diversity (Allendorf et al., 2010). Higher levels of heterozygosity indicate a greater number of genetic variants present, attributed to factors such as mutations, recombination, and migration.

In the context of disease genetics, heterozygosity plays a significant role in understanding inheritance patterns. For recessive genetic disorders, individuals must be homozygous for the disease-causing variant to manifest the condition. Heterozygosity at these loci indicates carrier status, where an individual carries one copy of the disease-causing allele but does not show symptoms (Gratten et al., 2008). This aids in studying the prevalence of carriers and understanding the inheritance patterns of genetic disorders.

Furthermore, heterozygosity is important in genetic association studies aimed at identifying genetic variants associated with diseases or complex traits. It contributes to the genetic variability that can influence susceptibility or protection against diseases, allowing for the detection of heterozygous genotypes that may confer a selective advantage or contribute to disease risk (Understanding Human Genetic Variation). Thus, assessing heterozygosity helps researchers characterise the genetic landscape of a population, identify regions of high or low diversity, and study patterns of genetic differentiation among populations.

Over the years, our Systems Genomics Lab has been conducting research on rare diseases and prostate cancer. We have developed and benchmarked pipelines for WES (Meena et al., 2018). These pipelines have been utilised in numerous studies aimed at enhancing our understanding of rare diseases, particularly anorectal malformations (ARM) (Mathur et al., 2018; Gupta et al., 2018, 2019, 2023). While our previous pipeline was capable of detecting potentially disease-causing variants, we identified the need for more stringent criteria in pathogenicity prediction, which laid the foundation for this current work.

Datasets

For our analysis, we used four datasets: one from the Genome in a Bottle (GIAB) project (NA12878, also known as HG001), two from the 1000 Genomes Project repository (HG00448, a sample of Han Chinese ancestry from southern China, and HG00284, a Finnish sample of European ancestry), and a prostate cancer (PCa) sample from our Systems Genomics Lab.

We downloaded rigorously validated 'gold standard' truth variant sets and high-confidence variant-calling regions for NA12878 (HG001) from the GIAB FTP repository (https://ftp-trace.ncbi.nlm.nih.gov/giab/ftp/release last accessed 1 August 2024). To analyse the distribution of pathogenic variants, we selected pathogenic and likely pathogenic variants with non-conflicting interpretations from ClinVar. The ClinVar database was downloaded as a variant call format (VCF) file (https://ftp.ncbi.nlm.nih.gov/pub/clinvar/vcf_GRCh38/archive_2.0/2023/clinvar_20230611.vcf.gz, last accessed 1 August 2024) to facilitate comparison. Our pipeline was then run on these widely accepted datasets in the scientific community.

The 1000 Genomes Project

The 1000 Genomes Project is a worldwide scientific initiative that aims to create a comprehensive catalogue of human genetic variants occurring at a frequency of at least 1% in the human population. Using NGS technology, the project, which ran from 2008 to 2015, examined the genomes of over 2,500 people from 26 different populations throughout the world. The resulting dataset, which is openly accessible for academic research, is now widely used in the study of human genetics and disease.

During the project period, high-coverage sequencing was prohibitively expensive for such a large number of samples. However, since each genomic region typically contains a limited number of haplotypes, the project adopted a strategic approach: sequencing many samples at low coverage (approximately 4× genome coverage per sample) and combining the data. While this depth of sequencing cannot identify every variant in each individual sample, the combined data from multiple samples enables the detection of variants present at frequencies as low as 1%. In the project's final phase, data from 2,504 samples were integrated using statistical methods including genotype imputation, allowing accurate genotype assignment for each sample across all identified variant sites.

Variant Callers

Numerous variant callers have been developed and combined into various pipelines to call (identify and characterise) variations in NGS data. A typical pipeline contains two main components: an aligner that maps the sequencing reads to a reference genome, and a variant caller that identifies variant sites and determines the subject's genotype(s). Extensive research has evaluated aligner effectiveness. Because alignment is essentially a string-matching procedure, Burrows–Wheeler Transform (BWT) based aligners are particularly efficient, as they exploit data compression features by building an index of the reference genome to aid string matching. These BWT-based aligners make a favourable trade-off between processing time, memory consumption, and precision (Lie et al., 2009). For our systematic comparison, we chose Bowtie2 as the common aligner for all variant callers and calling strategies implemented. In addition to the variant callers described below, we attempted to use SoapSNP and VarDict, but these proved incompatible with our pipeline (Lai et al., 2016). We used the following tools in our variant analysis pipeline:

Varscan: Varscan (https://dkoboldt.github.io/varscan/ last accessed 1 August 2024) is a platform-independent software tool developed at the Genome Institute at Washington University to detect variants in NGS data. It provides mutation calling capabilities for targeted, exome, and whole-genome sequencing data generated on Illumina, SOLiD, Life/PGM, Roche/454, and similar instruments. Most published variant callers for NGS data use a probabilistic framework, such as Bayesian statistics, to identify variants and rate their confidence. These methods are generally effective but can be confounded by several variables, including pooled samples, excessive read depths, and contaminated or impure samples. Varscan, in contrast, uses a robust heuristic/statistical technique that identifies variations meeting predetermined thresholds for read depth, base quality, variant allele frequency, and statistical significance. It takes SAMtools mpileup input to call variants. It also provides multiple options for variant analysis, including the following:

- Germline variants (SNP/indel) in individual samples or a pool of samples
- Multisample variants (shared or private) in multiple sample datasets
- Somatic mutations
- Somatic copy number alterations

FreeBayes: FreeBayes is a Bayesian variant caller that leverages population-wide information to accurately identify genetic variants from NGS data. It is written in C++ and implements a haplotype-based approach to detect SNVs, insertions, deletions, and complex variants. FreeBayes employs a probabilistic framework, utilising Bayes's theorem to estimate the likelihood of genetic variants given the observed sequencing data. This approach enables reliable variant detection, especially in regions of low coverage or complex genomic landscapes (Garrison and Marth, 2012).

SAMtools/BCFtools: SAMtools/BCFtools (https://www.htslib.org/ last accessed 1 August, 2024) is a suite of tools that can filter, query, convert, and validate variant calls in the VCF format and its binary equivalent, BCF. Variant calling with BCFtools typically involves two main steps. First, the '`bcftools mpileup`' command generates a pileup file, providing coverage information across each genomic position. Then, BCFtools processes these data to identify variants by applying a

Bayesian statistical framework and maximum likelihood estimation to assess the probability of variants at each site (Danecek et al., 2021).

vt: While not a variant caller but a toolset designed for post-processing the output of variant callers to support accurate variant discovery and analysis, vt (https://github.com/atks/vt last accessed 1 August, 2024) facilitates the manipulation, annotation, selection, simulation, and analysis of variants in NGS data. Working directly with VCF files, vt performs multiple processing steps, including indexing, sorting, normalisation (to ensure consistent representation of variants across datasets), and decomposition of both biallelic block substitutions and multiallelic variants into simpler components. This standardisation is critical for comparing variant call sets from different sources, enabling efficient and reliable downstream analyses (Adrian et al., 2015).

Bash and Linux

To accomplish this work, we used bash scripting extensively, and all of the pipeline development and related work was exclusively done on Linux servers employed in the Systems Genomics Lab. A typical and ideal configuration could be TB RAM with 64 core processors and 20 TB storage, depending on the samples.

Variants Called from Datasets

To evaluate the factors influencing the accuracy of our variant-calling pipeline, we used four WES datasets, as mentioned above: NA12878 (from GIAB), HG00284 (from 1000 Genomes), HG00448 (from 1000 Genomes), and PCa (a prostate cancer sample). All exome samples had high coverage (100–200×). We ran our workflow on each dataset and summarised the results in Table 17.1. While truth sets were available only for NA12878 (HG001) for direct accuracy assessment, the other datasets were used to optimise and validate the pipeline's performance.

As seen in Table 17.1, the final variant categories are pathogenic, likely pathogenic, pathogenic/likely pathogenic, variants of uncertain significance (VoUS), and variants of conflicting pathogenicity. All the categories are defined according to the tag found in the key 'CLNSIG' of the ClinVar database records.

From the results, it is evident that an average pipeline calls around 70–80K variants, out of which the final number of records after ClinVar mapping was found to be 20% of the consensus calls. For

TABLE 17.1
Variants Called from Our Pipeline Run on Corresponding Datasets

Cumulative Results of Pipeline Run

	Datasets			
Variant Categories	**HG00284**	**HG00448**	**PCa**	**NA12878**
Varscan	100,503	112,950	2,860,539	107,765
FreeBayes	407,388	368,739	6,485,069	580,300
SAMtools/BCFtools	835,042	773,214	1,318,728	796,993
vt	399,065	363,349	14,388,622	520,983
Consensus	73,988	81,156	68,552	74,159
ClinVar Validated	17,398	17,862	10,677	14,809
Pathogenic	3	3	11	3
Likely pathogenic	1	2	3	2
Pathogenic and likely pathogenic	1	0	2	0
Uncertain significance variants	28	33	109	79
Variants of conflicting pathogenicity	64	51	66	99

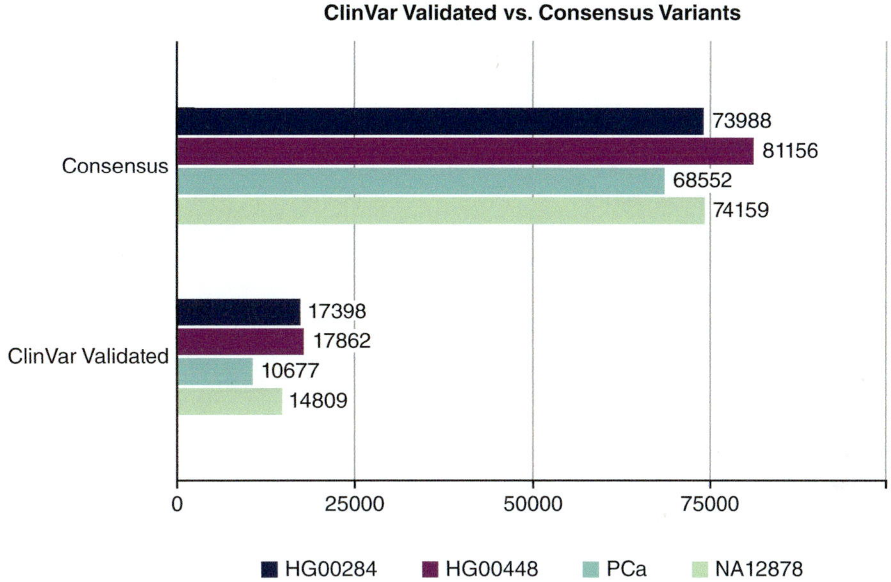

FIGURE 17.1 Comparison of consensus variants (identified by our pipeline) and ClinVar-validated variants across four datasets: HG00284 (Finnish), HG00448 (Han Chinese), PCa (prostate cancer), and NA12878 (GIAB reference). The bars show the total number of variants in each category.

ease of analysis, we included all tags containing 'pathogenic' as well as VoUS, as these indicate potential *de novo*, deleterious, or associated mutations (Figure 17.1).

From Figure 17.2, we can see that the number of uncertain variants consistently exceeds pathogenic variants, which is expected. Typically, loci with pathogenic attributes are rare, while most variant records are either conflicting or less significant. This pattern exists because variants from

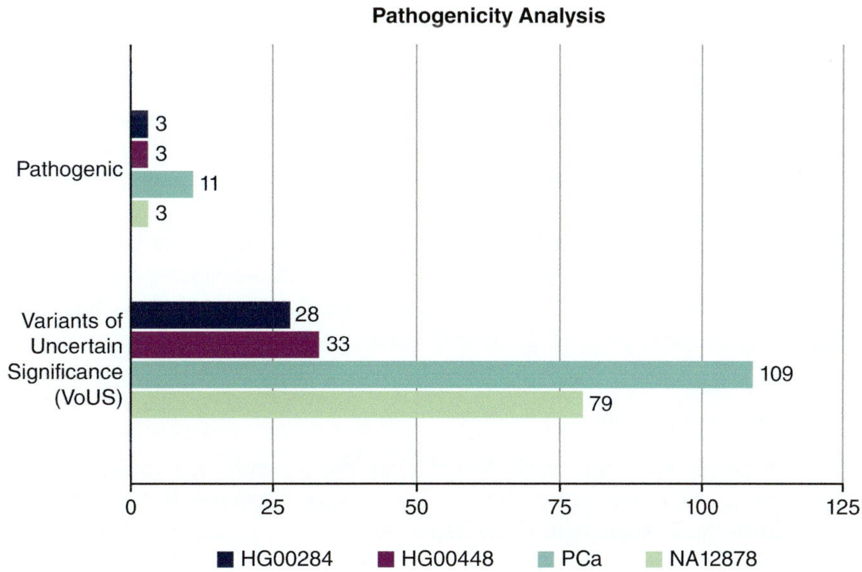

FIGURE 17.2 Distribution of pathogenic variants and VoUS identified across four datasets: HG00284 (Finnish), HG00448 (Han Chinese), PCa (prostate cancer), and NA12878 (GIAB truth set). PCa showed the highest number of both pathogenic variants (11) and VoUS (109), consistent with its disease status.

TABLE 17.2
Truth Set Analysis of NA12878 Yielding Stringent Pathogenic Variants

Truth Set Analysis of NA12878		
Variant Categories	**Pipeline**	**Truth Set**
Total SNPs	74,159	3,893,341
ClinVar validated	14,809	34,074
Pathogenic	5	9
Variants of uncertain significance	79	122
Variants of conflicting pathogenicity	99	116

NGS are difficult to interpret definitively, and our understanding of their phenotypic effects remains limited relative to the size of the human genome. The disease sample (PCa) shows a higher number of variants compared to the gold-standard reference datasets analysed in parallel, which is expected since these reference samples are not associated with any particular disease. However, there is a notable difference in the number of VoUS among datasets, with PCa showing the highest count. This can be attributed to its disease status, and these VoUS with conflicting pathogenicity may indicate *de novo* genomic changes that warrant further annotation.

Truth Set Analysis

We validated our pipeline using NA12878 as the truth set by comparing our results with the GIAB repository variants. We downloaded the benchmarked variants file from the GIAB repository and filtered out SNPs using the Genome Analysis Toolkit (GATK) SelectVariants function. This filtering was necessary as the file contained both SNPs and indels, while our study focused specifically on SNPs.

As shown in Table 17.2, our pipeline identified 74,159 total SNPs compared to 3,893,341 in the truth set. Of these, 14,809 variants were validated in ClinVar in our pipeline, compared to 34,074 in the truth set. Figure 17.3 provides a detailed comparison of pathogenic variants and VoUS between our pipeline and the truth set.

We then analysed the overlap between our results and the truth set variants (Figures 17.4 and 17.5). For pathogenic variants, our pipeline called five variants, while the truth set contained nine, with four variants common to both sets. For VoUS, our pipeline called 79 variants, and the truth set contained 122, with 73 variants found in both sets. Table 17.2 also shows that we identified 99 variants of conflicting pathogenicity compared to 116 in the truth set.

The sensitivity of our pipeline was calculated as the percentage of truth set variants that were correctly called. For pathogenic variants, our pipeline achieved 44% sensitivity (4/9 × 100), while for VoUS, it achieved 59% sensitivity (73/122 × 100). The average sensitivity across both variant types was 51.5%, suggesting that our pipeline calls approximately half of the variants present in the truth set.

Confusion Matrix and F_1 Score

In this study, we used the harmonic mean to calculate the F_1 score, a measure of a test's accuracy. The F_1 score is a value between 0 and 1, where a higher value indicates a better classifier. Specifically, the closer the F_1 score is to 1, the better the precision and recall of the classifier.

$$F_1 \text{ score} = 2 \times \frac{\text{precision} \times \text{recall}}{\text{precision} + \text{recall}}$$

Precision and recall are performance metrics used to evaluate the accuracy of the variant-calling algorithm. Precision, also known as positive predictive value (PPV), quantifies the number of true

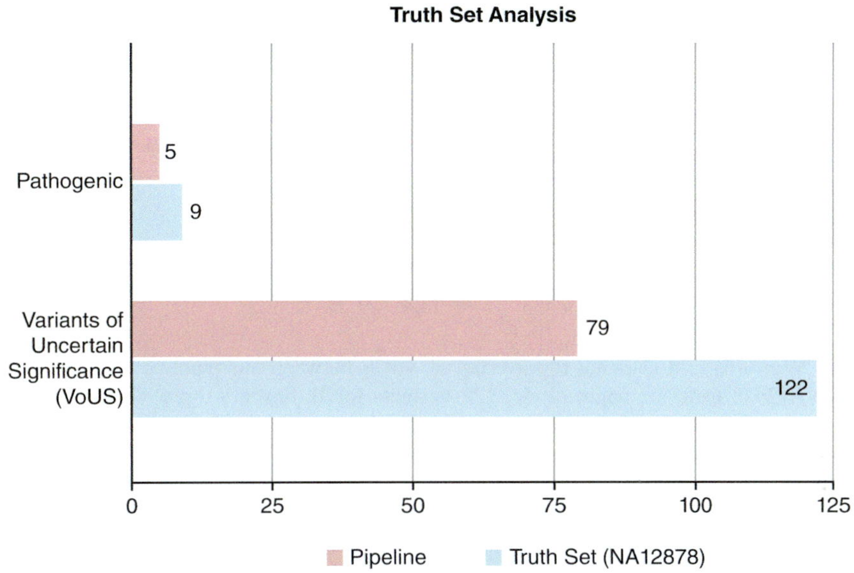

FIGURE 17.3 Comparison of pathogenic variants and VoUS between our pipeline results and the GIAB truth set for NA12878. Our pipeline identified fewer variants in both categories: 5 pathogenic variants and 79 VoUS compared to the validated set's 9 pathogenic variants and 122 VoUS.

positive predictions (i.e., variants correctly identified) out of all positive predictions made. Recall, also known as true positive rate (TPR), quantifies the number of true positive predictions out of all actual positive instances.

$$\text{precision} = \frac{\text{true positives (TP)}}{\text{true positives (TP)} + \text{false positives (FP)}}$$

$$\text{recall} = \frac{\text{true positives (TP)}}{\text{true positives (TP)} + \text{false negatives (FN)}}$$

In these formulas, true positives (TP) are the variants correctly identified by the algorithm, false positives (FP) are the non-variant positions incorrectly identified as variants, and false negatives (FN) are the actual variants that the algorithm failed to identify.

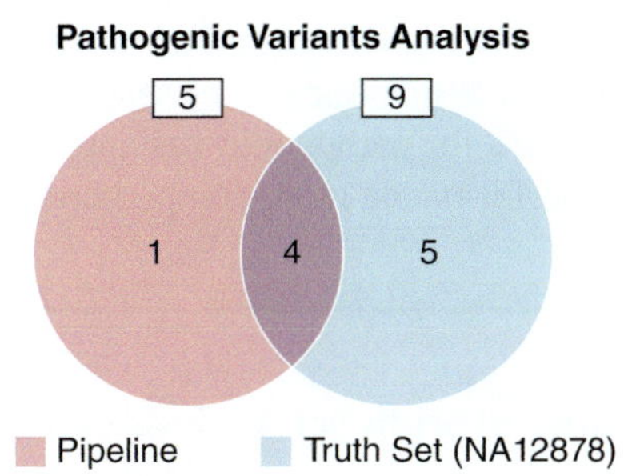

FIGURE 17.4 Venn diagram showing the overlap of pathogenic variants between our pipeline (left circle, 5 variants total) and the NA12878 truth set (right circle, 9 variants total). Four pathogenic variants were identified by both approaches, while one variant was unique to our pipeline and five were unique to the truth set. Distribution of variants: Left circle: 1 (unique to pipeline) + 4 (shared) = 5 total for pipeline; Right circle: 5 (unique to truth set) + 4 (shared) = 9 total for truth set; Overlap: 4 shared variants.

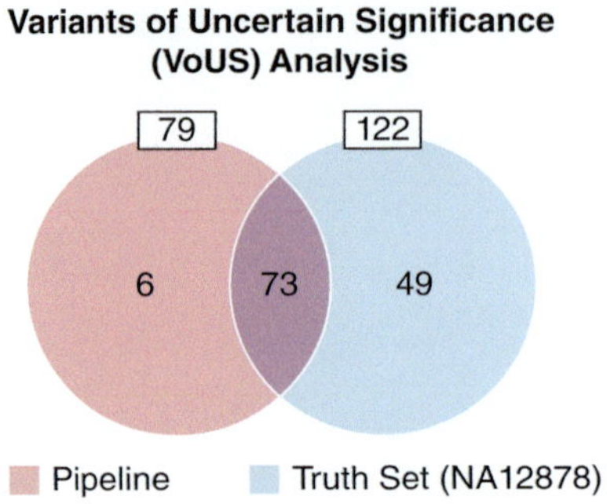

FIGURE 17.5 Venn diagram showing the overlap of VoUS between our pipeline (left circle, 79 variants total) and the NA12878 truth set (right circle, 122 variants total). Seventy-three VoUS were identified by both approaches, while six variants were unique to our pipeline and forty-nine were unique to the truth set. Distribution of variants: Left circle: 6 (unique to pipeline) + 73 (shared) = 79 total for pipeline; Right circle: 49 (unique to truth set) + 73 (shared) = 122 total for truth set; Overlap: 73 shared variants.

A perfect variant calling algorithm would have a precision and recall of 1, indicating that it correctly identified all variants and did not falsely identify any non-variant positions as variants.

We used a confusion matrix (Table 17.3) to compare the pathogenicity of the truth set against our pipeline's results. In this context, a positive value indicates that a variant was called, while a negative value means it was not. The total number of pathogenic variants in the ClinVar database is 236,034. For this sample, ten variants were called, leaving approximately 236,024 records uncalled by either method, resulting in a true negative (TN) result. Four variants were called by both methods, yielding a true positive (TP) result, while one variant was called by our pipeline but not according to the truth set, resulting in an FP.

$$F_1 \text{ score} = 2 * (0.8 * 0.5/0.8 + 0.5) = 0.6$$

$$\text{Where precision} = \text{TP}/(\text{TP} + \text{FP}) = 4/(4+1) = 0.8$$

$$\text{and recall} = \text{TP}/(\text{TP} + \text{FN}) = 4/(4+4) = 0.5$$

Our pipeline's performance can be assessed through several complementary metrics. Accuracy measures how often the pipeline is correct overall, considering both TP and TN. Precision measures the proportion of positive calls that are correct (TP / [TP + FP]), while sensitivity (or recall) measures the proportion of actual variants that are detected (TP / [TP + FN]). The F_1 score, calculated from the confusion matrix, was 0.6 (60%), providing a balanced measure of pipeline performance by combining precision and sensitivity. While our earlier sensitivity analysis showed how well the pipeline detected known variants (44% for pathogenic variants and 59% for VoUS), the F_1 score allows us to assess variant calling performance more comprehensively.

TABLE 17.3
Confusion Matrix of Truth Set Analysis on Pathogenicity

		Called by Our Pipeline	
		Negative	**Positive**
Present in Truth Set	**Negative**	236,024 (TN)	1 (FP)
	Positive	4 (FN)	4 (TP)

TABLE 17.4
Time Required for Each Run

Dataset	Size of the File	Time
HG00284	6.8 GB	7 h 15 m
HG00448	7.8 GB	10 h 00 m
PCa	10.0 GB	5 h 45 m
NA12878	7.4 GB	12 h 00 m

Run-time of the Pipeline

The pipeline takes around six to 12 hours for the complete run for a sample, averaging around 8.5 hours (Table 17.4). Pipeline execution time depends on multiple factors, including computational resources (CPU, memory, server load) and the size of the input data. The number of sequencing reads is particularly important, as it directly affects the size of alignment files (SAM format) that must be processed.

Heterozygosity Calls and Pathogenicity

Our exomic pipeline is designed to infer pathogenic variants, with a particular focus on heterozygous calls (typically denoted as 0/1). Analysis of heterozygous variants across our datasets (Table 17.5) revealed interesting patterns. The prostate cancer sample (PCa) showed notably higher heterozygosity, with 55,130 variants compared to approximately 45,500–46,000 variants in the other samples. This elevated heterozygosity in PCa is consistent with its disease status, as cancer samples often exhibit increased genetic variation. The three non-disease samples (HG00284, HG00448, and NA12878) showed remarkably consistent heterozygosity levels, ranging from 45,514 to 45,976 variants, suggesting our pipeline's reliability in detecting heterozygous variants across different populations.

Benchmarking NGS Pipelines

Numerous benchmarking studies have been conducted to compare the performance of different bioinformatics tools and establish quality metrics. Benchmarking NGS pipelines is essential to assess data quality and identify errors or biases introduced during the sequencing process. The accuracy and reproducibility of results obtained from NGS pipelines depend on the quality of the algorithms used for data processing, which in turn depends on the accuracy of the underlying assumptions and parameters.

Several methods are used for benchmarking NGS pipelines. Synthetic data, generated with known properties, can test pipeline accuracy and sensitivity. Real data assess pipeline performance in realistic settings. Reference datasets, such as those from the GIAB project developed by the

TABLE 17.5
Number of Heterozygous Variants from Our Analyses

Dataset	Heterozygous Variants
HG00284	45,514
HG00448	45,976
PCa	55,130
NA12878	45,628

National Institute of Standards and Technology (NIST), provide well-established standards for pipeline evaluation. The GIAB dataset, containing reference genomes and associated metadata, has been used to benchmark various NGS tools, including alignment tools, variant callers, and CNV callers.

Another approach uses benchmarking suites, such as the GATK Best Practices Pipeline, the BWA-MEM + GATK Best Practices Pipeline, and the Human Longevity, Inc. (HLI) WGS pipeline (Best Practices Workflows: https://gatk.broadinstitute.org/hc/en-us/sections/360007226651-Best-Practices-Workflows accessed 1 August 2024). These suites provide standardised tools and parameters for processing NGS data, enabling researchers to compare different pipelines using the same data.

In summary, benchmarking NGS pipelines through synthetic data, real data, reference datasets, and benchmarking suites ensures reliable and accurate results, enabling further advances in genomics research.

Consensus Strategy

We have developed an alternative approach to obtaining high-quality variants that focuses on the initial step of NGS data analysis. Traditionally, high-confidence variants require high-coverage sequencing, leading to increased cost per sample. Our approach instead leverages multiple variant callers to identify consensus variants, providing high-confidence calls even from lower-coverage data. Following this strategy, we developed a variant-calling pipeline that processes raw reads through four different tools: Varscan, FreeBayes, SAMtools/BCFtools, and vt. The resulting VCF files are then analysed by our custom script to identify consensus variants—those called by all four tools—producing a final list of high-confidence variants for downstream analysis (Figure 17.6).

NGS Workflow

In this analysis, we modified the third step (variant prioritisation) of the workflow described in a prior analysis of ours (Figure 17.7) (Meena et al., 2018). At that time, we performed variant calling using three different tools (Varscan, SAMtools/BCFtools, and vt). The resulting VCF files were passed on to the consensus calling script, which gave us a final consensus list of variants with

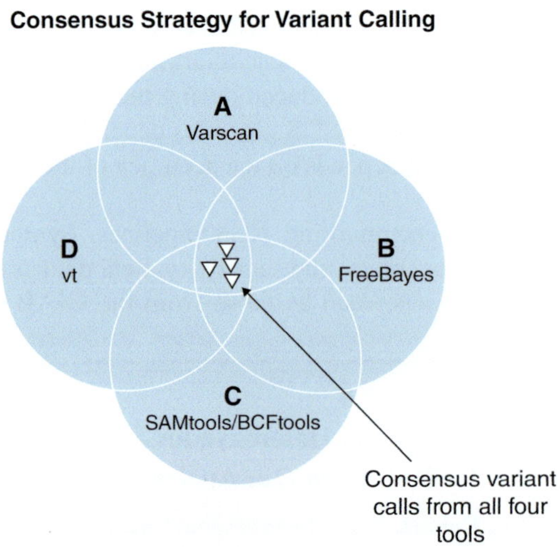

FIGURE 17.6 Consensus strategy for variant calling. The Venn diagram shows the intersection of variants called by four different tools (A: Varscan, B: FreeBayes, C: SAMtools/BCFtools, D: vt). The triangles in the central intersection represent high-confidence variants identified by all four tools, demonstrating how consensus calls can provide reliable variants even from lower-coverage sequencing data.

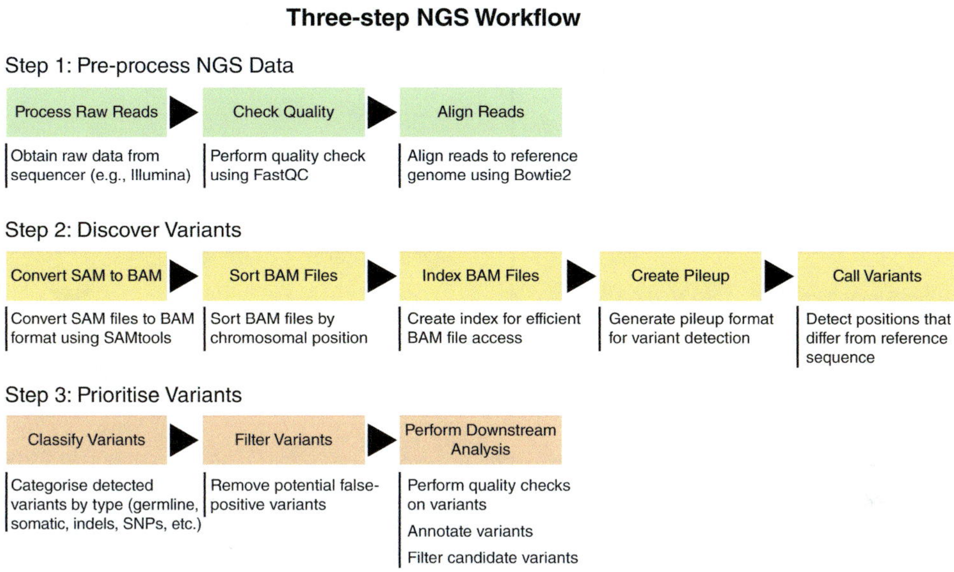

FIGURE 17.7 Three-step NGS workflow showing data pre-processing, variant discovery, and variant prioritisation stages. Adapted from Meena et al. (2018).

chromosome position, reference, and alternate allele. Downstream bioinformatics annotation can be supplemented to integrate the results with different pathway tools such as PANTHER, which assesses the ontology and pathways affecting the mutated genes. This can be further supplemented by using databases like ClinVar and dbSNP. In addition, global enrichment analysis and association networks using GeneMania would allow creation and visualisation of gene networks based on evidence found in pathways and gene–gene/protein–protein interactions (predicted and experimental).

Proposed Workflow

For this work, we developed an enhanced workflow that builds upon our previous pipeline by incorporating a consensus approach to variant calling. The workflow begins with raw FastQ reads, processes them through SAMtools, and then employs four variant analysis tools in parallel. The outputs are combined to find consensus variants, which are then validated against the ClinVar database to provide high-confidence variant calls with associated pathogenicity information. Figure 17.8 shows our proposed pipeline, which notably adds FreeBayes as a fourth tool and implements a consensus strategy for improving variant call confidence.

To accomplish the consensus pipeline, we developed a bash script that takes paired-end reads (forward and reverse) as input and produces variant classifications after mapping to the ClinVar variant database. These reads are first aligned to the reference sequence (hg38) using Bowtie2, an efficient tool for aligning reads to long reference sequences. Bowtie2 is particularly effective for reads ranging from 50 to 1,000 characters and uses an FM Index to minimise memory usage, requiring about 3.2 GB for the human genome. Gapped, local, and paired-end alignment modes are supported by Bowtie2, and the output is stored in a sequence alignment map (SAM) file, which is then converted to its compressed binary form (BAM). We utilised this version to make the file smaller and to enable indexing, which allows for quick random access to the data it contains.

For some historical context, the SAM format was developed by Heng Li, Bob Handsaker, and colleagues to store biological sequences aligned to a reference sequence. It emerged from the 1000 Genomes Project's need to move beyond the MAQ mapper format, with its tab-delimited structure influenced by BLAT's PSL format. The name 'SAM' originates from an earlier format developed by Gabor Marth that resembled BLAST output. It is commonly used for storing mapped data produced by NGS technologies,

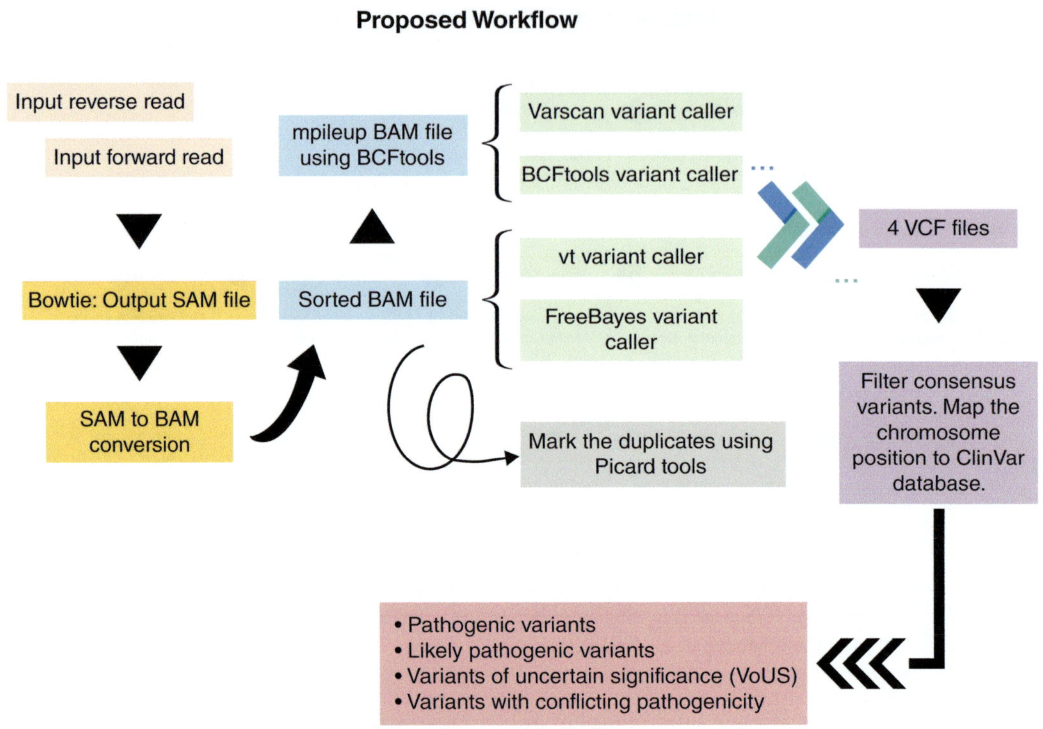

FIGURE 17.8 Proposed variant-calling pipeline incorporating a consensus approach. The workflow processes paired-end reads through alignment and format conversion steps, uses four parallel tools for variant calling (Varscan, FreeBayes, SAMtools/BCFtools, and vt), and implements consensus filtering before mapping variants to the ClinVar database for pathogenicity classification.

such as nucleotide sequences, and the standard has been expanded to allow unmapped data. The format has evolved to support both short and long reads (up to 128 Mbp) from various sequencing platforms.

The BAM file format includes an optional header describing data source, reference sequence, and alignment method, followed by the alignment section containing read mapping information in 11 required fields plus optional aligner-specific data. A sorted BAM has the data sorted by chromosomes/contigs/scaffolds—whatever is in your reference genome. To efficiently display/access data, the BAM file must be sorted before performing any analysis.

A text-based format called 'pileup' condenses the base calls of reads that have been aligned to a reference sequence, making SNP/indel calling and alignment more visible. For these format conversions, we use the SAMtools suite to convert between SAM/BAM and BAM/pileup formats. Our pipeline processes sequence information in both BAM and pileup formats as inputs for variant calling: Varscan and SAMtools/BCFtools use the mpileup format as input, while FreeBayes and vt directly use sorted BAM files.

The four tools each generate a VCF file, which are then filtered for only valid calls. These files are then processed through our consensus pipeline, which uses string manipulation and filtering in two steps: first, filtering for common chromosome positions, and second, identifying shared alternate alleles at these positions. This process yields a VCF file containing only variants that are called consistently by all four tools.

To ensure accurate detection of pathogenic variants, we used the ClinVar database to annotate variants with relevant information. ClinVar is a publicly accessible resource that provides reports on the associations between human variants and phenotypes, updated quarterly (https://www.ncbi.nlm.nih.gov/clinvar/docs/ftp_primer/, last accessed 1 August 2024). It includes germline and somatic variations of any size, type, or genomic position, contributed by clinical and research facilities, locus-specific databases, UniProt, expert panels, and practical guidelines. We downloaded the ClinVar database and matched chromosome numbers and positions from our consensus variant file against it using a string search. This process generated an output file containing chromosome positions, variant details, and associated pathogenicity tags. From this output, we filtered variants into categories based on five pathogenicity tags: *pathogenic*, *likely pathogenic*, *pathogenic/likely pathogenic*, *VoUS*, and *conflicting interpretations of pathogenicity*. Each category was saved in a separately named file.

CONCLUSION AND FUTURE PERSPECTIVES

The ongoing enhancement of variant-calling software necessitates regular, not only to evaluate performance but also to explore how available resources can be integrated to create more accurate and efficient variant-calling pipelines. Unlike previous studies, we adopted a systematic approach by evaluating the combined performance of multiple variant callers and employing a consensus strategy on WES samples. This approach allowed us to assess the relative importance of various factors in achieving accurate, reliable, and high-confidence variant discovery.

Our comparison showed that, despite some limitations, state-of-the-art variant callers such as Varscan, Freebayes, BCFtools, and vt offer accurate variant detection. Notably, combining these tools in a consensus approach enhances stringency and improves the identification of pathogenic variants. The pipeline's effectiveness was demonstrated through heterozygosity analysis, where it successfully detected expected differences between disease and reference samples while maintaining consistent detection levels across non-disease samples. Furthermore, we observed that variant callers have a greater influence on results than read aligners, underscoring the importance of selecting the appropriate tools for optimal results. By evaluating techniques such as Bayesian inference, heuristics, and maximum likelihood under different tools, we found that variant callers do not overfit to GIAB samples and perform comparably on non-GIAB samples. This conclusion was supported by an indirect evaluation using both GIAB and non-GIAB datasets (Barbitoff et al., 2022).

Currently, the pipeline is implemented using bash and awk scripting. In the future, we aim to enhance its flexibility and robustness by transitioning to Python and incorporating modern, advanced variant callers like VarDict and Clair3. These improvements are expected to enhance the pipeline's performance in terms of runtime, F_1 score, and the ability to capture pathogenic variants. In summary, we have demonstrated that even low-coverage sequencing can yield high-confidence variant calls when using a consensus-calling approach, and we have developed a pipeline to facilitate this process.

CODE AVAILABILITY

https://github.com/prashbio/Expipe

ACKNOWLEDGEMENTS

The authors gratefully acknowledge the chancellor, Amma-Mata Amritanandamayi and Bipin Nair, Dean, Amrita School of Biotechnology, for constant support.

REFERENCES

Adrian Tan, Gonçalo R. Abecasis and Hyun Min Kang. Unified representation of genetic variants. Bioinformatics. (2015) 31(13): 2202–4

Allendorf FW, Hohenlohe PA, Luikart G. Genomics and the future of conservation genetics. Nature Reviews Genetics. 2010. Available from: https://doi.org/10.1038/nrg2844.

Angerer P, Simon LM, Tritschler S, Wolf F, Fischer D, Theis FJ. Single cells make big data: New challenges and opportunities in transcriptomics. Current Opinion in Systems Biology. 2017. Available from: https://doi.org/10.1016/j.coisb.2017.07.004.

Barbitoff YA, Abasov R, Tvorogova VE, Glotov AS, Predeus AV. Systematic benchmark of state-of-the-art variant calling pipelines identifies major factors affecting accuracy of coding sequence variant discovery. BMC Genomics. 2022. Available from: https://doi.org/10.1186/s12864-022-08365-3.

Bartha Á, Győrffy B. Comprehensive outline of whole exome sequencing data analysis tools available in clinical oncology. Cancers. 2019. Available from: https://doi.org/10.3390/cancers11111725.

Danecek P, Bonfield JK, Liddle J, Marshall J, Ohan V, Pollard M, Whitwham A, Keane TM, McCarthy S, Davies RM, et al. Twelve years of SAMtools and BCFtools. GigaScience. 2021. Available from: https://doi.org/10.1093/gigascience/giab008.

Garrison E, Marth G. Haplotype-based variant detection from short-read sequencing. arXiv preprint arXiv:1207.3907 [q-bio.GN] 2012

Genome in a Bottle | NIST. NIST. 2023. Available from: https://www.nist.gov/programs-projects/genome-bottle.

Gratten J, Wilson AJ, McRae AF, Beraldi D, Visscher PM, Pemberton JM, Slate J. A localized negative genetic correlation constrains microevolution of coat color in wild sheep. Science. 2008. Available from: https://doi.org/10.1126/science.1151182.

Gupta S, Gupta N, Tiwari P, Menon S, Mathur P, Kothari SL, Nallapeta S, Medicherla KM, Suravajhala P. Lnc-EPB41-protein interactions associated with congenital pouch colon. Biomolecules. 2018. Available from: https://doi.org/10.3390/biom8030095.

Gupta S, Mathur P, Mishra AK, Medicherla KM, Bandapalli OR, Suravajhala P. Whole exome-trio analysis reveals rare variants associated with congenital pouch colon. Children (Basel). 2023. Available from: https://doi.org/10.3390/children10050902.

Gupta S, Tiwari P, Gupta N, Nunia V, Saxena AK, Simlot A, Kothari SL, Suravajhala P, Medicherla KM, Mathur P. Is pouch specific to colon and not ileum?. Current Pediatric Reviews. 2019. Available from: https://doi.org/10.2174/1573396315666190829155930.

Lai Z, Markovets A, Ahdesmäki MJ, Chapman B, Hofmann O, McEwen R, Johnson J, Dougherty B, Barrett JC, Dry JR. VarDict: A novel and versatile variant caller for next-generation sequencing in cancer research. Nucleic Acids Research. 2016. Available from: https://doi.org/10.1093/nar/gkw227.

Li H. Aligning sequence reads, clone sequences and assembly contigs with BWA-MEM. arXiv.org. 2013. Available from: https://arxiv.org/abs/1303.3997.

Li H, Durbin R. Fast and accurate short read alignment with Burrows-Wheeler transform. Bioinformatics. 2009;25(14):1754–60. doi: 10.1093/bioinformatics/btp324. Epub 2009 May 18. PMID: 19451168; PMCID: PMC2705234

Liu X, Han S, Wang Z, Gelernter J, Yang BZ. Variant callers for next-generation sequencing data: A comparison study. PLoS One. 2013. Available from: https://doi.org/10.1371/journal.pone.0075619.

Macaulay IC, Voet T. Single cell genomics: Advances and future perspectives. PLoS Genetics. 2014. Available from: https://doi.org/10.1371/journal.pgen.1004126.

Meena N, Mathur P, Medicherla K, Suravajhala P. A bioinformatics pipeline for whole exome sequencing: Overview of the processing and steps from raw data to downstream analysis. Bio-Protocol. 2018;8. doi: 10.21769/bioprotoc.2805.

Retterer K, Juusola J, Cho MT, Vitazka P, Millan F, Gibellini F, Vertino-Bell A, Smaoui N, Neidich J, Monaghan KG, et al. Clinical application of whole-exome sequencing across clinical indications. Genetics in Medicine. 2016;18(7):696–704. doi: 10.1038/gim.2015.148. Epub 2015 Dec 3. PMID: 26633542.

Seaby EG, Pengelly RJ, Ennis S. Exome sequencing explained: A practical guide to its clinical application. Briefings in Functional Genomics. 2015;15:374–384. doi: 10.1093/bfgp/elv054.

Slatko BE, Gardner AF, Ausubel FM. Overview of next-generation sequencing technologies. Current Protocols in Molecular Biology. 2018;122. doi: 10.1002/cpmb.59.

Van der Auwera GA, Carneiro MO, Hartl C, Poplin R, del Angel G, Levy-Moonshine A, Jordan T, Shakir K, Roazen D, Thibault J, et al. From FastQ data to high-confidence variant calls: The genome analysis toolkit best practices pipeline. Current Protocols in Bioinformatics. 2013;43. doi: 10.1002/0471250953.bi1110s43.

Wang K, Li M, Hakonarson H. ANNOVAR: Functional annotation of genetic variants from high-throughput sequencing data. Nucleic Acids Research. 2010;38:e164–e164. doi: 10.1093/nar/gkq603.

Wang Y, Navin NE. Advances and applications of single-cell sequencing technologies. Molecular Cell. 2015;58:598–609. doi: 10.1016/j.molcel.2015.05.005.

Yang Y, Muzny DM, Reid JG, Bainbridge MN, Willis A, Ward PA, Braxton A, Beuten J, Xia F, Niu Z, et al. Clinical whole-exome sequencing for the diagnosis of mendelian disorders. New England Journal of Medicine. 2013;369:1502–1511. doi: 10.1056/nejmoa1306555.

Index

Note: – *Italicized* page references refer to figures and **bold** references refer to tables.

D

E

T

U

V

W

X

Y

Z